AF506085

ADVANCES IN MALE MEDIATED DEVELOPMENTAL TOXICITY

ADVANCES IN EXPERIMENTAL MEDICINE AND BIOLOGY

ADVANCES IN MALE MEDIATED DEVELOPMENTAL TOXICITY

Edited by

Bernard Robaire and Barbara F. Hales

McGill University
Montreal, Quebec, Canada

Kluwer Academic / Plenum Publishers
New York, Boston, Dordrecht, London, Moscow

Library of Congress Cataloging-in-Publication Data

International Conference on Male Mediated Developmental Toxicity (2nd: 2001:
Montreal, Quebec)
    Advances in male mediated developmental toxicity/edited by Bernard Robaire and
Barbara F. Hales.
        p.   ; cm. — (Advances in experimental medicine and biology; v. 518)
    Includes bibliographical references and index.
    ISBN 0-306-47480-8
        1. Reproductive toxicology—Congresses.   2. Spermatozoa—Congresses.   3.
Mutagenesis—Congresses.   4. Developmental toxicology—Congresses.   I. Robaire, Bernard.
II. Hales, Barbara F.   III. Title.   IV. Series.
        [DNLM: 1. Spermatogenesis—drug effects—Congresses.   2.
Abnormalities—etiology—Congresses.   3. Environmental Exposure—adverse
effects—Congresses.   4. Pregnancy Outcome—Congresses.   WJ 834 I6035a 2003]
    RA1224.2 .I56 2001
    616.6'071—dc21

                                                                      2002040969

ISBN 0-306-47480-8

©2003 Kluwer Academic / Plenum Publishers, New York
233 Spring Street, New York, New York 10013

http://www.wkap.nl/

10 9 8 7 6 5 4 3 2 1

A C.I.P. record for this book is available from the Library of Congress

It is nearly a decade since the first Male Mediated Developmental Toxicity conference was held in Pittsburgh. The continuing public/scientific interest, growing amounts of animal data, introduction of innovative technologies, and increasing quantity of human epidemiological studies all suggest that male-mediated developmental toxicity is of major concern. A number of researchers concluded that a Second International Conference on Male Mediated Developmental Toxicity was necessary. The ensuing volume is particularly timely because it impacts on areas of special emphasis in many countries, with respect to children's and reproductive health, as well as to basic molecular mechanisms of environmental insult, and genetic susceptibility and predisposition.

The Programme and Local Organizing Committee, composed of Barbara Hales (Chair, McGill University), Bernard Robaire (McGill University), Daniel G. Cyr (INRS/Armand Frappier), Jacquetta M. Trasler (McGill University), Andrew F. Olshan (University of North Carolina), Sally Perreault Darney (US EPA), Donald R. Mattison (March of Dimes), and Jan M. Friedman (University of British Columbia), spent over two years identifying individuals who had made key contributions in this field over the past decade and planning various aspects of the meeting. The meeting was held in Montreal in June 2001. A total of 132 persons, coming from five continents and representing some 18 countries, took an active role in the proceedings. The conference was considered by all attendees to be a rousing success. Important discussions were held in the four break-out sessions, with a preliminary set of recommendations for action being presented by each panel.

We find throughout the volume an integration of information from laboratory and epidemiological studies, male reproduction, and teratology. Special emphasis on the need for translational research emerged as a common theme of the conference and is highlighted in several chapters. There is a need to develop a research agenda in this area which reflects our current understanding of biological plausibility.

In this volume, major contributions identify new developments and directions in the field over the past decade, as well as the challenges for the years ahead. The range of topics include: parental legacies and genomics; lifestyle, occupational and therapeutic paternal exposures and effects; effects on the gamete-packaging of human sperm; effects on fertilization and the embryo; role of DNA repair and germ cell apoptosis; stem cells, epigenetics and cloning; markers of sperm quality and relationship to progeny; model systems; implications to clinicians and genetic counsellors; and impact on reproductive technology and legislation.

We could not have had this meeting without the generous support provided by a number of governmental agencies, industry and charitable organizations. A full list of the conference's financial sponsors is provided in the volume. Sid Parkinson provided excellent administrative support throughout the planning and execution phases of the meeting as well as valuable support in editing this volume. We should also like to thank Jack Bishop for taking pictures during the meeting and kindly providing us with them. They provide a sense of the meeting for those that were not there and some memories for those that were.

Bernard Robaire
Barbara F. Hales

August 2002

LIST OF SPONSORS

Major Sponsor

U.S. National Institutes of Health (NIH)
    National Institute of Child Health and Development (NICHD)
    National Institute of Environmental Health Sciences (NIEHS)
    National Cancer Institute (NCI)

Supporters

Burroughs Wellcome Foundation
Canadian Institutes for Health Research (CIHR)
Fonds de la Recherche en Santé du Québec (FRSQ)
Hospital for Sick Children
March of Dimes
Pfizer
Schering Plough Research Institute
U.S. Environmental Protection Agency (EPA)

Donors

McGill University, Faculty of Medicine
Merck Frosst Canada
Wyeth Ayerst Research

Adedayo Adeeko
Dept of Pharmacology & Therapeutics
McGill University
3655 Prom. Sir-William-Osler
Montreal, QC, Canada, H3G 1Y6
Tel: 514-398-6241
Fax: 514-398-7120
Email: aadeeko@pharma.mcgill.ca

Adriana Aguilar
Dept of Pharmacology & Therapeutics
McGill University
3655 Prom. Sir-William-Osler
Montreal, QC, Canada, H3G 1Y6
Tel: 514-398-3634
Fax: 514-398-7120
Email: aaguilar@pharma.mcgill.ca

R. John Aitken
School of Biological & Chemical
   Sciences
University of Newcastle
Newcastle, NSW, Australia, 2308
Tel: 61-2-49-21-57-00
Fax: 61-2-4921-623
Email: jaitken@mail.newcastle.edu.au

Sarah Ali-Khan
Dept of Pharmacology & Therapeutics
McGill University
3655 Prom. Sir-William-Osler
Montreal, QC, Canada, H3G 1Y6
Tel: 514-398-3634
Fax: 514-398-7120
Email: salikhan@pharma.mcgill.ca

Diana Anderson
Dept of Biomedical Sciences
University of Bradford
Richmond Road
Bradford, West Yorkshire, UK,
   BD7 1DP
Tel: 44-1274-233569
Fax: 44-1274-309742
Email: d.anderson1@bradford.ac.uk

Lucy Anderson
National Cancer Institute
Bldg 538, FT Dietrich, FCRDC
Frederick, MD, USA, 21702
Tel: 301-846-5600
Fax: 301-846-5946
Email: andersol@mail.nciferf.gov

Jayaprakash Aravindakshan
INRS-IAF
245 Hymus Blvd
Pointe Claire, QC, Canada, H9R 1G6
Tel: 514-630-8837
Fax: 514-630-8850
Email: jparavindak@hotmail.com

Kelly Silveira e Athayde
School of Medicine
São Paulo University
Rua Jorge Faleiros, 234
São Paulo, Brasil, 04342-110
Tel: 55899115
Email: kesa@terra.com.br

Jacques Auger
Lab. Biol. Reprod.
Pavillion Cassini
Hôpital Cochin
University of Paris
123 Bd de Port Royal
Paris, France, 75014
Tel: 33-1-58-41-15-71
Fax: 33-1-58-41-15-65
Email: jacques.auger@cch.ap-hop-
  paris.fr

Janice Bailey
Dept Sciences Animales
Université Laval
Quebec, QC, Canada, G1K 7P4
Tel: 418-656-2131, x 3354
Fax: 418-656-3766
email: janice.bailey@crbr.ulaval.ca

HW Gordon Baker
Dept of Obstetrics and Gynaecology
Royal Women's Hospital
University of Melbourne
Melbourne, VC, Australia
Tel: 61-3-9344-2130
Fax: 61-3-9347-1761
Email: g.baker@unimelb.edu.au

Rodney L. Balhorn
Lawrence Livermore National
  Laboratory
PO Box 808
Livermore, CA, USA, 94550
Tel: 925-422-6284
Fax: 925-422-2282
Email: balhorn2@llnl.gov

Natan Bar-Chama
Mt Sinai Medical Center
One Gustave L. Levy Place, Box 1272
New York, NY, USA, 10029
Tel: 212-241-8711
Fax: 212-876-3246

Johanna Barthelemy
INRS-Sante Humaine
245 Hymus Blvd

Pointe-Claire, QC, Canada, H9R 1G6
Tel: 514-630-8837
Fax: 514-680-8850
Email: johanna.barthelemy@inrs-
  sante.uquebec.ca

Marisa S. Bartolomei
University of Pennsylvania School of
  Medicine
363 CRB
415 Curie Blvd
Philadelphia, PA, USA, 19063
Tel: 215-898-9063
Fax: 215-573-6434
Email: bartolom@mail.med.upenn.edu

Tara Barton
Dept of Pharmacology & Therapeutics
McGill University,
3655 Prom. Sir-William-Osler
Montreal, QC, Canada, H3G 1Y6
Tel: 514-398-3634
Fax: 514-398-7120
Email: tbarton@pharma.mcgill.ca

Jack B. Bishop
National Institute Environmental Health
  Sciences
Laboratory of Toxicology
Bldg. 101, Room B339, MD B3-05
Research Triangle Park, NC, USA,
  27709
Tel. 919-541-1876
Fax 919-541-4634
Email: bishop@niehs.nih.gov

Jens Peter Bonde
Dept of Occupational Medicine
Aarhus University Hospital
Norrebrogade 44
DK - 8000 Aarhus C, Denmark
Email: jpbon@aaa.dk

Riana Bornman
Dept of Urology
University of Pretoria
P/B N69 Pretoria
Pretoria, Gauteng, South Africa, 0001

Tel: 27-12-3541281
Fax: 27-12-3295152
Email: bornman@medic.up.ac.za

Michael Boubelik
Institute of Molecular Genetics
Videnska 1083
Prague 4, Czech Republic, 14220
Tel: 4202-4752568
Fax: 4202-4471707
Email: boubelik@biomed.cas.cz

Martin Brinkworth
Dept of Biomedical Sciences
University of Bradford
Bradford, UK, BD7 1DP
Tel: 44-1274-233584
Fax: 44-1274-390742
Email: m.h.brinkworth@bradford.ac.uk

Elizabeth Brown
School of Human Biosciences
La Trobe University
Victoria, Australia, 3086
Tel: 61-3-9479-5869
Fax: 61-3-9479-5784
Email: e.h.brown@latrobe.edu.au

Cecil Brownie
North Carolina State University
3309 Horton Street
Raleigh, NC, USA, 27607
Tel: 919-781-2903
Fax: 919-781-4877
Email: cecilandcadell@aol.com

Daniela Buckiova
Institute of Experimental Medicine
Videnska 1083
Prague 4, Czech Republic, 14220
Tel: 4202-4752698
Fax: 4202-4752782
Email: bucki@biomed.cas.cz

Louis Bujan
CECOS Midi-Pyrenes
Hopital La Grave
Toulouse, France, 31052

Tel: 33-56-17-77-841
Fax: 33-56-17-77-843
Email: bujan.l@chu-toulouse.fr

Lucrecia Calvo
Biologia de la Reproduccion
Ecuador 1465 - 2 B
Buenos Aires, Argentina, 1425
Tel: 5411-4825-5940
Fax: 5411-4988-0339
Email: lucrecalvo@pinar.com

Celine Campagna
Dept de Sciences Animales
Université Laval
Quebec, QC, Canada, G1K 7P4
Tel: 418-656-2131, x 6338
Fax: 418-656-3766
Email: multipass00@hotmail.com

Alastair Campbell
Centre for Reproductive Biology
MRC Human Reproductive Sciences
  Unit
Edinburgh, Scotland, UK, EH3 9ET
Tel: 44-131-229-2575
Fax: 44-131-536-4963
Email: a.campbell@hrsu.mrc.ac.uk

Robert Yuk Sing Cheng
National Cancer Institute
LCC, Bldg 538, Room 205E
Frederick, MD, USA, 21702
Tel: 301-846-7044
Fax: 301-846-5946
Email: rcheng@ncifcrf.gov

Moon Koo Chung
KRICT
PO Box 107, Yuseong
Daejeon, South Korea, 305-600
Tel: 82-42-860-7476
Fax: 82-42-860-7488
Email: mkchung@krist.re.kr

Alexis Codrington
Dept of Pharmacology & Therapeutics
McGill University

3655 Prom. Sir-William-Osler
Montreal, QC, Canada, H3G 1Y6
Tel: 514-398-3634
Fax: 514-398-7120
Email: acodring@pharma.mcgill.ca

Thomas Collins
Food & Drug Administration
6903 Ridgewood Avenue
Chevy Chase MD, USA, 20815
Tel: 301-951-8617
Email: tfc@chsan.fda.gov

Daniel G. Cyr
INRS-Institut Armand-Frappier
Centre de recherche en santé humaine
245 boulevard Hymus
Pointe Claire, QC H9R 1G6
Tel. 514-630-8833
Fax 514-630-8850
Email: Daniel.Cyr@INRS-
Sante.UQuebec.CA

Magdalena Cwikiel
Dept of Oncology
University Hospital
Lund, Sweden, 22185
Tel: 46-46-177520
Fax: 46-46-176080
Email: magdalena.cwikiel@onle.iu.se

James Dahlgren
University of California, Los Angeles
2811 Wilshire
Los Angeles, CA, USA, 90403
Tel: 310-449-5522, x 226
Fax: 310-449-5526
Email: dahlgren@envirotoxicology.com

Elaine Daniel
Pharmacia Corp.
7000 Portage Road
Kalamazoo, MI, USA, 49001
Tel: 616-833-0274
Fax: 616-833-9331
Email: elaine.m.daniel@pharmacia.com

Sally Perreault Darney
US EPA
MD-72
Research Triangle Park, NC, USA,
27711
Tel: 919-541-3826
Fax: 919-541-4017
Email: darney.sally@epa.gov

Tiaan De Jager
Dept des Sciences Animales
Université Laval
Sainte Foy, QC, Canada, G1K 7P4
Tel: 418-656-2131, x 7321
Fax: 418-656-3766
Email: tdejager@medic.up.ac.za

Kevin Denny
Argus Research
905 Sheehy Drive
Horsham, PA, USA, 19044
Tel: 215-443-8710
Fax: 215-443-8587
Email: kevin.denny@primedica.com

Lori Dostal
Pfizer Global R&D
2800 Plymouth Road
Ann Arbor, MI, USA, 48105
Tel: 734-622-7524
Fax: 734-622-3478
Email: lori.dostal@pfizer.com

Yuri Dubrova
Dept of Genetics
University of Leicester
Leicester, UK, LE1 7RH
Tel: 44-116-252-5654
Fax: 44-115-252-3378
Email: yed2@le.ac.uk

Florence Eustache
Lab. Biol. Reproved.
Pavillion Cassini
Hôpital Cochin
University of Paris
123 Bd de Port Royal
Paris, France, 75014

Tel: 33-1-58-41-15-71
Fax: 33-1-58-41-15-65
Email: florence.eustache@cch.ap-hop-
paris.fr

Don Evenson
Olson Biochemistry Lab
South Dakota State University
Brookings, SD, USA, 57007
Tel: 605-688-5474
Fax: 605-688-6296
Email:donald_evenson@sdstate.edu

Christopher Ford
University Div. of Ob. & Gyn
St Michael's Hospital
University of Bristol
Southwell Street
Bristol, UK, BS2 8EG
Tel: 44-117-928-5268
Fax: 44-117-928-5290
Email: chris.ford@bristol.ac.uk

Jan M. Friedman
Dept of Medical Genetics
University of British Columbia
C234 - 4500 Oak Street
Vancouver, BC, Canada, V6H 3N1
Tel: 604-875-2157
Fax: 604-875-2376
Email: jfriedman@cw.bc.ca

Anne Golden
Mt Sinai School of Medicine
One Gustave L. Levy Place, Box 1057
New York, NY, USA, 10029
Tel: 212-241-7866
Fax: 212-360-6965
Email: anne.golden@mssm.edu

John Graham
Cedars-Sinai Medical Center
444 S. San Vincente Blvd
Los Angeles, CA, USA, 90048
Tel: 310-423-9909
Fax: 310-423-9939
Email: john.graham@cshs.org

Christine Guillemette
Dept des Sciences Animales
Université Laval
Quebec, QC, Canada, G1K 7P4
Tel: 418-656-2131, x 6283
Fax: 418-656-3766
Email: chrisguillemette@hotmail.com

Lhanoo Gunawardhana
TAP Pharmaceutical Products Inc
675 Northfield Drive
Lake Forest, IL, USA, 60045
Tel: 847-236-2815
Fax: 847-236-2239
Email: lhanoo.gunawardhana@tap.com

Utpal Gupta
Pfizer Global R&D
Eastern Point Road, Bldg 274
Groton, CT, USA, 06340
Tel: 860-441-1864
Fax: 860-686-0433
Email: utpal_gupta@groton.pfizer.com

Barbara Hales
Dept Pharmacology & Therapeutics
McGill University
3655 Prom. Sir-William-Osler
Montreal, QC, Canada, H3G 1Y6
Tel. 514-398-3610
Fax 514-398-7120
Email bhales@pharma.mcgill.ca

Jung-Yeol Han
Samsung Cheil Hospital
1-19 Mookjung-Dong, Chung-Ku
Seoul, South Korea, 100-380
Tel: 82-2-2000-7175
Fax: 82-2-2278-4574
Email: prenatal@samsung.co.kr

Bryan Hardin
Bryan Hardin & Associates
33 Office Park Road, 4A PMB 355
Hilton Head Island, SC, USA, 29928
Tel: 843-363-9466
Fax: 843-363-9466
Email: bhardin@adelphia.net

Wafa Harrouk
Food & Drug Administration
5502 Besley Court
Rockville, MD, USA, 20851
Tel: 301-827-2876
Fax: 301-695-6775
Email: mxh4@cdrh.fda.gov

Russ Hauser
School of Public Health/Occupational
  Health
Harvard University
665 Huntington Ave, Bldg 1, Rm 1405.
Boston, MA, USA, 02115
Tel: 617-432-3326
Fax: 617-432-0219
Email: rhauser@hohp.harvard.edu

Keith Hazelden
Huntingdon Life Sciences
Eye, Suffolk, UK, IP23 7PX
Tel: 44-1379-672258
Fax: 44-1379-672358
Email: hazeldek@ukorg.huntingdon.com

Laura Hewitson
Oregon Health Sciences University
505 NW 185th Ave
Beaverton, OR, USA, 97006
Tel: 503-614-3713
Fax: 503-614-3725
Email: hewitson@ohsu.edu

Stephen Hooser
Animal Disease Diagnostic Laboratory
1175 ADDL
Purdue University
West Lafayette, IN, USA, 47907
Tel: 765-494-6831
Fax: 765-494-9181
Email: shooser@purdue.edu

Masao Horimoto
Pfizer Global R&D Nagoya
5-2 Taketoyo-cho
Chita, Aichi, Japan, 470-2393
Tel: 81-569-74-4604

Fax: 81-569-74-4767
Email: masao.horimoto@pfizer.com

Mark Hurtt
Pfizer Global R&D
Eastern Point Road, Bldg 274
Groton, CT, USA, 06340
Tel: 860-715-3118
Fax: 860-686-0433
Email: mark_e_hurtt@groton.pfizer.com

Gabor Huszar
Dept of Obstetrics & Gynaecology
Yale University School of Medicine
New Haven, CT, USA, 06520
Fax: 203-737-1200
Email: gabor.huszar@yale.edu

Stewart Irvine
Centre for Reproductive Biology
MRC Human Reproductive Sciences
  Unit
37 Chalmers Street
Edinburgh, Scotland, UK, EH3 9ET
Tel: 44-131-536-2574
Fax: 44-131-536-4963
Email:d.s.irvine@hrsu.mrc.ac.uk

Keith Jarvi
University of Toronto - Mount Sinai
  Hospital
600 University Ave, Ste 1525
Toronto, ON, Canada, M5G 1X5
Tel: 416-586-8867
Fax: 416-586-8354
Email: lpleshcan@mtsinai.on.ca

Michael Joffe
Dept of Epidemiology
Imperial College School of Medicine
St Mary's Campus
Norfolk Place
London, UK, W2 1PG
Tel: 44-20-7594-3338
Fax: 44-20-7402-2150
Email: m.joffe@ic.ac.uk

Tamara Kelly
Dept of Pharmacology & Therapeutics
McGill University,
3655 Prom. Sir-William-Osler
Montreal, QC, Canada, H3G 1Y6
Tel: 514-934-4400 x 5235
Fax: 514-934-4331
Email: tkelly1@po-box.mcgill.ca

Carole Kimmel
US EPA
NCEA (8623-D)/ORD
1200 Pennsylvania Ave, NW
Ariel Rios Bldg
Washington, DC, USA, 20460
Tel: 202-564-3307
Fax: 202-565-0078
Email: kimmel.carole@epa.gov

Gary Kimmel
US EPA
Ariel Rios Bldg (8623-D)
1200 Pennsylvania Ave, NW
Washington, DC, USA, 20004
Tel: 202-564-3308
Fax: 202-565-0078
Email: kimmel.gary@epa.gov

Sasha King
Centre for Reproductive Biology
MRC Human Reproductive Sciences
  Unit
37 Chalmers Street
Edinburgh, Scotland, UK, EH3 9ET
Tel: 44-131-229-2575
Fax: 44-131-228-5571
Email: s.king@hrus.mrc.ac.uk

Mary Lee
Duke University Med. Center
Box 3080, Pediatric Endocrine, Bell
  Building
Durham, NC, USA, 27707
Tel: 919-684-3772
Fax: 919-684-8613
Email: lee@mc.duke.edu

Michelle Leonard
Pfizer Centre de Recherche
Z.I. Poce/Cisse - BP 159
Amboise, France, 37601
Tel: 33-2-47-23-77-33
Fax: 33-2-67-23-79-39
Email: michele.leonard@pfizer.com

Daming Li
Dept of Pharmacology & Therapeutics
McGill University,
3655 Prom. Sir-William-Osler
Montreal, QC, Canada, H3G 1Y6
Tel: 514-398-6241
Fax: 514-398-7120
Email: dli@pharma.mcgill.ca

Dan Livy
Texas A&M University Health Science
  Center
228 Reynolds Medical Bldg
1114 TAMU
College Station, TX, USA, 77843-1114
Tel: 979-862-1153
Fax: 979-845-0790
Email: dlivy@medicine.tamu.edu

Michaela Luconi
Dept of Clinical Physiopathology
University of Florence
Viale Pieraccini G
Florence, Italy, IT-50139
Tel: 39-055-417-1370
Fax: 39-055-427-1371
Email: m.luconi@dfc.unifi.it

Susan E Maier
Texas A&M University System Health
  Science Center
228 Reynolds Medical Bldg
1114 TAMU
College Station, TX, USA, 77843-1114
Tel: 979-862-1153
Fax: 979-845-0790
Email: smaier@medicine.tamu.edu

Francesco Marchetti
Lawrence Livermore National Laboratory

7000 East Avenue L-448
Livermore, CA, USA, 94550
Tel: 925-423-6853
Fax: 925-424-3130
Email: marchetti2@llnl.gov

Renée H. Martin
Dept of Genetics
University of Calgary
Alberta Children's Hospital
1820 Richmond Road
Calgary, AB, Canada, T2T 5C7
Tel: 403-229-7369
Fax: 403-543-9100
Email:rhmartin@ucalgary.ca

Sue Marty
Dow Chemical Company
1803 Building
Midland, MI, USA, 48674
Tel: 989-636-6653
Fax: 989-638-9863
Email: mmarty@dow.com

Donald R. Mattison
March of Dimes
1275 Mamaroneck Ave
White Plains, NY 10605
Tel: 914-997-4649
Fax: 914-428-7849
E-mail: dmattison@modimes.org

Christine McCann
University of Manchester
G38, PPT, Stoppford Building
Oxford Road
Manchester, UK, MI3 0JH
Tel: 44-161-275-5727
Fax: 44-161-275-5600
Email: christine.mccann@man.ac.uk

Marvin Meistrich
MD Anderson Cancer Center
1515 Holcombe Blvd (Box 66)
Houston, TX, USA, 77030
Tel: 713-792-4866
Fax: 713-794-5369
Email: meistrich@manderson.org

Jacqueline Moline
Mount Sinai School of Medicine
One Gustave L. Levy Place, Box 1057
New York, NY, USA, 10029
Tel: 212-241-4792
Fax: 212-996-0407
Email:jacqueline.moline@mssm.edu

Ian Morris
Biological Sciences
University of Manchester
G38 Stoppford, Oxford Road
Manchester, UK, MI3 9PT
Tel: 44-161-275-5492
Fax: 44-161-275-5600
Email: i.morris@man.ac.uk

Rebecca Morris
US EPA - NHEERL
RTD, 6EEBB, MD-72
Research Triangle Park
NC, USA, 27711
Tel: 919-541-4204
Fax: 919-541-4017
Email: morris.rebecca@epa.gov

Wolfgang-Ulrich Muller
Inst fur Med. Strahlenbiologie
Uni-Klinikum Essen
Essen, Germany, D45122
Tel: 49-201-7234152
Fax: 49-201-7235966
Email:
    wolfgang.ulrich.mueller@uni.essen.de

John Mulvihill
University of Oklahoma
940 NE 13th Street, Room B2418
Oklahoma City, OK, USA, 73104
Tel: 405-271-8685
Fax: 405-271-8697
Email: john-mulvihill@ouhsc.edu

Eve Mylchreest
Dupont Haskell Laboratory
PO Box 50
Newark, DE, USA, 19714
Tel: 302-366-6543

Fax: 302-366-5003
Email: eve.mylchreest@usa.dupont.com

Veronique Nadeau
Dept Santé Environmentale et Santé au
  Travail
Université de Montréal
CP 6128 Succ Centre-Ville
Montreal, QC, Canada, H3C 3J7
Tel: 514-343-6111, x 2072
Fax: 514-343-2291
Email: nadeav@magellan.umontreal.ca

Tetsuji Nagao
Food & Drug Safety Center (FDSC)
729-5 Ochiai, Hadano
Kanagawa, Japan, 257-8523
Tel: 81-463-82-4751
Fax: 81 463 82-9627
Email: nagaot@kb3.so-net.ne.jp

Christopher Oakes
Dept of Pharmacology & Therapeutics
McGill University
3655 Prom. Sir-William-Osler
Montreal, QC, Canada, H3G 1Y6
Tel: 514-398-6241
Fax: 514-398-7120
Email: coakes@pharma.mcgill.ca

Andrew Olshan
Dept of Epidemiology, CB #7400
University of North Carolina
Chapel Hill, NC, USA, 27599-7400
Tel: 919-966-7424
Fax: 919-966-2089
Email: andy_olshan@unc.edu

Terrence Ozolins
Pfizer Global R&D
Eastern Point Road
Groton, CT, USA, 06340-8014
Tel: 860-715-2576
Fax: 860-441-0438
Email: ozolinstr@groton.pfizer.com

Kui Lea Park
NITR KFDA

5 Nokbun-dong, Eunpyung-gu
Seoul, South Korea, 122-704
Tel: 82-2-380-1788
Fax: 82-2-380-1791
Email: parkkl@kfda.go.kr

Kenneth Pavkov
Exxon Mobil Biomedical Sciences Inc.
1545 Route 22 East,
RM LC 370, PO Box 971
Annandale, NJ, USA, 08801
Tel: 908-730-1067
Fax: 908-730-1199
Email: klpavko@erenj.com

Jana Peknicova
Institute of Molecular Genetics
Videnska 1083
Prague 4, Czech Republic, 14220
Tel: 4202-4752642
Fax: 4202-44471707
Email: jpeknic@biomed.cas.cz

Lydia Zellers Philips
Thomas Jefferson University Hospital
1025 Walnut Street, Suite 605
Philadelphia, PA, UA, 19107-9775
Tel: 215-955-1936
Fax: 215-923-1420
Email: lydia.phillips@mail.tju.edu

Jacqui Piner
Gender & Reproductive Toxicology
Glaxo Smith Kline R&D
Park Road
Ware, Herts, UK, SG12 0DP
Tel: 44-1920-882828
Fax: 44-1920-882331
Email: jac8793@glaxowellcome.co.uk

Jianping Qiu
Cornell University Medical College
1300 York Avenue, LC428
New York, NY, USA, 10021
Tel: 212-746-6229
Fax: 212-746-8835
Email: jqiu@med.cornell.edu

Koon Rha
Dept of Urology
Yonsei University
CPO Box 8044
Seoul, South Korea, 120-751
Tel: 82-2-361-5805
Fax: 82-2-312-2538
Email:khrha@yumc.yonsei.ac.kr

Bernard Robaire
Dept. Pharmacology & Therapeutics
McGill University
3655 Prom. Sir-William-Osler
Montreal, QC H3G 1Y6
Tel. 514-398-3630
Fax 514-398-7120
Email brobaire@pharma.mcgill.ca

Wendie Robbins
University of California at Los Angeles
Room 5-254 Factor Building
Mailcode 956919
Los Angeles, CA, USA, 90095-6919
Tel: 310-825-8999
Fax: 210-206-3241
Email: wrobbins@sonnet.ucla.edu

Linda Roberts
Chevron Res. & Tech. Company
PO Box 1627, 100 Chevron Way
Richmond, CA, USA, 94802-0627
Tel: 510-242-7013
Fax: 510-242-7022
Email: lgro@chevron.com

Denny Sakkas
Dept of Obstetrics & Gynaecology
Yale University School of Medicine
333 Cedar Street
New Haven, CT, USA, 06520
Tel: 203-785-4005
Fax: 203-785-7134
Email: denny.sakkas@yale.edu

Nader Salama
Dept of Urology
Faculty of Medicine
University of Alexandria

Alexandria, Egypt
Fax: 20-3-4873076
Email: nadersalama@hotmail.com

David Savitz
Dept of Epidemiology, CB #7400 UNC
University of North Carolina
Chapel Hill, NC, USA, 27514
Tel: 919-966-7427
Fax: 919-966-2089
Email: david_savitz@unc.edu

Thomas Schmid
Dept of Biomedical Sciences
University of Bradford
Bradford, UK, BD7 1DP
Tel: 44-1274-233584
Fax: 44-1274-309742
Email: t.schmid@bradford.ac.uk

Charles Scriver
Dept of Pediatrics
McGill University
2300 Tupper Street
Montreal, QC H3H 1P3
Tel. 514-934-4418
Fax: 514-934-4331
Email: mc77@musica.mcgill.ca

Shayesta Seenundun
Dept of Pharmacology & Therapeutics
McGill University
3655 Prom. Sir-William-Osler
Montreal, QC, H3G 1Y6
Tel: 514-398-6241
Fax: 514-398-7120
Email: sseenund@pharma.mcgill.ca

Cynthia Shirley
MD Anderson Cancer
9238 Rockhurst
Houston, TX, USA, 77080
Tel: 713-462-8430
Fax: 713-794-5369
Email: cynthia.r.shirley@uth.tmc.edu

Shirley Siew
Michigan State University
East Fee Hall A 627
East Lansing, MI, USA, 48824
Tel: 517-353-9160 x 245
Fax: 517-432-1053
Email: shirley.siew@nt.msu.edu

Ellen Silbergeld
University of Maryland Medical School
10 S. Pine Street
Baltimore, MD, USA, 21201
Tel: 410-706-8707
Fax: 410-706-0727
Email: esilbergeld@epi.umaryland.edu

Marcello Spano
PCO-TOSS
ENEA CR Casaccia
Via Anguillarese 301
Rome, Italy, 00060
Tel: 39-06-3048-4737
Fax: 39-06-3048-6559
Email: spanomrc@casaccia.enea.it

Francois Spezia
Pfizer Global R&D
BP 159
Amboise, France, 37401
Tel: 33-2-47-23-77-04
Fax: 33-2-47-23-79-39
Email: francois.spezia@pfizer.com

Jeanne Stadler
Pfizer Centre de Recherche
Z.I. Poce/Cisse - BP 159
Amboise, France, 37601
Tel: 33-2-47-23-77-33
Fax: 33-2-67-23-79-39
Email: jeanne.stadler@pfizer.com

Robert Tardif
Université de Montréal
CP 6128 Succ Centre-Ville
Montreal, QC, Canada, H3C 3J7
Tel: 514-343-6111, x 1515
Fax: 514-343-2200
Email: robert.tardif@umontreal.ca

Michel Thabet
Procrea Quebec
1000 chemin Ste-Foy
Quebec, QC, Canada, G1S 2L6
Tel: 418-266-2876
Fax: 418-266-2879
Email:mthab@globetrotter.qc.ca

Jacquetta Trasler
Depts. Pediatrics, Human Genetics and
    Pharmacology & Therapeutics
Montreal Children's Hospital Research
    Institute
McGill University
2300 Tupper Street
Montreal, QC H3H 1P3
Tel: 514-934-4400 x 5235
Fax: 514-934-4331
Email: mdja@musica.mcgill.ca

Bennett Varsho
WIL Research Labs
1407 George Road
Ashland, OH, USA, 44805
Tel: 419-289-8700
Fax: 419-289-3650
Email: bvarsho@wilresearch.com

Larissa Vassilev
Dept of Biology
University of Utah
Biology Building, 257 S 14008
Salt Lake City, UT, USA, 84112
Tel: 801-585-5274
Fax: 801-581-4668
Email: vassilieva@biology.utah.edu

Robert Vinson
Dept of Pharmacology & Therapeutics
McGill University,
3655 Prom. Sir-William-Osler
Montreal, QC, Canada, H3G 1Y6
Tel: 514-398-3634
Fax: 514-398-7120
Email: rvinson@pharma.mcgill.ca

Ekkehart W. Vogel
Dept of Radiation Genetics & Chemical
  Mutagenesis
University of Leiden
Wassenaarseweg 72
Leiden, PO Box 9503, The Netherlands,
  2300 RA
Tel: 31-71-527-6147
Fax: 31-71-522-1615
Email: e.w.vogel@lumc.nl

Raphael Warshaw
CHSS
2811 Wilshire Blvd, Suite 510
Santa Monica, CA, USA, 90403
Tel: 909-579-0289
Fax: 909-579-0229
Email: ray@wdds.com

Tacey White
Sanofi-Synthelabo Research
9 Great Valley Parkway
Malvern, PA, USA, 19355
Tel: 610-889-8884
Fax: 610-889-6828
Email: tacey.white@sanofi-
  synthelabo.com

Andrew J. Wyrobek
Lawrence Livermore National
  Laboratory
7000 East Avenue L-448
Livermore, CA, USA, 94550
Tel: 925-422-6296
Fax: 925-424-3130
Email: wyrobek1@llnl.gov

Ryuzo Yanagimachi
The Institute for Biogenesis Research
University of Hawaii at Manoa
1960 East-West Rd, East Annex E124
Honolulu, HI, USA, 96822
Tel: 808-956-8746
Fax: 808-956-7316
Email:yana@hawaii.edu

Ashraf Youssef
TAP Pharmaceutical Products Inc
675 North Field Drive
Lake Forest, IL, USA, 60045
Tel: 847-317-3329
Fax: 847-236-2239
email: ashraf.youssef@tap.com

Katia Zubkova
Dept of Pharmacology & Therapeutics
McGill University
3655 Prom. Sir-William-Osler
Montreal, QC, H3G 1Y6
Tel: 514-398-6241
Fax: 514-398-7120
Email: kzubkova@pharma.mcgill.ca

Top row: Diana Anderson, Jack Bishop, Carole Kimmel
Middle row: Jan Friedman, Christopher Peters, Masao Horimoto

Top row: Renée Martin, Sally Perreault, Wendie Robbins, Bryan Hardin
Bottom row: Martin Brinkworth, Yuri Dubrova, Andrew Wyrobek, Bernard Robaire

Top row: Diana Lucifero, Kelly Silveira e Athayde, Robert Vinson,
Sarah Ali-Khan, Katia Zubkova
Middle row: Barbara Hales, Bernard Robaire, Marvin Meistrich, David Savitz

Top row; Franceso Marchetti, Renée Martin, Jens Peter Bonde
Middle row: Bernard Robaire, Diana Anderson, Jacquetta Trasler, Barbara Hales
Bottom row: Daniel Cyr, Francesco Marchetti, Sarah Ali-Khan, Diana Anderson, Jianping Qiu

CONTENTS

EXPOSURES AND EFFECTS: CAUSES OF CANCER AND
CONSEQUENCES OF TREATMENT

PREGNANCY OUTCOME

FEMALE-SPECIFIC REPRODUCTIVE TOXICITIES FOLLOWING PRECONCEPTION EXPOSURE TO XENOBIOTICS

Jack B. Bishop

NIEHS
111 TW Alexander Dr.
Research Triangle Park, NC 27709

INTRODUCTION

There are only a few testing strategies directed specifically at female reproductive toxicity. (Generoso et al, 1971; Generoso and Cosgove, 1973; Bishop et al., 1997). There are some elegant molecular studies of female germ cell and reproductive biology (Eichenlaub-Ritter, et al., 1988; Albertini, et al., 2001; Hunt et al., 1995) but few, if any, of these investigations involve toxicological perspectives. Future research efforts in female, as well as male, reproductive toxicology would benefit greatly by enhanced integration of knowledge and methodology from molecular developmental and cell biology. In this chapter, I will review some basic reproductive biology, emphasizing differences between females and males, describe tests for female reproductive toxicology used by the National Toxicology Program (NTP) at NIEHS, summarize NTP test data and discuss some of the dogmas of germ cell mutagenesis that have recently been dispelled by these data.

Features of female reproduction clearly differ from those of males even before birth (Figure 1). By the time a female is born, she has all of the oocytes she will ever possess, with most of those originally present in the ovary lost, or destined to be lost shortly, through apoptosis. All of those oocytes are arrested in the late meiotic, diplotene stage of oogenesis with each cell being diploid (2N) and containing 4 copies (4C) of the DNA. Meiotic division is normally completed only after ovulation and fertilization. From puberty onward males, on the other hand, have continuous cycles of spermatogenesis with mitotically dividing 2N-2C spermatogonia through meiotic 4N-2C spermatocytes to postmeiotic 1N-1C spermatids and sperm. In mice, the chromatin of the arrested diplotene stage oocyte is in a diffuse dictyate state in which the chromatin is relatively relaxed and accessible. The accessibility of the chromatin of male germ cells varies widely depending on the particular stage of spermatogenesis with the chromatin of mature sperm being highly compacted and bound by crosslinks in cystiene-rich protamine. Furthermore, oocytes as well as premeiotic, meiotic and early postmeiotic sperm are all DNA repair competent; late spermatids and spermatozoa, however, lack DNA repair enzymes and damage to their DNA can not be repaired until the sperm fertilizes an egg.

Advances in Male Mediated Developmental Toxicity, edited by Bernard Robaire and Barbara F. Hales.
Kluwer Academic/Plenum Publishers, 2003.

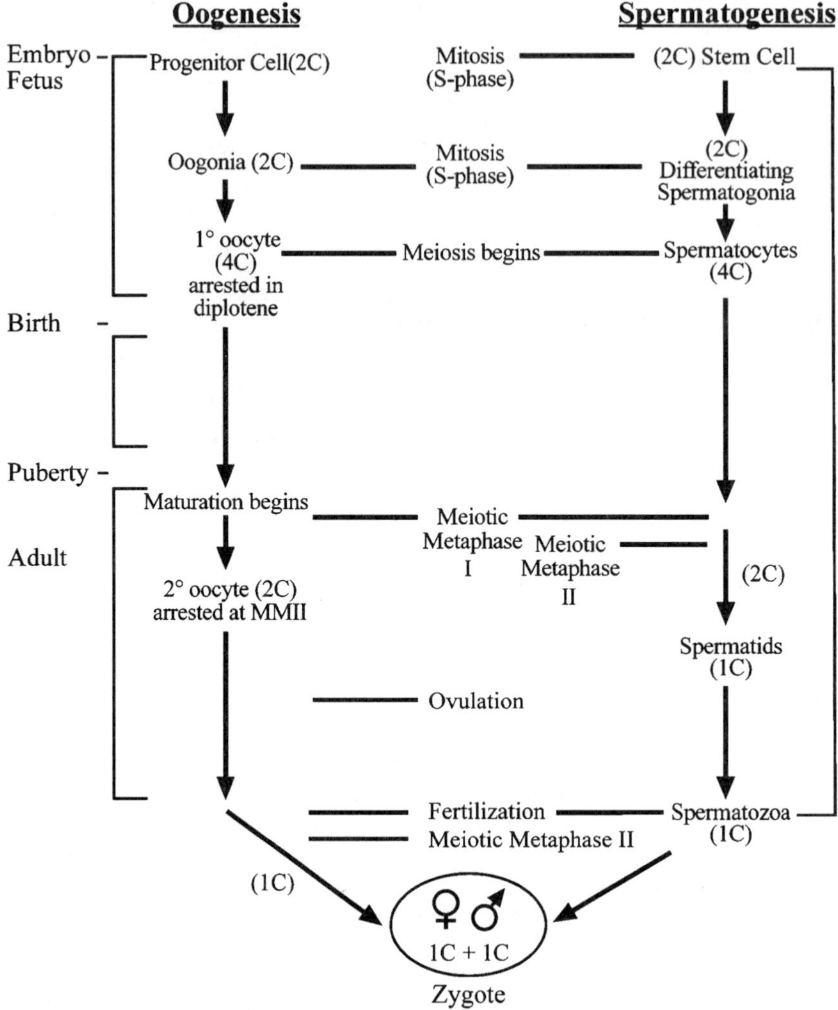

Figure 1. Cell division and DNA replication/reduction during oogenesis and spermatogenesis. Chromosme complement is indicated as 4C, 2C or 1C.

NTP ASSAYS FOR IDENTIFYING FEMALE REPRODUCTIVE TOXICANTS

There are three assays currently used by the NTP which allow us to detect the effect of toxicants upon female reproduction: reproductive assessment by continuous breeding (RACB); the total reproductive capacity test (TRCT); and the dominant lethal test (DLT).

Reproductive Assessment By Continuous Breeding (RACB)

The RACB is a multigeneration reproductive test and is the most routinely conducted test for reproductive toxicity used by the NTP. Both rats and mice are used with the species determined on the basis of the data needed. Over 125 chemicals have been tested in the RACB to date and ~5 additional chemicals are currently tested each year. The RACB test protocol (Figure 2) has evolved over the years. It currently involves daily, multi-dose chemical treatment (plus vehicle for controls) of F_0 males for four weeks (20

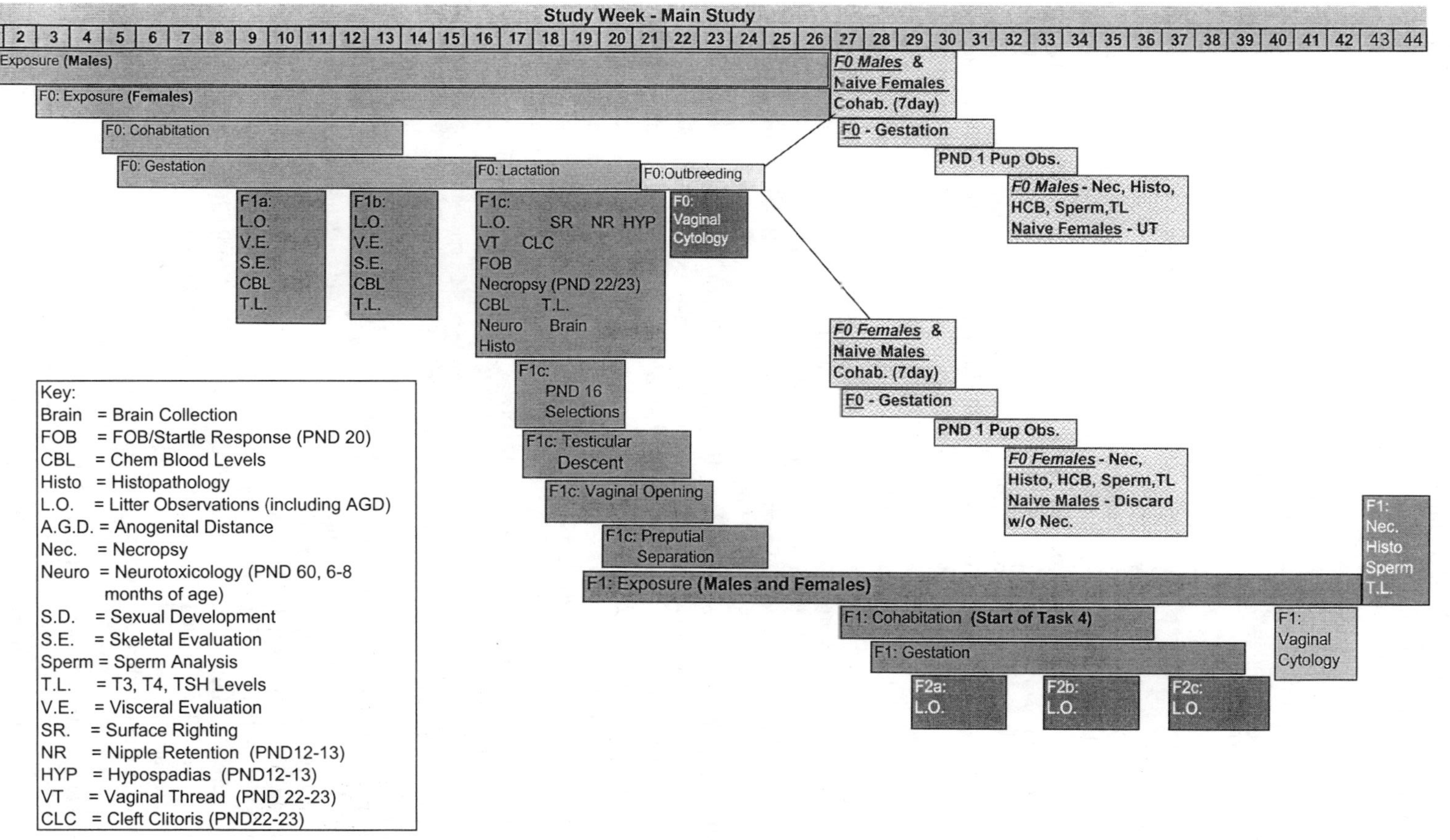

Figure 2. Reproductive assessment by continuous breeding (RACB) protocol.

per dose) and F_0 females for two weeks (20 per dose) prior to cohabitation. These F_0 animals are treated daily for an additional 18-19 weeks during which time they produce 3 F_1 litters. If an adverse reproductive effect is observed, an "outcross" test may be conducted to determine whether the effect is due to the male, female or both. In the "outcross" test, F_0 high dose males are mated to F_0 control females and F_0 high dose females are mated to F_0 control males, and their litters are evaluated as before. At necropsy, the reproductive tracts of F_0 animals are subjected to thorough histo-pathological examination, and standard vaginal cytology and sperm examinations (count, viability, motility) are conducted.

Two of the F_1 litters (generally the first 2) are discarded on post natal day 1 (PND1) after evaluation of numbers, pup weight and anogenital distance. A subset of these PND1 pups are subjected to examination for visceral malformations and their carcasses are preserved in 10% formalin for subsequent skeletal evaluation if desired. In some studies, pup blood and serum may be evaluated for chemical and hormone levels, respectively. One F_1 litter is kept (generally the third) and exposed through lactation until weaning and then treated daily after weaning. In addition to evaluation of weights and anogenital distance, these F_1 pups may also be evaluated for other developmental traits such as time of vaginal opening, preputial separation and testicular descent. Other special tests such as immunological competence, neurological function and/or brain examinations may also be included. Twenty F_1 males and 20 F_1 females from each dose group are selected for cohabitation to produce three litters of an F_2 generation. The F_2 pups are discarded after recording numbers, pup weight and anogenital distance. When the F_1 animals are necropsied, reproductive tract histo-pathological examination, and vaginal cytology and sperm evaluations are performed as they were with the F_0 animals.

The adverse reproductive effects most often seen in RACB studies include decreased numbers of pups per litter, decreased numbers of litters, decreased pup weights and altered sex ratios. Decreased pups per litter is the parameter most likely to relate to direct germ cell damage and/or untoward developmental effects in utero. However, changes in any number of the endpoints measured in the RACB can reflect a variety of non-germ cell effects such as general maternal and/or uterine toxicity, pituitary-gonadal or other hormonal alterations in parents as well as in offspring, etc.. Of 125 chemicals studied in the RACB over the past ~20 years, 68 were found to induce some form of reproductive toxicity in females. About 1/3 of those were thought to cause reproductive toxicities in females but not males and about 1/2 involved reproductive toxicity that included a reduction in the number of pups per litter.

Total Reproductive Capacity Test

The TRCT (Bishop et al., 1997; Generoso et al., 1971; Generoso and Cosgove, 1973) has only been conducted with mice. It involves an acute (generally a single up to a maximum of five) intraperitoneal injection of a chemical at relatively high maximum tolerated (MTD) and/or half MTD doses. Treated females are then single-pair mated with an untreated male for their reproductive life-span (at least 18 19-day intervals covering 347 days; Figure 3). Their litters are recorded and discarded at birth. Effects are measured as a significant reduction in the total number of litters per female, the total numbers of offspring per female, and/or the number of offspring per female in a mating interval. A significant reduction in the number of offspring per female in the first and/or second mating intervals is often associated with dominant lethal effects (Bishop et al., 1997). Positive TRCT results have been further characterized through follow-up female DLT studies as well as via histological examination of ovarian sections to determine whether there are reduced numbers of healthy primordial oocytes and/or early to late growing follicles.

The long-term reproductive assessment feature of the TRCT procedure is its most unique feature. It provides a capacity for detecting a range of toxic insults upon reproduction, from follicular toxicities to dominant lethality. A total of 46 chemicals have been tested in the TRCT to date (Bishop et al., 1997 and unpublished results; Generoso et al., 1971; Generoso and Cosgove, 1973; Sudman et al., 1991). While many of the chemicals tested have been shown to affect the reproductive performance of females through mutagenic effects on oocytes and/or cytotoxic effects on follicles, no causative mechanism could be identified for the observed reduction in female reproductive performance for some of them.

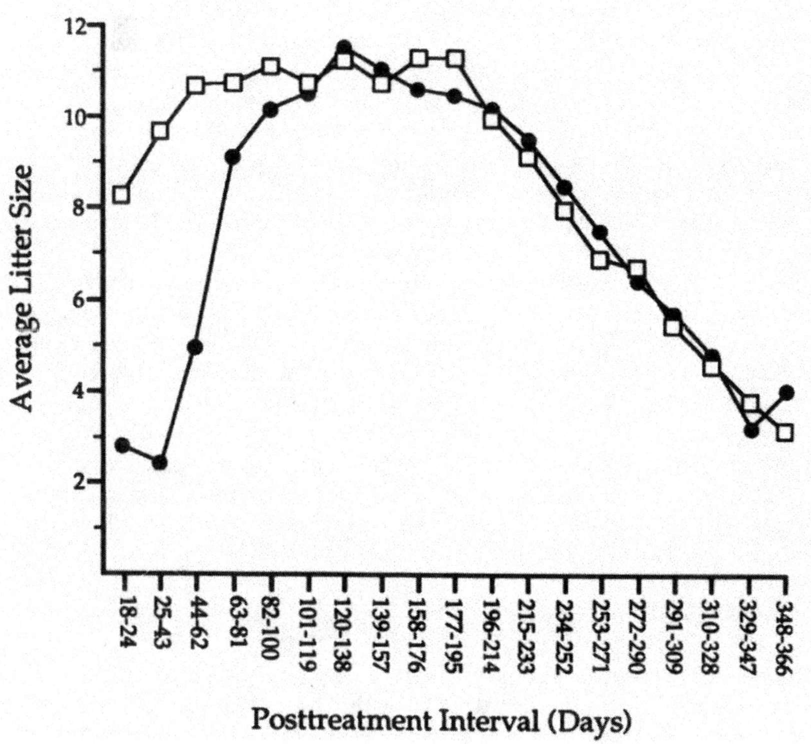

Figure 3. Effects of bleomycin (BLM) treatment on reproductive performance of treated (SECxC57BL) F1 females mated to (C3H/R1xC57BL) F1 males. The first interval represents litters conceived within the first 4.5 days post-treatment; the remaining intervals represent data pooled from successive 19-day intervals.

Dominant Lethal Tests (DLTs)

(DLTs, which are intended to measure mutagenic damage in the germ cell, have been conducted most often with male mice. However, some DLTs have been conducted with male rats, and likewise, with female mice. Treatments often involve acute dosing, frequently by intraperitoneal injection; but a variety of routes have been used including oral gavage, drinking water, dermal, inhalation, and subcutaneous and intravenous injections. Subchronic exposures from 3 days to 8 weeks are also reported. The treated males (or females) are cohabitated with untreated females (or males) over 1- to 5-day mating intervals to sample specific exposed germ cell stages. When subchronic exposures are employed, only one or two 5-day matings are required, and thus fewer animals are needed; but information about spermatogenic stage sensitivity is obviously lost. Pregnant females are humanely sacrificed between mid-gestation (~day 14) and near term (~day 17 or 18), and their uterine contents examined for total number of implantations, live fetuses (both normal and abnormal), late dead fetuses, late embryonic death, and early embryonic death and resorption moles (Lockhart et al., 1987). When untreated females were mated to males treated with the alkylating chemical methylmethane sulfonate, the frequencies of resorption moles in utero at mid-gestation and unstable chromosome aberrations (such as terminal deletions, dicentrics and acentric fragments) observed in first-cleavage embryos, were found to be highly correlated (Brewen et al., 1975). This suggests that resorption moles representing "dominant lethal effects" result from DNA lesions induced in the male germ cells that are fixed in the zygote as unstable chromosome aberrations which lead to postimplantation early embryonic death. Recent data from analysis of chromosome damage with chromosome painting support this result (Marchetti et al., 1997, 2001).

Space requirement is probably the greatest deterrent to conducting rat DLTs. At least 30 pregnant females (~300 total implantations) per dose and germ cell stage sampling time are required for an adequate dominant lethal test. A thorough examination of all spermatogenic stages at just the MTD and control doses may require more than 4000 animals; such studies with rats are obviously more costly than similar studies with mice. Nevertheless, certain chemicals like dibromochloropropane have been shown to induce dominant lethals in male rats but not male mice (Teramoto et al., 1980), and rats remain an important model for such studies.

For female DLTs, there are other scientific and technological challenges. The most obvious is the logistical issue of the number of conceptus evaluated per treated female; while a treated male can be mated to multiple females for analysis of effects upon his germ cells, the analysis of female germ cells is restricted to those individuals treated. Further, there can be confounding effects of maternal toxicity upon the measure of direct germ cell toxicity. To help sort out such confounding effects, cytogenetic analysis of oocytes or first-cleavage embryos and/or reciprocal egg-transfer experiments may have to be conducted. These types of studies raise the level of complexity for a thoroughly conducted female dominant lethal test much higher than that of the male dominant lethal test.

GERM CELL MUTAGENESIS DOGMA

There have been a number of dogmas in germ cell mutagenesis dispelled by dominant lethal tests conducted during the last decade of the Twentieth Century. Three of these dogmas ate illustrated by the patterns of in vivo genetic toxicities shown in Table 1. Most germ cell tests have been predicated on the observation of positive results in somatic cell tests, especially in vivo bone marrow (BM) chromosome aberration (ABS) or micronucleus (MN) tests. Thus, it may not be surprising that the DLT positives are also BM/MN/ABS positive. The first of the 3 illustrated dogmas that all germ cell mutagens are somatic cell mutagens, which was recently affirmed by Tinwell et al. (2001) and Ashby and Tinwell (2001), has more recently been disproven by Witt et al. (2002) in a male mouse DLT study of N-hydroxymethyl acrylamide. Doses of 360-720 ppm N-hydroxymethyl acrylamide in drinking water induced significant, dose-related (30-50%) increases in dead implants, with no indication of any increase of MN in either the bone marrow or peripheral blood of the treated males.

Two additional, related dogmas are: all morphological specific locus test (MSLT) mutagens are dominant lethal test (DLT) mutagens; and all male pre-meiotic/gonia/stem cell mutagens are male post-meiotic mutagens. The pattern of MSLT-to-DLT mutagenicity is influenced by the predominant practice of only testing DLT positive chemicals in the MSLT, just as the relationship of DLT-to-MN results is influenced by MN positive chemicals being selected for testing in the DLT. Further, most chemical mutagens tested to

Table 1: Patterns of In Vivo Genetic Toxicity in Rodents[a]

CHEMICALS	BM MN/ABS	DLT	HTT	MSLT
4-AAF	-	-		
Pyrene	-	-		
THC	-	-	-	
Caffeine	-	-	-	
EDB	-	-		
DBCP	-	-		
2-AAF	+	-		
Urethane	+	-		
Vincristine	+	-		
Adriamycin	+	-[b]		
Platinol	+	-[b]		
HEMPA	+	-[b]		
Hycanthone		-[b]		
B(a)P	+	+	-	
6-MP	+	+	-	
Myleran	+	+		
Ethylene oxide	+	+	+	
Acrylamide	+	+	+	
MethylenebisACR	+	+	+	
EthylenebisACR		+	+	
EMS	+	+	+	+/-
Cyclophosphamide	+	+	+	+/-
Chlorambucil	+	+	+	+/-
Melphalan		+		+/-
Diethylsulfate		+		+/-
Mitomycin C	+	+	+	?/+
Ethylnitrosourea	+	+	+	+/+
Methylnitrosourea		+	+	+/+
Procarbazine	+	+	+	+/+
TEM	+	+	+/+	+/+

[a] Heritable Translocation Test (HTT) and Morphological Specific Locus Test (MSLT) double entry responses are indicated as post-meiotic/premeiotic; single entries for the Dominant Lethal Test (DLT) and HTT) are for post-meiotic germ cell stages.

[b] Positive DLT response in female mice were demonstrated to be genetic via egg transfer experiments and/or analysis of first cleavage metaphase **chromosomes.**

date have been found to be positive in the DLT and/or MSLT in post-meiotic stages of spermatogenesis with only a few found to be positive in pre-meiotic stages. Until recently, all of the chemicals found to be positive in pre-meiotic stages of spermatogenesis were also found to be positive in post-meiotic stages; with the single exception of the DNA base analogue, anti-cancer drug, 6-mercaptopurine, which exhibited a somewhat unusual pattern of dominant lethality affecting only a narrow window in preleptotene spermatocytes (Generoso et al. 1975).

Both of these dogma were contradicted by the Bleomycin MSLT results of Russell et al. (2000b). Sudman et al. (1992) had previously tested the anti-cancer drug Bleomycin in a male mouse DLT and found it negative. Russell et al. (2000b), however, found that Bleomycin significantly increased the frequency of MSLT mutations recoverd from exposed spermatogonia; they also reconfirmed the lack of Bleomycin mutagenicity in post-meiotic stages for both the DLT and the MSLT. The topoisomerase inhibitor etoposide has also been shown in recent studies to induce dominant lethal (Shelby et al., 2001) and specific locus mutations (Russell et al., 1998) in premeiotic primary spermatocytes but not in post-meiotic stages of spermatogenesis. Etoposide has also been shown to induce chromosome aberrations and aneuploidy in exposed primary spermatocytes, as detected through analysis of 1st cleavage metaphase chromosome by fluorescence in situ hybridization chromosome painting techniques (Marchetti et al., 2001). Russell et al (2000a) also demonstrated that etoposide exposure of primary spermatocytes alters the meiotic recombination frequency.

One final dogma in germ cell mutagenesis was that female germ cells are much less sensitive to mutagenic damage than male germ cells. Most chemical mutagens which were positive in male germ cells either produced a lesser response in female germ cells or were negative in females. All chemical mutagens found to induce dominant lethals in female germ cells had also been found to induce dominant lethals in male germ cells. However, Generoso and colleagues (Katoh et al., 1990; Sudman and Generoso, 1991; Sudman et al., 1992) have subsequently identified several environmental chemicals that were apparent female specific dominant lethal mutagens.

SUMMARY AND CONCLUSIONS

Females do differ from males in their germ cell and general reproductive responses to toxicants. Chromatin structure is perhaps one factor that contributes to sex differences in germ cell response to toxicants. Differences in available targets likely contribute to response differences between sexes as well as to within sex differences between germ cell stages. It is important to consider these differences when conducting reproductive toxicity studies and interpreting the results.

REFERENCES

Albertini, D.F., Combelles, C.M., Benecchi, E., Carabatsos, M.J., 2001, Cellular basis for paracrine regulation of ovarian follicle development. *Reproduction.* 121:647-653.

Ashby, J., and Tinwell, H., 2001, Continuing ability of the rodent bone marrow micronucleus assay to act as a predictor of the possible germ cell mutagenicity of chemicals. *Mutat Res.* 478:211-213.

Bishop, J.B., Morris, R.W., Seely, J.C., Hughes, L.A., Cain, K.T., Generoso, W.M., 1997, Alterations in the reproductive patterns of female mice exposed to xenobiotics. *Fundamental and Applied Toxicology.* 40:191-204.

Brewen, J.G., Payne, H.S., Jones, K.P., Preston, R.J., 1975, Studies on chemically induced dominant lethality: I. The cytogenetic basis of MMS-induced dominant lethality in postmeiotic male germ cells. *Mutat Res.* 33:239-249.

Eichenlaub-Ritter, U., Chandley, A.C., Gosden, R.G.., 1988, The CBA mouse as a model for age-related aneuploidy in man: studies of oocyte maturation, spindle formation and chromosome alignment during meiosis. *Chromosoma.* 96:220-226.

Generoso, W.M., Stout, S.K., Huff, S.W., 1971, Effects of alkylating chemicals on reproductive capacity of adult female mice. *Mutat Res.* 13:171-184.

Generoso, W.M., and Cosgove, G.E., 1973, Total reproductive capacity procedure in female mice. In Chemical Mutagenesis - Principles and Methods for Their Detection (A Hollaender ed) pp. 241-258. Plenum, New York.

Generoso, W.M., Preston, R.J., Brewen, J.G., 1975, 6-mercaptopurine, an inducer of cytogenetic and dominant-lethal effects in premeiotic and early meiotic germ cells of male mice. *Mutat Res.* 28:437-447.

Hunt, P., LeMaire, R., Embury, P., Sheean, L., Mroz, K., 1995, Analysis of chromosome behavior in intact mammalian oocytes: monitoring the segregation of a univalent chromosome during female meiosis. *Hum Mol Genet.* 4:2007-2012.

Katoh, M.A., Cain, K.T., Hughes, L.A., Foxworth, L.B., Bishop, J.B., Generoso, W.M., 1990, Female-specific dominant lethal effects in mice. *Mutat Res.* 230:205-217.

Lockhart, A.C., Bishop, J.B., Piegorsch, W.W., 1991, Issues regarding data acquisition and analysis in the dominant lethal assay. Proceedings of the Biopharmaceutical Section of the American Statistical Association, pp. 234-237.

Marchetti, F., Bishop, J.B., Lowe, X., Generoso, W.M., Hozier, J., Wyrobek, A.J., 2001, Etoposide induces heritable chromosomal aberrations and aneuploidy during male meiosis in the mouse. *Proc Natl Acad Sci USA.* 98:3952-3957.

Marchetti, F., Lowe, X., Bishop, J. and Wyrobek, A.J., 1997, Induction of chromosomal aberrations in mouse zygotes by acrylamide treatment of male germ cells and their correlation with dominant lethality and heritable translocations. *Environ Mol Mutagen.* 30:410-417.

Russell, L.B., Hunsicker, P.R., Johnson, D.K., Shelby, M.D., 1998, Unlike other chemicals, etoposide (a topoisomerase-II inhibitor) produces peak mutagenicity in primary spermatocytes of the mouse. *Mutat Res.* 400:279-286.

Russell, L.B., Hunsicker, P.R., Hack, A.M., Ashley, T., 2000a, Effect of the topoisomerase-II inhibitor etoposide on meiotic recombination in male mice. *Mutat Res.* 464:201-212.

Russell, L.B., Hunsicker, P.R., Kerley, M.K., Johnson, D.K., Shelby, M.D., 2000b, Bleomycin, unlike other male-mouse mutagens, is most effective in spermatogonia, inducing primarily deletions. *Mutat Res.* 469:95-105.

Shelby, M.D., Bishop, J.B., Hughes, L.A., Morris, R.W., Generoso, W.M., 2001, Primary spermatocytes are the main targets for induction of etoposide dominant lethals and heritable translocations in male mice. *Mutat Res.* (in press)

Sudman, P.D., and Generoso, W.M., 1991, Female-specific mutagenic response of mice to hycanthone. *Mutat Res.* 246:31-43.

Sudman, P.D., Rutledge, J.C., Bishop, J.B., Generoso, W.M., 1992, Bleomycin: female-specific dominant lethal effects in mice. *Mutat Res.* 296:143-156.

Teramoto, S., Saito, R., Aoyama, H., Shirasu, Y., 1980, Dominant lethal mutation induced in male rats by 1,2-dibromo-3chloropropane (DBCP). *Mutat Res.* 77:71-78.

Tinwell, H., Brinkworth, M.H., Ashby, J., 2001, Further evidence for the rodent bone marrow micronucleus assay acting as a sensitive predictor of the possible germ cell mutagenicity of chemicals. *Mutat Res.* 473:259-261.

Witt, K.L., Hughes, L.A., Burka, L.T., Bishop, J.B., 2002, The mouse micronucleus assay is not a predictor of the N-hydroxymethyl acrylamide induced germ cell mutagenicity seen in the mouse dominant lethal assay. *Mutat Res.* (in press)

OVERVIEW OF MALE-MEDIATED DEVELOPMENTAL TOXICITY

Diana Anderson

Department of Biomedical Sciences
University of Bradford
Bradford
West Yorkshire, BD7 IDP, U.K.

INTRODUCTION

In recent years, the public has become more aware that exposure of males to certain agents can adversely affect their offspring; for example, fathers who smoke appear to give rise to tumours in the F_1 generation (Sorahan et al., 1997a and b; Ji et al., 1997). Savitz (1994) has reported an increased incidence of miscarriages after potential exposure to a variety of agents. Also, Lefebvre et al. (1998) have shown that the paternally transmitted and paternally imprinted gene, MEST, is involved in normal maternal behaviour. MEST-deficient females show abnormal behaviour and intrauterine and postnatal growth retardation of progeny. This is even more evidence of how important the male is to the successful development of the future generation.

Current guidelines for regulatory testing require that only the female is tested for teratogenic effects. However, since the male contributes half of the genetic information of the genome to developing offspring, then males could also be examined for induced "teratogenic" effects (congenital malformations). Transplacental carcinogenesis is recognised in the female, but carcinogenesis mediated through the male germ cells is not so well appreciated and understood. Congenital malformations and cancer could arise after exposure of males to both chemicals and physical agents, such as radiation.

Radiation Exposure

The hazards associated with the dangers of exposure to ionising radiation have been recognised for nearly a century, but interest was aroused when a cluster of leukaemia cases was identified in young children living in Seascale, close to the nuclear processing plant at Sellafield in West Cumbria (Black, 1984). Clusters were sought and found in the vicinity of other nuclear establishments in the UK and other countries (Roman et al., 1987, Cook-Mozaffari et al., 1994, Bithell et al., 1994). Leukaemia clusters are not found exclusively in the vicinity of nuclear reactors (Buckley et al., 1989). The report of Gardner et al. (1990 a,b.), however, suggested that occupational exposure of men at Sellafield might be linked to increased susceptibility to leukaemia or non-Hodgkin's lymphoma in children. COMARE (1996), the Committee on Medical Aspects of Radiation in the Environment appointed by the UK Department of Health, could find no epidemiologial evidence from studies in other locations, elsewhere in Cumbria (Wakeford and Parker, 1996), near to a similar plant at Dounreay in Scotland (Urquhart et al., 1991), around the Aldermaston and Burghfield nuclear weapons establishments in England (Roman et al., 1993) or from Ontario, Canada (Kinlen et al., 1993). In spite of these negative surveys, the Gardner report was used in a civil court case on behalf of two of the alleged victims of paternal irradiation at Seascale against British Nuclear Fuels (Wakeford and Tawn, 1994). The case foundered on "the balance of probabilities" (Doll et al., 1994). A survey by Parker et al. (1993) and parallel studies carried out by the HSE (1993, 1994) corroborated this report – see also Draper et al. (1997). Gardner acknowledged that possible exposure to internally incorporated radionuclides had not been taken into account and such sources could possibly explain their findings. The relative risk factor for leukaemia and non–Hodgkin's lymphoma for all surveys other than Seascale were not significantly different from 1.0 (Little et al., 1994b and 1996), however, for children born to exposed residents of Seascale, the risk was 36 times higher than the control level.

In 1993, Roman and his colleagues, in a small case–control study among children living in West Berkshire/North Hampshire also came to the conclusion that pre-conception, paternal irradiation might lead to an increased risk of cancer. As early as 1966, Graham et al. had come to the same conclusion from a study following diagnostic X-irradiations, but after maternal exposure. However, The Oxford Survey (Kneale and Stewart, 1980) found no association in 4542 children who died of cancer with parental exposure. Studies of the F_I population after the atomic bombs in Hiroshima and Nagasaki revealed no increase in malignancy with increasing parental gonadal dose (Yoshimoto, 1990; Yoshimoto et al., 1991, Little et al., 1994a). In 1800 offspring from male cancer patients who received radiotherapy in the UK (Hawkins et al., 1989) and the USA (Li et al., 1979; Mulvihill et al., 1987), only heritable retinoblastomas were found. Dubrova (1996) reported that radiation exposure at Chernobyl had induced heritable mutations in the male germ line.

There was evidence for the transmission of abnormalities and susceptibility after paternal exposure from experimental X-ray studies in mice (Nomura, 1982; Kirk and Lyon, 1984). In Nomura's study, the tumours were clearly heritable as shown by F2 transmission. There were increases in leukaemia up to 18-fold for one strain, when radiation was given at the spermatogonial stage (Nomura, 1991). These findings have been criticised (Selby 1990; Cox 1992), because of a lack of simultaneous controls, a seasonal variation in tumourigenicity, a small level of pneumonia in the experimental groups other than the control group, and a possible stimulation of pre-existing tumour pre-disposing mutations. Cattanach et al. (1995), in experiments which mimicked Nomura's, but not using the same strain, could not confirm his findings, and Cosgrove et al. (1993) found that offspring of males whose germ cells were irradiated when they were spermatogonia lived normal lifespans. Takakhashi et al. (1992) reported inheritance by first generation male offsring, but not female, of a cancer–prone genetic trait after irradiation of the fathers with californium-252 neutrons. Nomura et al. (1983) found that a subsequent challenge to the offspring with urethane, known to induce lung tumours, stimulated large clusters of tumour nodules. Cattenach et al. (1998), again mimicking Nomura's work, failed to corroborate these findings. However, similar results to those of Nomura et al. (1983) were obtained by Vorobotsova and Kitaev (1988) and Vorobtsova et al. (1993) who showed that a second insult could promote lung and skin cancers, respectively.

Lord et al. (1998 a,b) treated male mice with plutonium-239 and generated offspring in which haemopoiesis was assayed. For the secondary insult, female mice were treated with either methyl nitrosourea (MNU - a leukemia-inducing drug) or a leukemia-inducing sub-lethal dose of 3.3 Gy γ-rays. To overcome criticisms of earlier work, parallel control

Table 1. Responses in different studies in rats and mice

Compound	Species	Treatment			Endpoints		F₁ karyotype		F₁ Tumours	Refs
		Acute	Sub-Chronic	Chronic	F_0 Dominant lethal mutations	F_1 Congenital malformations	Foetus	Adult	Adult	
Cyclophos-phamide	Mouse[a]	♂			+	+	−	ND	ND	1
	Rat[b]			♂	+	+	+	+	±	2,3
1,3-butadiene	Mouse[a]	♂			+	−	−	ND	−	4
	Mouse[a]		♂		+	+/−	−	ND	−	5,6
	Rat[b]		♂		−	ND	ND	ND	ND	4
Urethane	Mouse[a]	♂			−	−	−	−	+[c]	7
			♂		−	−	−	−	−	7

[a]CD-1 mice; [b]Sprague-Dawley rats; [c]Males only.

*+=statistically significant increase above untreated controls; − = no statistically significant increase; ± = equivocal response; +/− = one study positive, one negative; ND = not done.

Refs: 1-Jenkinson et al, 1987, 2-Jenkinson and Anderson, 1990, 3-Francis et al., 1990, 4-Anderson et al., 1998, 5-Anderson et al., 1996, 6-Brinkworth et al., 1998, 7-Edwards et al., 1999.

groups and two different mouse strains with a spontaneous leukemia incidence rate of zero were used at two different centres. Each mouse could be considered as an individual. The inter-animal variation was larger than usual and the balance of haemopoiesis in offspring was disturbed in a significant number of mice. These mice also showed a significant trend to higher levels of chromosomal aberrations in bone marrow cells. In mice secondarily insulted with MNU or sub-lethal radiation, there was a significant increase in the rates of lympho-haemopoietic malignancy and a change in the disease patterns, e.g., myeloid leukemia now predominated, whereas in MNU-treated mice it normally only developed secondarily to thymic tumours. Thus, on balance, experimental evidence from radiation studies does suggest the possibility of transmission of paternally-mediated congenital effects.

Chemical Exposure

Certainly, there is evidence from chemical exposure for such paternally-mediated effects. Such effects have been found in experimental animal models after exposure of the males to drugs such as cyclophosphamide. (Trasler et al., 1985, 1986, 1987, Trasler and Robaire, 1988; Qiu et al., 1992, 1995; Robaire and Hales, 1994; Hales and Robaire, 1994, Brinkworth, 2000) or to environmental chemicals such as lead or dibromochloropropane (Uzych, 1985; Whorton et al., 1979). In man, various studies (Buffler et al., 1982; Fried et al., 1987; Potashnik and Yanai-Inbar, 1987; Sentura et al., 1985) have failed to show an increase in foetal malformations in children fathered by men exposed to chemicals, whilst other studies have indicated such effects (Gulati et al., 1986; Cohen et al., 1974; Infante et al., 1976). Olshan and Schnitzer (1994) listed various birth defects after occupational exposure of fathers in various industries. Savitz (1994) reported an increased incidence of miscarriage after fathers' occupational exposure.

Congenital malformations and tumours can be studied after exposure of male rodents in an extended dominant-lethal assay where untreated females mated to treated males are examined the day before term, as opposed to mid-term in the conventional study (Knudsen et al., 1977). At this stage, congenital malformations, such as hydrocephaly, exencephaly, cleft palate, open eye, runts (dwarfs), oedema, anasarca and gastroschisis can be determined. Some of these abnormalities have similar manifestations in humans. The foetuses can also be examined for skeletal malformations by using alizarin staining. If the F_0 treated and control males are mated with more than one female, then in the F_1 generation, litters of the extra female(s) can be examined for the same effects in live-born offspring, confirming the original observation. Litters can also be allowed to develop to adulthood where tumours can be observed and karyotype analysis can be performed on fetuses and adult offspring to determine if induced genetic damage can be transmitted.

By using this type of study design, Anderson and co-workers have examined cyclophosphamide, 1,3-butadiene and urethane, using chronic and acute exposure. A summary of the findings is shown in table 1.

Cyclophosphamide was positive in the rat after chronic gavage exposure for 33 weeks, for endpoints of dominant lethal mutation plateauing at 75% after week 7 (Jenkinson et al., 1990, Francis et al., 1990), congenital malformations (Table 2) and tumours (Table 3). F_1 karyotype analysis, both in the foetus (Table 4) and adults, was carried out where chromosome abnormalities were found in all cells of two of the adults, confirming transmission of induced damage through the male germ line (Francis et al., 1990). Such effects with cyclophosphamide have also been shown by other workers (Trasler et al., 1985) and led to the belief that chronic exposure might be a more realistic model than acute exposure, since in the workplace and environmentally, man is chronically exposed.

Table 2. Characterisation of Sprague-Dawley rat foetal abnormalities after cyclophosphamide and allyl alcohol treatment of F_0 males

Abnormality	Control	Allyl alcohol[a]	Cyclophosphamide[b]
Anasarca		1	13 (7)[c]
Anasarca + craniofacial abnormality			3 (2)
Anasarca + skeletal abnormality			4 (2)
Exencephaly		1	
Hydrocephaly			6
Craniofacial abnormality			6
Craniofacial and skeletal abnormality		1	2
Anaemia			4
Gastroschisis			2 (1)
Abnormal placenta			1
Growth retarded foetuses (Runts)	13	13	61
Total	13	16	102

[a]25 mg/kg body weight; [b]3.5 mg/kg body weight for 4 weeks and 5.1 mg/kg body weight from weeks 5-33 subsequently. [c]Numbers in parentheses indicate dead foetuses. Allyl alcohol is a metabolite of cyclophosphamide.

Table 3. Tumours and hydronephrosis identified macroscopically at post-mortem in female offspring from cyclophosphamide-treated and control male Sprague-Dawley rats[*]

Abnormalities identified macroscopically at post mortem	Histological findings	Paternal treatment	Age at post mortem (weeks)					
			Up to 53	54-66	67-79	80-91	92-104	Total
Hydronephrosis	–	CP	2[#]	2c[##]	1d[#]	3[###]	2e[##]	10
		Control	1	0	1	1b[#]	1a	4
Liver tumour	Fibrosarcoma	CP	0	0	0	0	0	0
		Control	0	0	0	0	1a	0
Lung tumour	Fibrosarcoma	CP	0	0	0	0	0	0
		Control	0	0	0	0	1a	1
Pituitary tumour	Adenoma	CP	0	0	0	0	2e[#]	2
		Control	0	1	0	1b	1a	3
Lymph node tumour	Adenofibroma	CP	0	0	0	0	0	0
		Control	0	0	1[#]	0	0	1
Vaginal tumour	Fibrosarcoma	CP	0	1	0	0	0	1
		Control	0	0	0	1	0	1
Ovarian tumour	Fibrosarcoma	CP	0	0	0	0	0	0
		Control	0	0	0	0	1a	1
Mammary tumour	Total (including tumours not examined)	CP	1	5	5	4	4	19
		Control	1	2	8	8	4	23
Uterine tumour	Adenofibroma	CP	–	1	1	1	–	3
		Control	–	–	–	1	2	3
	Total (including tumours not examined)	CP	0	1	1d	1	1[#]	4f
		Control	0	0	0	0	0	0
	Sarcoma	CP	0	1c	–	0	–	1
		Control	0	0	0	0	0	0
	Carcinoma	CP	0	0	–	1[#]	–	1
		Control	0	0	0	0	0	0
Total number of female offspring		CP	10	14	8	9	6	47
		Control	9	19	18	14	7	67

[*]Some tumours were examined histologically, and the findings are shown in the table. [#]Indicates each animal from which a karyotype was analysed (note that some animals had more than one macroscopic abnormality; see a–e). [a]All these abnormalities found in one animal; similarly, [b,c,d,e]; CP = cyclophosphamide; – = not examined histologically, [f]=borderline significance (p=0.051) by comparison with controls; doses as in Table 2.

In mice, 1,3-butadiene was positive for endpoints of dominant lethal mutation and congenital malformations after an exposure of 10 weeks (Table 5), even when compared to the historical control congenital malformation data (Table 6). There were significant effects in one study (Anderson et al., 1996) and not in another (Brinkworth et al., 1998) with no increase in tumours after sub-chronic inhalation exposure (Anderson et al., 1996). In the rat, no dominant lethality was observed after 10 weeks' exposure and there were no increases in congenital malformations in mice after 4 weeks' exposure (Table 7) (Anderson et al., 1998).

For urethane, however, there were negative results in mice for dominant lethality and congenital malformations after sub-chronic exposure in the drinking water (Table 8), although there was an increase in tumours in males after acute intraperitoneal (ip) treatment (Edwards et al., 1999). (Table 9). A study in ICR mice (Nomura, 1983) after acute ip treatment with urethane, also obtained negative dominant lethal mutation results, confirming results by other workers, but showed an increase in congenital malformations, tumours in the F_1 generation and transmitted tumours in the F_2 and F_3 generation.

Since tumours can be manifest without dominant lethal mutations, as is the case for urethane (Edwards et al., 1999) (Tables 1 and 9) the different endpoints may be independent genetic (germ cell transmissible) events and might be animal species and/or strain dependent (see mice versus rats after 1,3-butadiene exposure in Table 1). The question of acute versus chronic exposure might also be agent/compound dependent (Table 1).

Table 4.Results of analysis of abnormal Sprague-Dawley rat foetus karyotypes after cyclophosphamide and allyl alcohol treatment of F_0 males

Foetus No.	Abnormality	Centromeres (No.)	Karyotype abnormality
AA 42	Runt	43 (46)[a]	Tris. + 3 Fragments
AA 48	Anasarca/Runt	43	Tris. Ch. not identified
AA 94	Craniofacial	43	Tris. Ch. not identified
CP 29	Runt	42	Trans. intra Ch. 1
CP 31	Runt	42	Trans. Ch. 6 → Ch. 2
CP 58	Runt	42	Trans. Chs. not identified
CP 60	Runt	41	Monos. + Trans. Ch. 2 → Ch. 3
CP 64	Runt	42	Trans. Acrocentric Ch. → Ch. 2
CP 82	Runt	41	Monos. + Trans. Ch. 4/5 → Ch. 3
CP 90	Runt	42	Trans. Ch. 1 → Ch. 2
CP 91	Runt	42	Trans. Ch. 1 → Ch. 2
CP 106	Runt	42	Deletion Ch. 1
CP 51	Anaemic/Runt	42	Trans. Ch. 3 → Ch. 2
CP 70	Anaemic	41	Monos. Ch. 2 + Trans. Ch. 17/2 → Ch. 1
CP 120	Anaemic	41	Monos.
CP 63	Anasarca/Runt	42	Trans. Ch. 3 → Ch. 13
CP 100	Anasarca	42	Trans. Ch. 5 → Ch. 1
CP 59	Craniofacial	43	Tris. Ch. 4/5
CP 110	Craniofacial	42	Trans. Ch. 10 → Ch. 2
CP 52	Hydroceph./Runt	42	Trans. Chs. not identified
CP 101	Skeletal	41	Monos. + Trans. Chs. not identified
CP 66	Abnormal placenta	42	Trans. Ch. 4/5 → Ch. 2

Ch. = Chromosome; Trans. = Translocation; Monos. = Monosomy; Tris. = Trisomy; [a]3 small fragments were present in every metaphase and may have been centromeric; CP = cyclophosphamide; AA = allyl alcohol; doses as in Table 2.

Table 5. Effect of 1,3-butadiene on reproductive outcome in CD–1 mice after subchronic (6 h/day, 5 days/week, 10 weeks) exposure of males

Treatment	No. of males	Males mated to at least 1 female		No. of females[a]	Pregnant females		No. of pregnant females used in assay for dominant lethal mutation[b]
		n	%		n	%	
Control	25	23	92	50	41	82	23
12.5 ppm	25	25	100	50	45	90	24
1250 ppm	48[c]	43	90	96	74	77	38

	Implantations		Early deaths		Late deaths		Late deaths including dead foetuses		Abnormal foetuses	
	n	Mean ± S.D.	n	Mean ± S.D.	n	Mean[d] ± S.D.	n	Mean[d] ± S.D.	n	Mean[d] ± S.D.
Control	278	12.09±1.28	13	0.050±0.059	0		2	0.007±0.022	0	
12.5 ppm	306	12.75±2.51	16	0.053±0.058	7	0.023[**]±0.038	8	0.026[*]±0.042	7[e]	0.024[*]±0.062
1250 ppm	406	10.68[**]±3.10	87	0.204[***]±0.161	6	0.014[***]±0.032	7	0.016[*]±0.034	3[f]	0.011[**]±0.044

Significantly different from control at [*]$P<0.05$; [**]$P<0.01$; [***]$P<0.001$ (by analysis of variance and least significance test on arc-sine transformed data).

[a]Each male housed with two females for up to 1 week; [b]Not more than one for each male; the other females were allowed to litter; [c]2/50 males died after treatment (one at week 5, one at week 1); [d]Per implantation per pregnancy; [e]Four exencephalies (three in one litter), two runts (70% and 60% of mean body weight of others in litter, total litter sizes, 7 and 9, respectively), one with blood in amniotic sac but no obvious gross malformation (significance of difference not altered if this foetus is excluded); [f]One hydrocephaly, two runts (71% and 75% of mean body weight of others in litter; total litter sizes, 2 and 11, respectively)

Table 6. Results of assays for dominant lethal mutations and congenital malformations in CD–1 mice in previous studies (i.e. historical control data)

Study	No. of foetuses	No. of live foetuses	No. of abnormal foetuses	Abnormalities	No. of late deaths	No. of early deaths
Jenkinson *et al.* (1987) Experiment 1	720	671	8	(1 Exencephaly, 1 open eye, 6 runts)	7	34
Jenkinson *et al.* (1987) Experiment 2	1212	1125	6	(1 Exencephaly, 1 open eye, 4 runts)	19	62
Anderson *et al.* (1987)	–	680	2	(1 Exencephaly, 1 open eye)	–	–
Anderson *et al.* (unpublished) Experiment 1	254	234	5	(1 Hydrocephaly, 4 runts)	3	12
Anderson *et al.* (unpublished) Experiment 2	279	259	1	(1 Bent hind limb)	3	16
Totals	2465	2969	22		32	124
Percentage of number of implants					1.30%	5.03%[a]
Percentage of number of live foetuses			0.74%			

[a]A similar value (5.37%) for the percentage of early deaths in the CD–1 mouse strain was obtained in the 1970s and 1980s in over 700 foetuses [21].

Table 8. The effect of urethane in an extended dominant lethal assay in CD–1 mice after acute (intraperitoneal) or sub-chronic (drinking water) exposure of males

Treatment	Number of pregnant females[a]	Implantations (per pregnant female)[b,c]	Early deaths[b,d]	Late deaths[b,d]	Abnormal foetuses[b,d,e]
Acute					
Control	24	11.58±2.39	0.069±0.10	0.011±0.03	0.003±0.01
1.25 g/kg bwt	18	12.06±3.04	0.018±0.03	0.008±0.02	0.014±0.06
1.75 g/kg bwt	15	11.80±2.54	0.049±0.08	0.012±0.03	0.005±0.02
Sub-chronic					
Control	20	12.50±3.38	0.093±0.23	0.012±0.03	0.028±0.05
1.25 mg/ml for 10 weeks	23	12.91±1.73	0.014±0.03	0.023±0.04	0.006±0.02
3.75 mg/ml for 9 weeks	20	11.25±2.34	0.028±0.05	0.015±0.04	0

[a]Approximately half the pregnant females were allowed to litter; [b]Mean and standard deviation shown; [c]Tested using two-sample t-test (no significant differences found); [d]Data expressed per implantation per female. Transformed data [22] tested using two-sample t-test (no significant differences found); [e]See text for description of abnormal foetuses.

It is assumed that such effects as shown above could only be produced as a result of alterations induced in the exposed males' germ cells, because of their heritability through the generations.

Table 7.   Summary of tumour incidence for F_1 adult offspring from CD-1 male mice treated subchronically with butadiene

		F_1 animals			Treatment group					
		Control			12.5 ppm Butadiene			1250 ppm Butadiene		
		Age at necropsy (weeks)			Age at necropsy (weeks)			Age at necropsy (weeks)		
Tumour site	Sex	No. of animals	<74[a]	74–75[b]	No. of animals	<74[a]	74–75[b]	No. of animals	<74[a]	74–75[b]
Liver	Male	11	1 (71)	10	22	2 (65,70)	20[d]	25	3 (49,54,1)	22[h]
	Female	1	1 (54)	0	1	0	1	1	0	
Kidney	Male	0	0	0	0	0	0	1	1 (49)	0
	Female	1	1 (49)	0	0	0	0	1	1 (66)	0
Lung	Male	1	0	1	2	0	2	2	1 (69)	1[l]
	Female	0	0	0	0	0	0	4	2 (54,71)	1
Limb	Male	0	0	0	0	0	0	2	2 (45,52)	0
	Female	2	2 (45,52)	0	0	0	0	0	0	0
Urogenital	Male	0	0	0	0	0	0	1	1 (48)	0
	Female	1	1 (48)	0	0	0	0	0	0	0
Stomach	Male	0	0	0	1	0	1[e]	0	0	0
	Female	0	0	0	0	0	0	0	0	0
Pancreas	Male	0	0	0	1	0	1[f]	0	0	0
	Female	0	0	0	0	0	0	0	0	0
Abdominal	Male	0	0	0	1	1 (51)	0	0	0	0
	Female	0	0	0	0	0	0	1	1 (70)	0
Total with at least one tumour	Male	12	1	11	25[g]	3	22	30[k]	8	22
	Female	5[c]	5	0	1	0	1	7[i]	4	3
No tumours	Male	85	12	73	72[g]	9	63	68[k]	4	64
	Female	114[c]	14	100	106	26	80	93[i]	21	72
Total	Male	97	13	84	102[g]	22	85	100[k]	12	86
	Female	121[c]	19	100	107	26	81	101[j]	25	75

[a]Number in parentheses indicate age (weeks) at which animal was humanely killed or died due to sickness.
[b]All surviving animals were humanely killed for necropsy at 74–75 weeks.
[c]No data available for two F_1 females.
[d]Also stomach and pancreas; [e]Also liver and pancreas; [f]Also liver and stomach; [g]No data available for five F_1 males; [h]Also lung; [i]Also liver;
[j]No data available for one F_1 female; [k]No data available for two F_1 males.

Table 9.The effect of urethane on tumour incidence in the offspring of CD–1 mice after acute (intraperitoneal) exposure of males (number of F_1 animals shown)

Paternal treatment (acute):			Control			1.75 g/kg bwt	
Tumour site	Sex	Total	45–82 weeks old	83–84 weeks old	Total	45–82 weeks old	83–84 weeks old
Kidney	M	1	1	0	0	0	0
	F	1	0	1	1	1	0
Limb	M	1	1	0	2	2	0
	F	5	4	1	1	1	0
Liver	M	10	1	9	23*	6	17
	F	1	0	1	0	0	0
Lung	M	5	1	4	6	2	4
	F	1	1	0	2	1	1
Pancreas	M	0	0	0	0	0	0
	F	0	0	0	2	0	2
Thymus	M	0	0	0	1	0	1
	F	0	0	0	0	0	0
Urogenital	M	0	0	0	0	0	0
	F	1	0	1	0	0	0
Uterus/ovaries	F	4	1	3	4	4	0
Total with at least one tumour	M	17	4	13	27	8	19
	F	13	6	7	10	7	3
Total with no tumours	M	81	18	63	83	27	56
	F	86	23	63	87	12	75
Total no. of F_1 animals	M	99	22	76	111	35	75
	F	100	29	70	97	19	78

*Statistically significant from controls p=0.026.

The exact time of mating, within the week after treatment, and local husbandry conditions can have an effect on observed responses. In order to obtain sufficient numbers of offspring for analysis, there is a delicate balance between death through dominant lethality and survival of normal and malformed offspring, creating a "window" for detection of effects. As with any toxicological model, careful control of parameters is required. However, it is a useful model for examining inherited congenital malformations and tumours which can be attributed to exposure of the male and could be useful for predicting possible effects in man.

In the human situation, e.g. after chemotherapy, males can be advised not to indulge in sexual practices for a few months, to allow progression of newly formed germ cells through the next spermatogenic cycle, and often males do not feel well enough to mate. Thus, the animal models might not exactly reflect the human situation because animals are constantly mating.

However, there can also be difficulties in detecting reproductive effects directly in humans e.g. when interviewing for reproductive outcome. This can be illustrated by reproductive studies with vinyl chloride. Personal interviews and/or questionnaires are a primary source of data for monitoring programmes. In gathering information covering reproductive events, studies based on husbands' indirect reports yielded considerably lower figures for pregnancy loss (Infante et al., 1976) than those based on interviews with wives (Buffler et al., 1982). Individuals clearly have a much better recall for events in their own lives, and the circumstances of pregnancy are far more significant for a woman than a man; gathering information directly from the wives of employees would be a valuable technique in industrial male reproductive monitoring programmes. Therefore, even human models are not perfect and animals models can provide useful information for man.

CONCLUSION

Due to difficulties in conducting controlled reproductive studies in humans, it is important to use model systems in rodents to try to understand how paternal exposure could result in congenital abnormalities in offspring of man and/or produce a predisposition to cancer. Further work is desirable using these model systems with other chemicals to try to understand how predictive they are for man.

REFERENCES

Anderson, D., Edwards, A. J., Brinkworth, M.H., and Hughes, J.A., 1996, Male-mediated F_1 effects in mice exposed to 1,3-butadiene. *Toxicology.* 113:120-127.

Anderson, D., Hughes, J.A., Edwards, A.J., and Brinkworth, M.H., 1998, A comparison of male-mediated effects in rats and mice exposed to 1,3-butadiene. *Mutat Res.* 397:77-84.

Bithell, J.F., Dutton, S.J., Draper, G.J., and Neary, N.M., 1994, Distribution of childhood leukaemias and non-Hodgkin's lymphomas near nuclear installation in England and Wales. *Br Med J.* 309:501-505.

Black, D., 1984, Investigations of the possible increased incidence of cancer in West Cumbria, Report of the Independent Advisory Group (London: HMSO).

Brinkworth, M.H., 2000, Paternal transmission of genetic damage: findings in animals and humans. *Int J Andrology.* 23:123-135.

Brinkworth, M.H., Anderson, D., Hughes, J.A., Jackson, LI ,Yu, T-W., and Neischlag, E., 1998, Genetic effects of 1,3-butadiene on mouse testis. *Mutat Res.* 397:67-75.

Buckley, J.D., Robison, L.L., Swothinsky, R., Garabrant, D.H., Lebeau, M., Manchester, P., Nesbit, M.E., Odom, L., Peters, J M , and Woods, E.G , 1989, Occupational exposures of parents of children with acute nonlymphocytic leukemia: a report from the Children's Cancer Study Group. *Cancer Res.* 49:4030-4037.

Buffler, P.A., and Aase, J.M., 1982, Genetic risk and environmental surveillance. *Jnl Occupa Med.* 24:305-314.

Cattanach, B.M., Papworth, D., Patrick, G., Goodhead, D.T., Hacker, T., Cobb, L., and Whitehill, E., 1998, Investigation of lung tumour induction of C3H/H mice, with and without tumour promotion with urethane, following paternal X-irradiation. *Mutat Res.* 403:1-12.

Cattanach, B.M., Patrick, G., Papworth, D., Goodhead, D.T., Hacker, T., Cobb, L., and Whitehill, E., 1995, Investigation of lung tumour induction in BALB/C^J mice following paternal X-irradiation. *Int Jnl Rad Biol.* 67:606-616.

Cohen, E.N., 1974, Occupational disease among operating room personnel: a national study. *Anaesthesiology.* 41:321-340.

Cohen, E.N., 1975, A survey of anaesthetic health hazards among dentists. J Am Dent Assoc. 90:1291-1296.

Comare, 1996, Possible effects of paternal preconception irradiation in cancer, Committee on Medical Aspects of Radiation in the Environment, 4th report (London: Department of Health), pp.82-96.

Cook-Mozaffari, P.J., Darby, S.C., Doll, R., Forman, D., Hermon, C., Pike, M.C., and Vincent, T., 1989, Geographical variation in mortality from leukaemia and other cancers in England and Wales in relation to proximity to nuclear installations, 1968-1978. *Br J Cancer.* 59:476-485.

Cosgrove, G.E., Selby, P.B., Upton, A.C., Mitchell, T.J., Steele, M.H., and Russell, W.L., 1993, Lifespan and autopsy findings in the first generation offspring of X-irradiated male mice. *Mutat Res.* 319:71-79.

Cox, R., 1992, Transgeneration carcinogenesis: are there genes that break the rules?, *Radiat Prot Bull.* 129:15-23.

Doll, R., Evans H.J., and Darby, S.C., 1994, Paternal exposure not to blame. *Nature.* 367:678–680.

Draper, G.J., Little, M.P., Sorahan, T., Kinlen, L.J., Bunch,.K.J., Conquest, A.J., Kendall, G.M., Kneale, G.W., Lancashire, R.J., Muirhead, C.R., O'Connor, C.M., and Vincent, T.J., 1997, Cancer in the offspring of radiation workers: a record linkage study. *Br Med Jnl.* 315:1181-1188.

Dubrova, Y.E., Nesterov, V.N., Krouchinsky, N.G., Ostapenko, V.A., Neumann, R., Neil D.I., and Jefferys, A.J., 1996, Human minisatellite mutation rate after the Chernobyl accident. *Nature.* 380:683-686.

Edwards, A.J., Anderson D. Brinkworth, M.H., Myers, B., and Parry, J.M., 1999, An investigation of male-mediated F_1 effects in mice treated acutely and sub-chronically with urethane. *Teratog Carcinog Mutag.* 19:87-103.

Francis,A.J. Anderson, D., Evans J.G., Jenkinson, P.C., and Godbert P., 1990, Tumours and malformations in the adult offspring of cyclophosphamide-treated and control male rats in preliminary communication. *Mutat Res.* 229:239-246.

Fried, P., Steinfield, R., Casileth, B., and Steinfeld, A., 1987, Incidence of developmental handicaps among the offspring of men treated for testicular seminoma. *Int J Androl.* 10:385-387.

Gardner, M.J., Snee, M.P., Hall, A.J, Powell, C.A, Downes S., and Terrell, J.D., 1990a, Results of case control study of leukaemia and lymphoma among young people near Sellafield Nuclear Plant in West Cumbria *Br Med Jnl.* 300:423-429.

Gardner, M.J, Hall, A.J., Snee, M.P., Downes, S., Powell, C.A., and Terrell, J.D., 1990b, Methods and basic data of case control study of leukaemia and lymphoma among young people near Sellafield Nuclear Plant in West Cumbria. *Br Med Jnl.* 300:429-434.

Graham, S., Levin, M.L., Lilienfield, A.M., Schuman, L.M., Gibson, R., Dowd, J.E., and Hemplemann, L., 1966, Preconception, intrauterine and postnatal irradiation as related to leukaemia. *Nal Cancer Inst Monographs.* 19:347-371.

Gulati, S.C., Vega, R., Gee, T., Kosiner, B., and Clarkson, B., 1986, Growth and development of children born to patients after cancer therapy. Cancer invest. 4:197-205.

Olshan, A.F. and Schnitzer, P.G., Paternal occupation and birth defects, in: *Male-Mediated Developmental Toxicity*, A F.Olshan and D. R. Mattison., eds., Plenum Press, New York, pp 153-167S

Hales, B., and Robaire, B., The male mediated developmental toxicity of cyclophosphamide, in: *Male-Mediated Developmental Toxicity*, A. F.Olshan and D. R. Mattison., eds., Plenum Press, New York, pp. 105-116

Hawkins, M.M., Draper, G.J., and Smith, R.A., 1989, Cancer among 1,348 offspring of survivors of childhood cancer. *Int J Cancer.* 43:975-978.

Health and Safety Executive, 1993, HSE investigation of leukaemia and other cancers in the children of male workers at Sellafield (London: HSE).

Health and Safety Executive, 1994, HSE investigation of leukaemia and other cancers in the children of male workers at Sellafield: review of the results published in October 1993 (London: HSE).

Infante, P.A., Wagoner, J.K., McMichael, A.J., and Waxweiller., 1976, Genetic risks of vinyl chloride. *Lancet.* 1476:734-735.

Jenkinson, P.C., Anderson, D., and Gangolli, S.D., 1987, Increased incidence of abnormal foetuses in the offspring of cyclophosphamide and allyl alcohol-treated male mice. *Mutat Res.* 188:57-62.

Jenkinson, P.C., and Anderson, D., 1990, Malformed foetuses and karyotype abnormalities in the offspring of cyclophosphamide and allyl alcohol-treated male rats. *Mutat Res.* 229:173-184.

Ji, B.T., Shu, X.O., Linet, M.S., Zheng, W., Wacholder S, Gao, Y.T., Ying, D.M., and Jin, F., 1997, Paternal cigarette smoking and the risk of childhood cancer among offspring of nonsmoking mothers. *J Natl Cancer Inst.* 89:238-244.

Kinlen, L.J., 1993b, Can paternal preconceptional radiation account for the increase of leukaemia and non-Hodgkin's lymphoma in Seascale?. *Br Med J.* 307:444-445.

Kinlen, L.J., Clar, K., and Balkwill, A., 1993a, Paternal preconceptional radiation exposure in the nuclear industry and leukaemia and non-Hodgkin's lymphoma in young people in Scotland. *Br Med J.* 306:1153-1158.

Kirk, M.K., and Lyon, M.F., 1984, Induction of congenital malformations in the offspring of male mice treated with X-rays at pre-meiotic and post-meiotic stages. *Mutat Res.* 125:75-85.

Kneale, G.W., and Stewart, A.M., 1980, Pre-conception X-rays and childhood cancers. *Br J Cancer.* 41:222-226.

Knudsen, I., Hansen, E.V., Meyer, A.O., and Poulsen, E., 1977, A proposed method for the simultaneous detection of germ-cell mutations leading to foetal death (dominant lethality) and of malformations (male teratogenicity) in mammals. *Mutat Res.* 478:267-270.

Lefebvre, L.,Viville, S., Barton, S.C., Ishino, F., Keverne E.B., and Surani, M.A., 1998, Abnormal maternal behaviour and growth retardation associated with loss of the imprinted gene. *Nat Genet.* 20:163-167.

Li, F.P., Fine, W., Jaffe, N., Holmes, G.E., and Holmes, F.F., 1979, Offspring of patients treated for cancer in childhood. *J Natl Cancer Inst.* 62:1193-1197.

Little, M.P., Wakeford, R., and Charles, M.W., 1994a, An analysis of leukaemia, lymphoma and other malignancies together with certain categories of non-cancer mortality in the first generation of offspring (F-1) of the Japanese bomb survivors. *J Radiol Prot.* 14:203-218.

Little, M.P., Wakeford, R., and Charles, M.W., 1994b, A comparison of the risks of leukaemia in the offspring of the Sellafield workforce born in Seascale and those born elsewhere in West Cumbria with the risks in the offspring of the Ontario and Scottish worforces and the Japanese bomb survivors. *J Radiol Protect.* 14:187-201.

Little, M.P., Wakeford, R., Charles, M.W., and Andersson, M., 1996, A comparison of the risks of leukaemia and NHL in the first generation offspring (F1) of the Danish Thorotrast patients with those observed in other studies of parental preconception irradiation. *J Radiol Prot.* 16:1-9.

Lord, B.I., Woolford, L.B., Wang, L., Stones, V.A., McDonald, D., Lorimore, S.A., Papworth, D., Wright, E.G., and Scott, D., 1998a, Tumour induction by methyl-nitroso-urea following preconceptional paternal contamination with plutonium-239. *Brit J Cancer*. 78:301-311.

Lord, B.I., Woolford, L.B., Wang, L., Stones, V.A., McDonald, D., Lorimore, S.A., Papworth, D., Wright, E.G., and Scott, D., 1998b, Induction of lympho-haemopoietic malignancy: impact of preconception paternal irradiation. *Int Jnl Rad Biol*. 74:721-728.

Mulvihill, J.J., Myers, M.J., Connelly, R.R., Byrne, J., Austin, D.F., Bragg, K., Cook, J.W., Hassinger, D.D., Holmes, F.F., Holmes, G.F., Krause, M.R., Latourette, H.B., Meigs, J.W., Naughton, M.D., Steinhorn, S.C., Strong, L.C., Teta, M.J., and Weyer, P.J., 1987, Cancer in offspring of long-term survivors of childhood and adolescent cancer. *Lancet*. 2:813-817.

Nomura, T., 1982, Parental exposure to X-rays and chemicals induces heritable tumours and anomalies in mice. *Nature*. 296:575-577.

Nomura, T., 1991, Paternal exposure to radiation and offspring cancer in mice: reanalysis and new evidences. *J Radiat Res*. 2:64-72.

Nomura, T., 1983, X-ray induced germ line mutation leading to tumours and its manifestation in mice given urethane post-natally. *Mutat Res*. 121:59-66.

Olshan, AF. and Schnitzer, Paternal occupation and birth defects, in: *Male-Mediated Developmental Toxicity*," A F.Olshan and D R. Mattison., eds., Plenum Press. New York, pp. 153-167

Parker, L., Craft, AW., Smith, J., Dickenson, H., Wakeford, R., Binks, K., McElvenny, D., Scott, L., and Slovak, A., 1993, Geographical distributions of preconceptional radiation doses to fathers employed at the Sellafield nuclear installation, West Cumbria. *Br Med J*. 307:966-971.

Potashnik,G., and Yanai-Inbar, I., 1987, Dibromochloropropane (DBCP): an 8 year reevaluation of testicular function and reproductive performance. *Fertil Steril*. 47:317-323.

Qiu, J., Hales, B. F. and Robaire, B., 1992, Adverse effects of cyclosphamide on progeny outcome can be mediated through post-testicular mechanisms in the rat. *Biol Reprod*. 46:926-931.

Qiu, J., Hales, B. F. and Robaire, B., 1995, Damage to rat spermatozoal DNA after chronic cyclophosphamide exposure. *Biol Repro*. 53:1465-1473.

Robaire, B and Hales, B.F., 1994, Post testicular mechanisms of male-mediated developmental toxicity, in: *Male-Mediated Developmental Toxicity*, A. F.Olshan and D. R. Mattison., eds., Plenum Press. New York, pp. 93-103

Roman, E., Watson, A., Beral, V., Buckle, S., Bull, SD., Baker, K., Ryder, H., and Barton, C., 1993, Case-control study of leukaemia and non-Hodgkin's lymphoma among children aged 0-4 years living in West Berkshire and North Hampshire health districts. *Br Med J*. 306:615-621.

Savitz, D.A., 1994, Paternal exposures and pregnancy outcome: miscarriage, stillbirth, low birth weight, preterm delivery, in: *Male-mediated Developmental Toxicit*, A.F. Olshan, D.R. Mattison, ed., New York: Plenum Press; pp.177-196.

Selby, P.B., 1990, Experimental induction of dominant mutations in mammals by ionizing radiations and chemicals. *Issues Rev Terat*. 5:181-253.

Sentura, Y.D., Peckman, C.S., and Peckman, M,J., 1985 Children fathered by men treated for testicular cancer. Lancet. 2:766-769.

Sorahan T., Prior, P., Lancashire, R.J., Faux, S.P., Hulten, M.A., Peck, I.M., and Stewart, I.M., 1997, Childhood cancer and parental use of tobacco : deaths from 1971 to 1976. *Br J Cancer*. 76:1525-1531.

Sorahan, T., Lancashire, R.J., Hulten, M.A., Peck, I., and Stewart, A.M., 1997, Childhood cancer and parental use of tobacco:deaths from 1953 to 1955. *Br J Cancer*. 75:124-138.

Takahashi, T., Watenabem H., Dohi, K., and Ito, A., 1992, 252cf relative biological effectiveness and inheritable effect of fission neutrons in mouse liver tumorigenesis. *Cancer Res*. 52:1948-1953.

Trasler, J.M., Hales, B.F., and Robaire, B., 1985, Paternal cyclophosphamide treatment of rats causing fetal loss and malformations without affecting male fertility. *Nature*. 316:144-146.

Trasler, J.M., Hales, B. F., and Robaire,B., 1986, Chronic low dose cyclophosphamide treatment of adult male rats: effect on fertility, pregnancy outcome and progeny. *Biol Reprod*. 34:275-283.

Trasler, J.M., Hales, B. F., and Robaire,B., 1987, A time course study of chronic paternal cyclophosphamide treatment in rats: effects on pregnancy outcome and the male reproductive and haematologic systems. *Biol Reprod*. 37:317-326.

Trasler, J.M., and Robaire, B., 1988, Effects of cyclophosphamide on selected cytosolic and mitochondrial enzymes in the epididymis of the rat. *J Androl*. 9:143-152.

Urquhart, J.H., Black, R.J., Muirhead, M.J., Sharp, L., Maxwell, M., Eden, O.B., and Jones, D.A., 1991, Case-control study of leukaemia and non-Hodgkin's lymphoma in children in Caithness near the Dounreay nuclear installation. *Br Med J*. 302:687-692.

Uzych,L., 1985, Teratogenesis and mutagenesis associated with the exposure of human males to lead: a review. *Yale J Biol Med*. 58:9-17.

Vorobtsova, I.E., Aliyakparova, C.M., and Anisimov, V.N., 1993, Promotion of skin cancers by 12-0-tetradecanooylphorbol-13-acetate in two generations of descendants of male mice exposed to X-irradiation. *Mutat Res.* 287:207-216.

Vorobtsova, I.E., and Kitaev, E.M., 1988, Urethane-induced lung adenomas in the first generation progeny of irradiated male mice. *Carcinogenesis.* 9:1931-1934.

Wakeford, R., and Parker, L., 1996, Leukaemia and non-Hodgkin's lymphoma in young persons resident in small areas of West Cumbria in relation to paternal preconceptional irradiation. *Brit J Cancer.* 73:673-679.

Wakeford, R., and Tawn, E.J., 1994, Childhood leukaemia and Sellafield: the legal cases. *J Radiol Prot.* 14:293-316.

Whorton, M. D., Milby, T.H., Krauss, R.M., and Stubbs, H.A., 1979, Testicular function in DBCP exposed pesticide workers. J Occup Med. 21:161-166.

Yoshimoto, Y., 1990, Cancer risk among children of atomic bomb survivors. A review of RERF epidemiologic studies. Radiation Effects Research Foundation. *JAMA.* 264:622-623.

Yoshimoto, Y., Neel, J.V., Schull, W.J., Kato, H., Soda, M., Eto, R., and Mabuchi, K., 1990, Malignant tumours during the first 2 decades of life in the offspring of atomic bomb survivors. *Amer J Hum Genetics.* 46:1041-1052.

EPIDEMIOLOGIC EVIDENCE ON BIOLOGICAL AND ENVIRONMENTAL MALE FACTORS IN EMBRYONIC LOSS

Jens Peter Bonde,[1] Henrik I. Hjøllund,[1] Tine B. Henriksen,[2] Tina K. Jensen[3], Marcello Spanò[4], Henrik Kolstad[1], A. Giwercman[5] , Lone Storgaard[1] , Erik Ernst[6] and Jørn Olsen[7] .

[1] Department of Occupational Medicine, Aarhus University Hospital, Denmark
[2] Perinatal Epidemiological Research Unit, Department of Obstetrics and Gynaccology, Aarhus University Hospital, Denmark
[3] Department of Growth and Reproduction, The National Hospital, Denmark
[4] Section of Toxicology and Biomedical Sciences, Department of the Environment,   ENEA CR Casaccia, Rome, Italy
[5] Department of Andrology, University of Lund, Sweden
[6] Fertility Clinic, Aarhus University Hospital, Denmark
[7] The Danish Epidemiology Science Center, Aarhus University, Denmark

INTRODUCTION

Epidemiological studies on male mediated developmental toxicity may focus on links between environmental exposure of the male and pregnancy failures or disorders in the offspring. Alternatively, the focus may be on biological markers of male fecundity and developmental disorders. The study design only rarely allows for a combination of these two approaches. Demonstration of effects of male exposure on the male reproductive system, and subsequent demonstration of links between observed abnormalities of the male system with abnormalities of foetal and childhood development in the offspring would provide strong evidence for causal associations between exposure and outcomes.

The Danish First Pregnancy Planner Study represents a major Danish research effort into couple fertility. In a prospective design we collected semen samples and urine specimens in young couples who, for the first time, tested their reproductive capability following discontinuation of contraception (Bonde et al., 1998b). This study provides the options of studying effects of male occupational and lifestyle factors on the male reproductive system, the time taken to conceive and occurrence of embryonic loss, as detected by human chorionic gonadotrophin (hCG) in urine. Following a brief presentation of the Danish First Pregnancy Planner Study, this paper will summarize and review findings which are of interest from the male-mediated developmental toxicity point of view. Some results have been published (Bonde et al., 1998a; Jensen et al., 1998b and c;

Advances in Male Mediated Developmental Toxicity, edited by Bernard Robaire and Barbara F. Hales.
Kluwer Academic/Plenum Publishers, 2003.

Hjollund et al., 2000) but several preliminary findings from ongoing analyses are also presented.

THE DANISH FIRST PREGNANCY PLANNER STUDY

The Danish First Pregnancy Planner Study is a follow-up study of environmental and biological determinants of fertility in 430 Danish couples who discontinued contraception to have their first child (Bonde et al., 1998b). The study population was sampled from four Danish trade unions to increase exposure prevalence and thus the ability to examine specific occupational hypotheses. In 1992-94 we wrote letters to 52,225 union members (metal workers, office workers, nurses and day care workers) ages 20-35 years, who lived with a partner and had no children. In order to counteract preferential participation of infertile couples, partners were ineligible if they had ever tried to have a child whether this attempt was successful or not. Women were thus not eligible if they ever had a miscarriage, an induced abortion, had given birth or if the man had fathered a child or an abortive pregnancy. The exact number of eligible couples in the source population is unknown but an average participation rate of 16% of all eligible subjects was estimated (Bonde et al., 1998b). Follow-up began when a couple stopped using contraception and was completed after 6 menstrual cycles or if a pregnancy was recognised by the general practitioner within that period. At enrolment, each man provided a semen sample and both partners completed a questionnaire on their demographic, medical and reproductive background, their occupations and lifestyle. During follow-up, each women kept a diary of menstrual bleeding and sexual intercourse. We also collected questionnaires each month on lifestyle factors during the week proceeding day 14-21 from the last menstrual bleeding.

Each woman collected first morning urine specimens during 10 consecutive days from the first day of bleeding in each menstrual cycle. Early pregnancy loss was considered present if one immune fluorometric value of hCG was above 1 IU/L followed by a decline (Bonde et al., 1998b). Further information on the outcome of clinically recognised pregnancies which were diagnosed during the six cycles of follow-up was collected by questionnaire and telephone interview.

We obtained complete data on 183 pregnancies during the 6 months of follow-up. Of these 133 resulted in delivery of a child and 55 (30.4%) were spontaneously aborted [34 embryonic losses before six completed weeks of pregnancy were detected by hCG measurements (62%) and 21 were clinically recognised between 6 and 15 weeks of gestation (38%)].

SEMEN QUALITY AND EMBRYONIC LOSS

It is often argued that chemical exposure of the human male is unlikely to be related to adverse outcomes such as miscarriage, birth defects, growth retardation and cancer, because millions of sperm cells are competing to reach and fertilise the ovum. Impaired sperm cells may have small chances of winning this competition. This may be true for chemicals interfering with sperm motility or sperm capacitation but does not necessarily apply to spermatozoa having an altered genome. Thus, we know that spermatozoa with genetic defects are capable of fertilisation and of subsequently causing abnormalities in offspring, e.g., Down's Syndrome (Colie, 1993), retinoblastoma and Prader-Willi Syndrome (Olshan and Faustman, 1993). Nevertheless, if crude markers of poor semen quality such as sperm density and morphology are related to foetal survival, an increased awareness of male mediated effects would be warranted - not to speak about concern about IVF treatment, in particular intracytoplasmic sperm injection [ICSI]. Several substances encountered in the work environment have impact on semen quality (Bonde and Giwercman, 1995) and sperm concentration and sperm morphology are strong and independent predictors of time taken to conceive (Bonde et al., 1998a). A case-referral study from Stockholm done as early as 1962 found a high prevalence of morphologically abnormal sperm in men whose wives had a clinically recognised spontaneous abortion when compared with men fathering a child (Furuhjelm et al., 1962). Several studies have examined abnormal sperm as a possible factor in recurrent abortion but results are conflict-

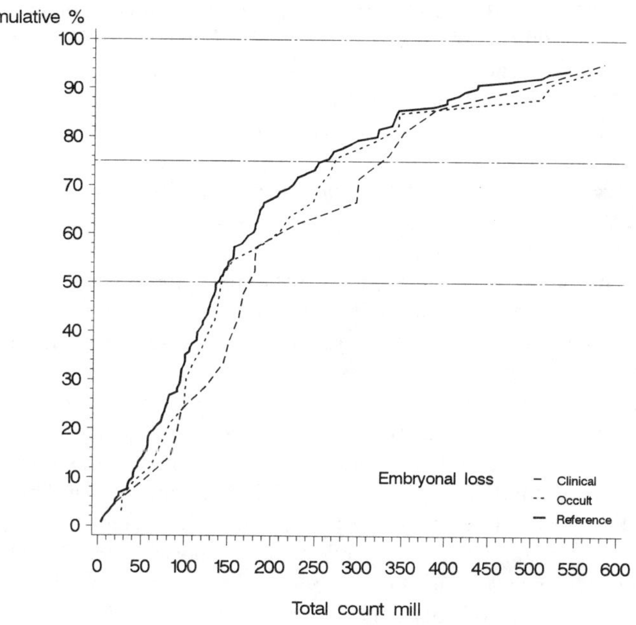

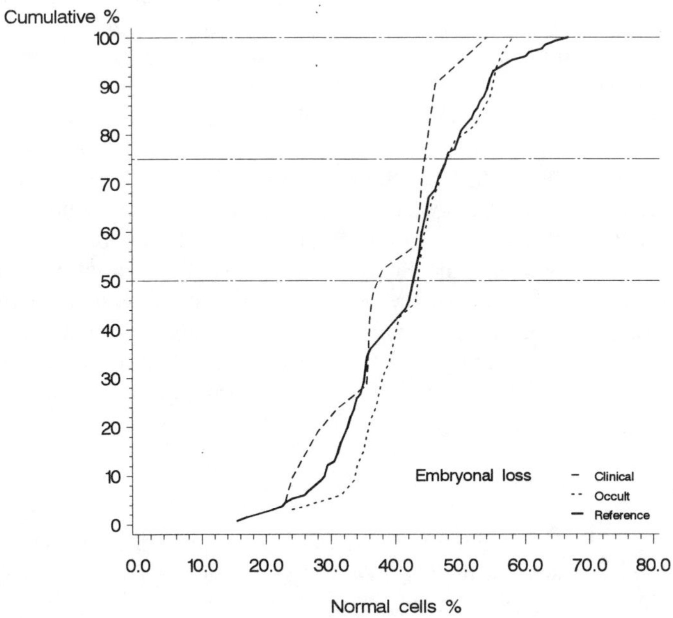

Figure 1A-1B. Cumulative distribution of total sperm count and morphology by pregnancy survival. The Danish First Pregnancy Planner Study of 430 couples (Bonde et al., 1998b).

ing (Homonnai et al., 1980; Sbracia et al., 1996; Kiefer et al., 1997; Katsoff et al., 1999). Moreover, studies of pregnancy outcomes following in-vitro fertilisation compare the rate of spontaneous abortion between different types of IVF treatments, but we lack comparisons with the background rate in the population (Coulam et al., 1996; Govaerts et al., 1998; Westergaard et al., 2000). The Danish First Pregnancy Planner Study offers good options to address the question in a controlled design. The distributions of total sperm count and percentage of abnormal sperm cells, as determined by the WHO classification, were very similar in the spouses of women with, respectively, a normal delivery, a clinically recognised abortion, and a subclinical abortion detected by hCG (Figure 1A and 1B). Similar findings were obtained for sperm cell concentration. It seems from these data that the conventional measures of semen quantity and quality are not strongly related to embryonic loss, but the limited number of abortions in this study should be acknowledged.

The sperm concentration is fluctuating widely in repeated samples from the same man within short time spans. A semen specimen obtained one to 6 months before conception may not accurately reflect the ejaculate which actually results in fertilisation. This would dilute the effect on embryonic loss - if any. On the other hand, the sperm morphology is rather stable within a man and is perhaps, a priori, of greater interest when considering male factors in embryonic loss. We found no indication whatsoever of an effect on either sperm concentration or sperm morphology (Figure 1A and 1B).

The sperm chromatin structure assay (SCSA) indicates the susceptibility of sperm chromatin DNA to acid induced denaturation in situ. Flow cytometry is used to measure the metachromatic shift of acridine-orange fluorescence from green (natural double stranded DNA) to red (denatured single stranded DNA) (Evenson et al., 1991). The main parameters include the mean-αT which is the average ratio between red and total (green + red) fluorescence, the percentage of cells outside the main population (COMP) in a sample of some 10,000 sperm cells in an ejaculate. Abnormal parameters are believed to reflect abnormal nuclear condensation of sperm DNA and single strand breakage (Sailer et al., 1995). The SCSA is feasible to use in epidemiological studies (Spano et al., 1998), is susceptible to the action of several chemicals in rodents, and predicts human fertility independently of sperm concentration and morphology (Spano et al., 2000). The Danish First Pregnancy Planner Study indicates, with borderline significance, that a high percentage of cells with abnormal chromatin (percentage COMP) was associated with an increased rate of early embryonic loss before 6 completed gestational weeks but not with embryonic loss at a later stage (Figure 2B). These findings are in agreement with results from a recent US study (Evenson et al., 1999; Larson et al., 2000). The other parameters of abnormal sperm chromatin were not associated with early or later embryonic loss in the Danish First Pregnancy Planner Study (mean-αT: Figure 2A; other parameters not shown). As is the case for sperm morphology, the SCSA values show little variation in repeated samples from any given man. Therefore, the delay of 1-6 menstrual cycles between the measured and the fertilising ejaculate is not expected to dilute or bias the results. Obviously, there is a need to corroborate or refute these findings and to investigate direct measures of genetic sperm abnormalities in relation to foetal survival in humans.

WELDING OF STAINLESS STEEL AND EMBRYONIC LOSS

Stainless steel is an alloy of chromium, nickel and iron while mild steel essentially is iron with minute amounts of several metals. Welding of stainless steel, but not of mild steel, may confer a high pulmonary uptake of hexavalent chromium from the welding particulates. Following uptake, chromium is distributed to all compartments of the body

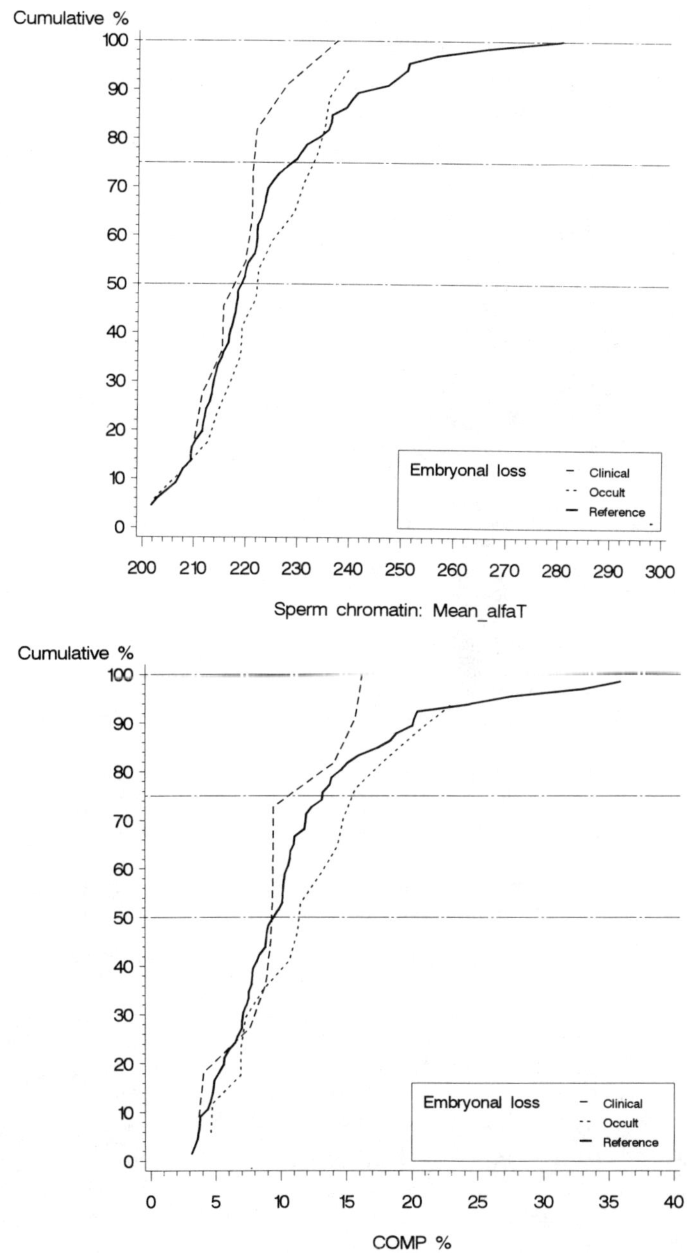

Figure 2A and 2B. Cumulative distribution of sperm chromatin parameters by pregnancy survival. The Danish First Pregnancy Planner Study of 430 couples (Bonde et al., 1998b).

including the testis. Hexavalent chromium is known to have mutagenic effects in both somatic cells and germ cells. Experimental studies indicate that hexavalent chromium administered to male rodents impairs the viability of embryos fathered by exposed males (Anderson et al., 1994). A nationwide cohort study of adverse pregnancy outcomes in metal workers found an increased risk of self reported spontaneous abortions in spouses of stainless steel but not of mild steel welders [relative risks of 2.0 (1.1-3.5) and 1.0 (0.5-2.0), respectively] (Bonde et al., 1992). The risk estimates were higher in high-level exposed manual metal arc stainless steel welders than in low-level exposed tungsten inert gas stainless steel welders, but no excess risk was found in the same cohort in analyses based on hospital records of spontaneous abortion (Hjollund et al., 1995). In general, the rate of embryonic loss declines rapidly during the first 6-12 weeks of gestation (Figure 3).

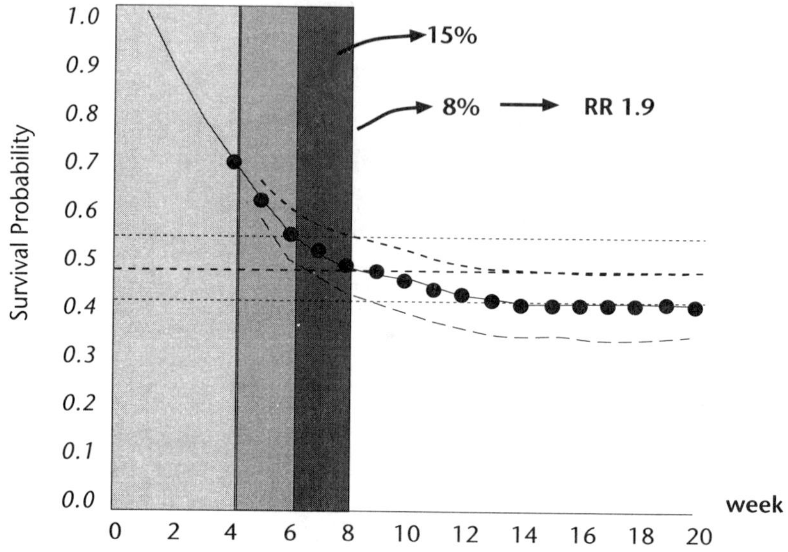

Figure 3. Detection bias in epidemiological studies based on clinical abortions. Pregnancy survival in the Danish First Pregnancy Planner Study (Bonde et al., 1998b).

Small shifts in recognition and reporting of spontaneous abortion between compared groups may result in strongly biased risk estimates. This is a potentially severe methodological problem in all studies of clinically recognised spontaneous abortion and could be the explanation of the conflicting finding in the above 2 studies. Using a standardised objective measure of early embryonic loss, the Danish First Pregnancy Planner Study bypassed detection bias. In this study, the risk of spontaneous abortion was increased in spouses of stainless steel welders (RR 3.5, 95% CI1.3-9.1) but not in spouses of mild steel welders (RR 1.0, 95% CI 0.5-2.1) (Hjollund et al., 2000). In stainless steel welders the risk of early hCG detected loss and of clinically recognised loss was at the same level (RR 3.0, 95% CI1.1-8.0 and RR 3.2, 95% CI 1.1-9.8, respectively; Figure 4).

The level of chromium in urine samples taken at the end of a work shift was not markedly increased in stainless steel welders, but the urine levels do not reflect accumulation of chromium in the tissues. Therefore, it is of interest that the maternal age adjusted risk of pregnancy loss increased as the number of years of stainless steel welding increased (Figure 5). Toxicokinetic studies have indicated a very slow elimination of chromium from some compartments of the body. This seems to result in a gradual built up of chromium in tissues during long term exposure. However, fertility effects on the male reproductive system were not present for sperm concentration, morphology or motility (Hjollund et al., 1998). On the other hand, as shown earlier in this paper, these biological markers of male fecundity seem not to be related to pregnancy loss in general. If the increased risk of spontaneous abortion is mediated by stainless steel welding exposure, more subtle and specific effects on sperm cell function should be identified. As early as 1985, it was shown that cyclophosphamide, an alkylating anticancer drug, given to male

Figure 4. The risk of embryonic loss in stainless steel and mild steel welders. The Danish First Pregnancy Planner Study of 430 couples (Hjollund et al., 2000).

rodents before mating to untreated females caused very high rates of pre-implantation as well as post-implantation embryonic loss at levels not causing reduced sperm count (Trasler et al., 1985). Local intra-uterine toxicity by agents transmitted by the seminal fluid (Hales et al., 1986) is unlikely since levels of chromium in urine and semen were very low.

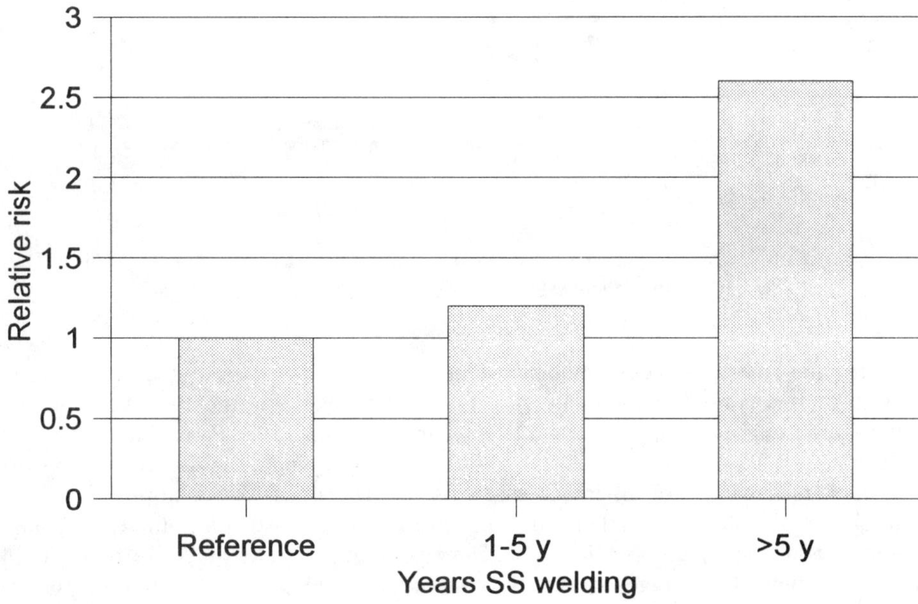

Figure 5. The risk of pregnancy loss in partners of long term stainless steel welders. The Danish First Pregnancy Planner Study of 430 couples (Hjollund et al., 2000).

LIFESTYLE

Male smokers seem not to have reduced fecundity (Jensen et al., 1998a; Juul et al., 1999), and the effect of smoking on sperm count and morphology is small, if at all present (Vine et al., 1993, 1997). A meta-analysis of published studies on sperm count in smokers indicates an average 10-15% reduction in sperm count, but an effect this size hardly translates into delayed conception or increased frequency of infertility (Bonde et al., 1999). The sperm chromatin seems remarkably resistant to the effects of toxicants in tobacco smoke (Spano et al., 1998). However, recently two studies have demonstrated an increased rate of sperm chromosomal aberrations in smokers (Rubes et al., 1998; Harkonen et al., 1999). Using the fluorescence in-situ hybridisation technique in decondensed human sperm nuclei, Rubes et al. reported increases in disomic sperm for chromosomes 8, X and Y in teenage smokers. Lahdetie et al. found a higher rate of disomy and diploidy for chromosome 1 and 7 in farmers who smoke (Harkonen et al., 1999). Although the frequency of disomy of each individual sperm chromosome is low, in the range of 0-50 per 10,000 sperm in healthy men, the proportion of sperm with at least one abnormal chromosome could be higher. If sperm with abnormal chromosomes are not selected against at fertilisation (Marchetti et al., 1999), an increased risk of embryonic loss would be expected. The Danish First Pregnancy Planner Study does not, however, provide clear evidence of increased rate of embryonic loss in spouses of male smokers (Figure 6). Differences between non-smokers and moderate smokers were small through all gestational ages and heavy smokers had the lowest rate of spontaneous abortions. These findings are in accordance with several other studies (Coste et al., 1991; Windham et al., 1992b, 1999; Chatenoud et al., 1998).

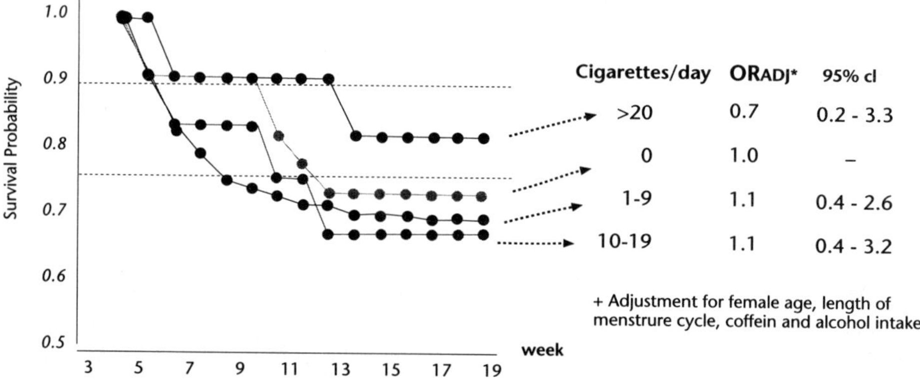

Figure 6. Survival of pregnancies by level of male smoking during the week surrounding fertilization. The Danish First Pregnancy Planner Study of 430 couples (Bonde et al., 1998b).

During the past three years, evidence that caffeinated beverage consumption during the first trimester confers a moderate risk for spontaneous abortion has been found in studies of good quality (Dlugosz et al., 1996; Fenster et al., 1997; Parazzini et al., 1998; Robertson et al., 1998; Cnattingius et al., 2000). One study indicates that risk is confined to non-smokers and only result in increased risk of abortions with a normal karyotype (Cnattingius et al., 2000). Another study indicates that only women with severe nausea during the first trimester are at risk. It is of interest that fecundity may also be related to intake of caffeinated beverages in female non-smokers (Jensen et al., 1998b), possibly because smoking increases the rate of caffeine metabolism (Brown et al., 1988). Knowledge on male mediated pregnancy loss related to intake of caffeine is limited (Jensen et al., 1998b). In one study the fecundity in non-smoking men, but not in smoking men, decreased with increasing level of caffeine intake, but findings might be due to chance and have not been corroborated (Jensen et al., 1998b).

The Danish First Pregnancy Planner Study provides some indication of increased risk of embryonic loss related to male intake of caffeine during the last cycle of spermatogenesis prior to conception (Figure 7), but possible mechanisms remain obscure and would probably be different from mechanisms operating in women.

Adverse embryonic and offspring effects of male alcoholic beverage consumption may be important to study. The concentration of ethanol in semen is in the same range as that found in blood, and direct local toxicity cannot be completely ignored. Moreover, the dominant lethal test for alcohol is positive in both rats and mice, and other experimental studies show decreased survival of foetuses following alcohol administration to males before mating (Abel, 1995). However, epidemiological studies are few (Windham et al., 1992a; Parazzini et al., 1990). Results from the First Pregnancy Planner Study are in the process of publication elsewhere.

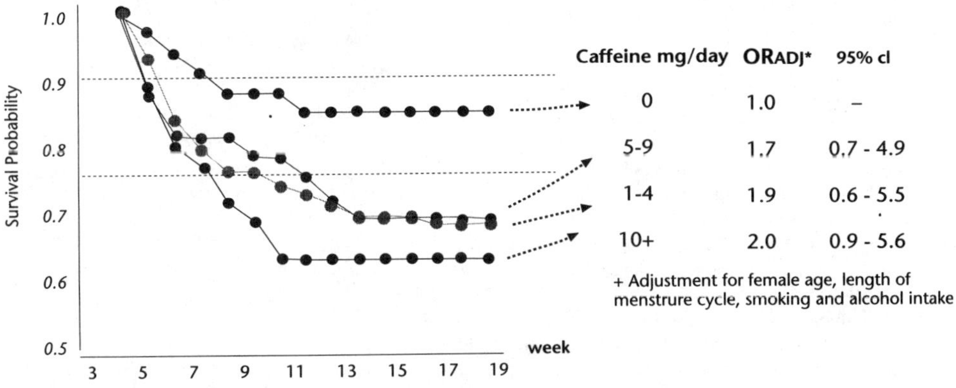

Figure 7. Survival of pregnancies by level of paternal caffeine intake during the week surrounding fertilisation. The Danish First Pregnancy Planner Study of 430 couples (Bonde et al., 1998b).

CONCLUSION

Studies of miscarriage may be a feasible and sensitive approach to increase knowledge on male-mediated developmental toxicity, in particular if the problem of detection bias in observational studies is accounted for. Couples treated for infertility with IVF techniques offer new possibilities for cost effective studies of male-mediated developmental toxicity, unless it turns out that female factors and the treatment per se have a strong impact on the rate of spontaneous abortion. While conventional measures of semen quality are not related to embryonic loss, suggestive evidence links abnormal sperm chromatin and sperm chromosome anomalies to environmental exposures. This emphasises the need to corroborate or refute male mediated effects of several occupational and lifestyle exposures.

REFERENCES

Abel, E.L., 1995, A surprising effect of paternal alcohol treatment on rat fetuses. *Alcohol.* 12:1-6.
Anderson, L., Kasprzak, K.S., Rice, J.M., 1994, Preconception exposure of males and neoplasia in their progeny:effects of metals and consideration of mechanisms. In: Olshan AF, Mattison DR, editors. Male-mediated developmental toxixity. New York and London: Plenum Press.

Bonde, J.P., Olsen, J.H., and Hansen K.S., 1992, Adverse pregnancy outcome and childhood malignancy with reference to paternal welding exposure. *Scand J Work Environ Health.* 18:169-177.

Bonde, J.P., Giwercman, A., 1995, Occupational hazards to male fecundity. *Reprod Med Rev.* 4:59-73.

Bonde, J.P., Ernst, E., Jensen, T.K., Hjollund, N.H., Kolstad, H., Henriksen, T.B. et al., 1998a, Relation between semen quality and fertility: a population-based study of 430 first-pregnancy planners. *Lancet.* 352:1172-1177.

Bonde, J.P.E., Hjollund, N.H.I., Jensen, T.K., Ernst, E., Kolstad, H., Henriksen, T.B. et al., 1998b, A follow-up study of environmental and biological determinants of fertility among 430 Danish first pregnancy planners: Design and methods. *Reprod Toxicol.* 12:19-27.

Bonde, J.P., Hjollund, N.H., Kolstad, H.A., Abell, A., Larsen, S.B., 1999, Environmental semen studies--is infertility increased by a decline in sperm count? *Scand J Work Environ Health.* 25 Suppl 1:12-16.

Brown, C.R., Jacob, P., III, Wilson, M., Benowitz, N.L., 1988, Changes in rate and pattern of caffeine metabolism after cigarette abstinence. *Clin Pharmacol Ther.* 43:488-491

Chatenoud, L., Parazzini, F., di Cintio, E., Zanconato, G., Benzi, G., Bortolus, R. et al., 1998, Paternal and maternal smoking habits before conception and during the first trimester: relation to spontaneous abortion. *Ann Epidemiol.* 8:520-526.

Cnattingius, S., Signorello, L.B., Anneren, G., Clausson, B., Ekbom, A., Ljunger, E. et al., 2000, Caffeine intake and the risk of first-trimester spontaneous abortion. *N Engl J Med.* 343:1839-1845.

Colie, C.F., 1993, Male mediated teratogenesis. *Reprod Toxicol.* 7:3-9.

Coste, J., Job-Spira, N., Fernandez, H., 1991, Risk factors for spontaneous abortion: a case-control study in France. *Hum Reprod.* 6:1332-1337.

Coulam, C.B., Opsahl, M.S., Sherins, R.J., Thorsell, L.P., Dorfmann, A., Krysa, L. et al., 1996, Comparisons of pregnancy loss patterns after intracytoplasmic sperm injection and other assisted reproductive technologies. *Fertil Steril.* 65:1157-1162.

Dlugosz, L., Belanger, K., Hellenbrand, K., Holford, T.R., Leaderer, B., Bracken, M.B., 1996, Maternal caffeine consumption and spontaneous abortion: a prospective cohort study. *Epidemiology.* 7:250-255.

Evenson, D.P., Jost, L.K., Baer, R.K., Turner, T.W., Schrader, S.M., 1991, Individuality of DNA denaturation patterns in human sperm as Measured by the sperm chromatin structure assay. *Reprod Toxicol.* 5:115-125.

Evenson, D.P., Jost, L.K., Marshall, D., Zinaman, M.J., Clegg, E., Purvis, K. et al., 1999, Utility of the sperm chromatin structure assay as a diagnostic and prognostic tool in the human fertility clinic. *Hum Reprod.* 14:1039-1049.

Fenster, L., Hubbard, A.E., Swan, S.H., Windham, G.C., Waller, K., Hiatt, R.A. et al., 1997, Caffeinated beverages, decaffeinated coffee, and spontaneous abortion. *Epidemiology.* 8:515-523.

Furuhjelm, M., Jonson, B., Lagergren, C.G., 1962, The quality of human semen in spontanous abortion. *Int J Fertil.* 7:17-21.

Govaerts, I., Devreker, F., Koenig, I., Place, I., Van den, B.M., Englert, Y., 1998, Comparison of pregnancy outcome after intracytoplasmic sperm injection and in-vitro fertilization. *Hum Reprod.* 13:1514-1518.

Hales, B.F., Smith, S., Robaire, B., 1986, Cyclophosphamide in the seminal fluid of treated males: transmission to females by mating and effect on pregnancy outcome. *Toxicol Appl Pharmacol.* 84:423-430.

Harkonen, K., Viitanen, T., Larsen, S.B., Bonde, J.P., Lahdetie, J., 1999, Aneuploidy in sperm and exposure to fungicides and lifestyle factors. ASCLEPIOS. A European Concerted Action on Occupational Hazards to Male Reproductive Capability. *Environ Mol Mutagen.* 34:39-46.

Hjollund, N.H., Bonde, J.P., Hansen, K.S., 1995, Male-mediated risk of spontaneous abortion with reference to stainless steel welding. *Scand J Work Environ Health.* 21:272-276.

Hjollund, N.H., Bonde, J.P., Jensen, T.K., Ernst, E., Henriksen, T.B., Kolstad, H.A. et al., 1998, Semen quality and sex hormones with reference to metal welding. *Reprod Toxicol.* 12:91-95.

Hjollund, N.H., Bonde, J.P., Jensen, T.K., Henriksen, T.B., Andersson, A.M., Kolstad, H.A. et al., 2000, Male-mediated spontaneous abortion among spouses of stainless steel welders. *Scand J Work Environ Health.* 26:187-192.

Homonnai, Z.T., Paz, G.F., Weiss, J.N., David, M.P., 1980, Relation between semen quality and fate of pregnancy: retrospective study on 534 pregnancies. *Int J Androl.* 3:574-584.

Jensen, T.K., Henriksen, T.B., Hjollund, N.H., Scheike, T., Kolstad, H., Giwercman, A. et al., 1998a, Adult and prenatal exposures to tobacco smoke as risk indicators of fertility among 430 Danish couples. *Am J Epidemiol.* 148:992-997

Jensen, T.K., Henriksen, T.B., Hjollund, N.H., Scheike, T., Kolstad, H., Giwercman, A. et al., 1998b, Caffeine intake and fecundability: a follow-up study among 430 Danish couples planning their first pregnancy. *Reprod Toxicol.* 12:289-295.

Jensen, T.K., Hjollund, N.H., Henriksen, T.B., Scheike, T., Kolstad, H., Giwercman, A. et al., 1998c, Does moderate alcohol consumption affect fertility? Follow up study among couples planning first pregnancy. *BMJ.* 317:505-510.

Juul, S., Karmaus, W., Olsen, J., 1999, Regional differences in waiting time to pregnancy: pregnancy-based surveys from Denmark, France, Germany, Italy and Sweden. The European Infertility and Subfecundity Study Group. *Hum Reprod.* 14:1250-1254.

Katsoff, D., Kiefer, D., Check, M., Check, J.H., 1999, Oligoasthenozoospermia is not associated with spontaneous abortions following in vivo pregnancies contrasting with in vitro fertilization data. *Arch Androl.* 43:203-205.

Kiefer, D., Check, J.H., Katsoff, D., 1997, Evidence that oligoasthenozoospermia may be an etiologic factor for spontaneous abortion after in vitro fertilization-embryo transfer. *Fertil Steril.* 68:545-548.

Larson, K.L., DeJonge, C.J., Barnes, A.M., Jost, L.K., Evenson, D.P., 2000, Sperm chromatin structure assay parameters as predictors of failed pregnancy following assisted reproductive techniques. *Hum Reprod.* 15:1717-1722.

Marchetti, F., Lowe, X., Bishop, J., Wyrobek, A.J., 1999, Absence of selection against aneuploid mouse sperm at fertilization. *Biol Reprod.* 61:948-954.

Olshan, A.F., Faustman, E.M., 1993, Male-mediated developmental toxicity. *Reprod Toxicol.* 7:191-202.

Parazzini, F., Bocciolone, L., Vecchia, L., Negri, E., Fedele, L., 1990, Maternal and paternal moderate daily alcohol consumption and unexplained miscarriages. *Br J Obstet Gynaecol.* 97:618-622.

Parazzini, F., Chatenoud, L., di Cintio, E., Mezzopane, R., Surace, M., Zanconato, G. et al., 1998, Coffee consumption and risk of hospitalized miscarriage before 12 weeks of gestation. *Hum Reprod.* 13:2286-2291.

Robertson, J., Feldkamp, M., Jennings, J., Leen-Mitchell, M., Martinez, L., Carey, J.C., 1998, Maternal alcohol and caffeine consumption and spontaneous abortion. *Epidemiology.* 9:583-584.

Rubes, J., Lowe, X., Moore, D., Perreault, S., Slott, V., Evenson, D. et al., 1998, Smoking cigarettes is associated with increased sperm disomy in teenage men. *Fertil Steril.* 70:715-723.

Sailer, B.L., Jost, L.K., Evenson, D.P., 1995, Mammalian sperm DNA susceptibility to in situ denaturation associated with the presence of DNA strand breaks as measured by the terminal deoxynucleotidyl transferase assay. *J Androl.* 16:80-87.

Sbracia, S., Cozza, G., Grasso, J.A., Mastrone, M., Scarpellini, F., 1996, Semen parameters and sperm morphology in men in unexplained recurrent spontaneous abortion, before and during a 3 year follow-up period. *Hum Reprod.* 11:117-120.

Spano, M., Kolstad, A.H., Larsen, S.B., Cordelli, E., Leter, G., Giwercman, A. et al., 1998, The applicability of the flow cytometric sperm chromatin structure assay in epidemiological studies. Asclepios. *Hum Reprod.* 13:2495-2505.

Spano, M., Bonde, J.P., Hjollund, H.I., Kolstad, H.A., Cordelli, E., Leter, G., 2000, Sperm chromatin damage impairs human fertility. The Danish First Pregnancy Planner Study Team. *Fertil Steril.* 73:43-50.

Trasler, J.M., Hales, B.F., Robaire, B., 1985, Paternal cyclophosphamide treatment of rats causes fetal loss and malformations without affecting male fertility. *Nature.* 316:144-146.

Vine, M.F., Hulka, B.S., Margolin, B.H., Truong, Y.K., Hu, P.C., Schramm, M.M. et al., 1993, Cotinine concentrations in semen, urine, and blood of smokers and nonsmokers. *Am J Public Health.* 83:1335-1338.

Vine, M.F., Setzer, R.W., Jr., Everson, R.B., Wyrobek, A.J., 1997, Human sperm morphometry and smoking, caffeine, and alcohol consumption. *Reprod Toxicol.* 11:179-184.

Westergaard, H.B., Johansen, A.M., Erb, K., Andersen, A.N., 2000, Danish National IVF Registry 1994 and 1995. Treatment, pregnancy outcome and complications during pregnancy. *Acta Obstet Gynecol Scand.* 79:384-389.

Windham, G.C., Fenster, L., Swan, S.H., 1992a, Moderate maternal and paternal alcohol consumption and the risk of spontaneous abortion. *Epidemiology.* 3:364-70.

Windham, G.C., Swan, S.H., Fenster, L., 1992b, Parental cigarette smoking and the risk of spontaneous abortion. *Am J Epidemiol.* 135:1394-1403.

Windham, G.C., Von Behren, J., Waller, K., Fenster, L., 1999, Exposure to environmental and mainstream tobacco smoke and risk of spontaneous abortion. *Am J Epidemiol.* 149:243-247.

MECHANISMS OF MALE MEDIATED DEVELOPMENTAL TOXICITY INDUCED BY LEAD

Ellen K. Silbergeld[1], Betzabet Quintanilla-Vega[2], and Robin E. Gandley[3]

[1]University of Maryland Baltimore, Department of Epidemiology & Preventive Medicine, 10 South Pine Street, MSTF 9-34, Baltimore, MD 21201
[2]Toxicology Section, CINVESTAV-IPN, PO Box 14-740, Mexico City, 07000, Mexico
[3]Magee-Women's Research Institute, 204 Craft Avenue, Room 630, Pittsburgh, PA 15213

INTRODUCTION

Lead remains one of the most significant occupational and environmental hazards world-wide, despite major efforts to ban its use in gasoline and paints. While environmental exposures have generally fallen in countries such as Canada and the US, occupational exposures remain significant. Workers are exposed in major industries, such as mining, battery manufacture, and electronics; environmental exposures result from emissions to air from stationary sources such as smelters and incinerators, contamination of drinking water from lead plumbing, contact with lead based paint, and leaching of lead from ceramics and glassware. Over the past 10 years, many countries have adopted public health guidance to prevent lead poisoning in children, using current epidemiological and toxicological information to set a blood lead level of 10 mcg/dL as an indicator of potentially toxic exposures. However, occupational guidelines and standards adopted to prevent adult lead toxicity have not been changed for over 20 years. In most developed countries, occupational exposures are set to prevent blood lead elevations above 40 or 50 mcg/dL. These exposures are clearly unsafe since effects on neurological, renal, and nervous system functions have been documented in adults with these levels of lead in blood (WHO, 1990).

One of the original criteria for setting the occupational lead standard in the US was to protect men against reproductive toxicity (see Silbergeld et al., 1991). Data available at the time, mainly those published by Lancranjan et al (1975), indicated that the rate of spermatogenic abnormalities was increased among lead exposed men whose blood lead levels were in the range of 70 to 90 mcg/dL. More recently, in a prospective study of Danish lead workers, Viskum et al (1999) reported decreases in sperm motility and impaired performance in *in vitro* tests of sperm function in men with blood lead levels in

Advances in Male Mediated Developmental Toxicity, edited by Bernard Robaire and Barbara F. Hales. Kluwer Academic/Plenum Publishers, 2003.

the range of 30-40 mcg/dL. However, it has been generally assumed that the clinical outcome of such lead-induced spermatogenic changes would be a reduction in male fertility. In reproductive toxicology, even the effects of chromosomal or genetic damage in sperm are generally assumed to have the potential of being expressed as decreased fertility, as a consequence of intrauterine death or failure of early pre-implantation development, or abnormal progeny. In toxicology, the major test for male-mediated developmental toxicity remains the dominant lethal test, which measures embryonic loss as an endpoint.

The possibility that paternal lead exposures could impair the development of children has not been considered, although there is epidemiological literature suggesting such an association (reviewed by Olshan et al., 1991 and Faustman and Olshan, 1993). These epidemiological data, while still only suggestive, are limited by two important factors. First, it is practically impossible to eliminate co-exposures of the mother in studies of neonatal outcomes hypothesized to be related to paternal lead exposure. Studies of men rarely include measurements of lead exposure in women (Irgens et al., 2000; Kristensen et al., 1993), and studies of pregnant women have never included measurements of lead exposure in men. The environmental studies obviously include men and women, even if exposure assessment is only carried out on one parent, while for occupational studies, the female partners of the men in the study can be exposed by releases from the factory (very likely in cases of smelter workers) or by take-home exposures to lead (e.g., McDiarmid and Weaver, 1993). Second, the interpretation of these studies has been limited by lack of information to support the biological plausibility of observed associations. This limitation has also affected the interpretation of the few experimental studies on the associations of paternal lead exposure and offspring development (reviewed below). Until recently, there was relatively little understanding of the biological mechanisms by which paternal exposures could affect offspring development. Recent advances in developmental biology have provided significant increases in our understanding of the paternal role in development. Thus, it is appropriate to re-evaluate both the epidemiological and toxicological data on paternal lead exposure and developmental toxicity.

Experimental Studies of Male Mediated Effects of Lead on Development

Lead exposure to both men and experimental animals impairs reproductive capacity, including testicular toxicity, effects on spermatogenesis, and altered androgen metabolism (Winder, 1989; Alexander et al., 1996). Chronic low level exposures may compromise fertility in the absence of demonstrable effects on endocrine function and semen quality (Foster et al., 1996; Johansson and Wide, 1986). Lead is also a transplacental carcinogen (Waalkes et al., 1995). Associations between paternal exposure to low levels of lead and offspring development, in the absence of infertility or overt testicular damage, have been reported in two key studies published in the literature by Stowe and Goyer (1971) and by Brady et al. (1975). These authors reported effects on reproductive and learning behavior in the F_1 generation of rodents whose fathers were exposed to lead. Stowe and Goyer (1971) also reported persistent reproductive deficits in the F_1 males sired by exposed F_0 males, which they associated with impaired mating behavior. Brady et al. (1975) demonstrated deficits in learning and retention in the F_1 males and females, similar to findings reported in rodents exposed in utero via exposure of the mother. We followed up on these studies by examining the effects of paternal lead exposure on neurobiological development of Sprague Dawley rat pups. These studies were followed by investigations of early events in the pre-implantation embryo, and by mechanistic studies of lead effects on sperm.

Effects of Paternal Lead Exposure on Fertility and Neurodevelopment of F_1 and F_2 Offspring

We initiated our studies of paternal lead exposures by studying neurological development in rats. In all studies, adult male Sprague Dawley rats were exposed for five weeks to lead added to drinking water as lead acetate. This exposure encompassed the postmeiotic stages of spermatogenesis, which we selected on purpose. After five weeks, the males were placed with unexposed females and allowed to mate. Fertility was assessed by determining success in breeding (vaginal plug), production of live born offspring, and litter size.

Neurodevelopment was studied by electrophysiological and neuromorphological methods, using organotypic cultures of hippocampal cells derived from GD18 fetuses and PND1 neonates. Some of the F_1 offspring were kept through adulthood and then bred to unexposed animals. The F_2 offspring were then studied for transmitted effects of the exposures of F_0 males.

In all studies, adult male Sprague Dawley rats (10-12 weeks of age) were exposed for 35 days prior to breeding, by providing drinking water *ad libitum* to which lead acetate was added (0, 10, 25, or 250 ppm as lead). Blood lead levels were determined at specific times, using flameless atomic absorption spectrophotometry, by the Clinical Pathology Laboratory of the University of Maryland Medical Center, which has participated in the CDC proficiency testing system. Over the seven weeks of exposure, blood lead levels rose from a control range of 0.2-1.6 μg/dL to an average of 10.0 μg/dL in the 10 ppm group, 18.8 μg/dL in the 25 ppm group, and 60.2 μg/dL in the 250 ppm group. These exposures also increased concentrations of lead in testis and epididymis significantly. At the highest dose, lead levels in these tissues were elevated to approximately 2 - 3 times control levels (it should be noted that the concentrations of lead in "control" drinking water were measured to be 5.6 ppm).

Table 1. Effects of Paternal Lead Exposure on Fertility in Rats.

Dose (Generation)	% Breeding Success[a]	Litter Size	Superovulation[b]
Male-Female			
C-C (F_0)	83% (n=30)	15± 4	84% (n=31)
10-C (F_0)	71% (7)	17± 1	Not Done
25-C (F_0)	85% (20)	14± 5	81% (44)
250-C (F_0)	65% (20)	15± 2	77% (48)
C-C/C-C (F_1)	83% (12)	16± 4	80% (7)
250-C/C-C (F_1)	71% (14)	13± 6	19% (15)
C-C/250-C (F_1)	60% (10)	13± 4	46% (7)

[a]Fertility using free breeding was measured as the number of litters produced from the given breeding pairs.
[b]Fertility using superovulated control females was measured as 2-cell embryos produced verses the total number of oocytes available.

All males appeared to successfully mate within 3 - 5 days as evidenced by the presence of a vaginal plug. However, fertility (percent of pairs producing litters) was reduced in the 250 ppm males directly exposed to lead and also in their F1 male and female offspring, as shown in Table 1. No effect was observed on litter size. These results are consistent with those reported by Stowe and Goyer (1971), and indicate that paternal lead exposure may affect fertility in both male and female offspring. Stowe and Goyer (1971) suggested that these impairments were associated with behavioral changes; we did not measure mating behavior in our study.

Neurodevelopmental analyses were performed on hippocampal cells obtained from neonates. The hippocampus was selected because of its well-described sensitivity to lead during development (Silbergeld 1992; Kuhlman et al., 1997). In addition, the neurobehavioral toxicity, described by Brady et al (1975), in rats fathered by lead-exposed rats is consistent with hippocampal toxicity. For our studies, three to five hippocampi per litter from one-day old pups or fetuses were pooled after microdissection. The tissues were minced, trypsinized, and triturated, and then plated in modified Eagle's medium (with 10% horse serum and 10% fetal calf serum) at a density of 7.5×10^5 cells/2 ml. After 7 days, the hippocampal cultures consisted mainly of pyramidal cells; these cells were utilized for neuromorphological analyses using Neurolucida methods to measure area of cell bodies and dendritic arborization. In separate cultures, biochemical assessments of total protein synthesis were also done after one hour labeling with ^{35}S-methionine in methionine-deficient medium. Newly synthesized proteins were quantitatively measured by counting total incorporated radioactivity and also qualitatively evaluated with 2-dimensional gel electrophoresis.

Paternal lead exposure did not cause major changes in hippocampal cultures. Neural cultures from all litters grew to similar density by seven days, and total hippocampal protein synthesis, measured in TCA-precipitated proteins, was not affected by paternal lead exposure. These results indicated that paternal lead exposure did not have major effects on cell viability or general parameters of development *in vitro*. Using 2-dimensional gel electrophoresis, 10 major proteins were visualized by autoradiography. There were no significant differences in the relative labeling of these separated proteins in any litters.

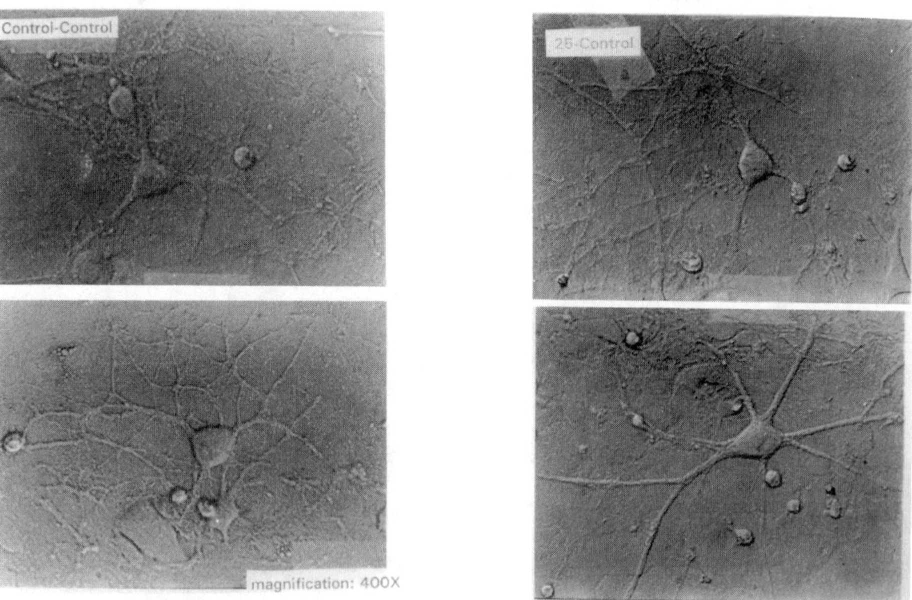

Figure 1A and 1B. Scanning photomicrographs of hippocampal cell cultures (400x) taken at 28 days in culture. 1A: cultures derived from fetuses fathered by control males. 1B: cultures derived from fetuses fathered by males exposed to 25-ppm lead in drinking water.

The morphometric analyses demonstrated differences between cultures from controls and from paternally lead-treated litters. As shown in figure 1, there were qualitative differences in the cultures, depending on paternal treatment. These differences were assessed quantitatively using Neurolucida (Fig. 2) imaging of individual neurons and computer-based assessment of cell body size and degree of branching or arborization. In comparison to cultures from control F_1 offspring, cultures from offspring of lead-treated parents (male, female, or both) tended to have more cells with larger cell body areas (Fig. 3). Analyses of dendritic branching indicated that either paternal or maternal exposures increased higher order branching (defined as the number of branches along a single axon, going from the cell body). In cultures derived from litters with both parents exposed to 250 ppm, the order of branching was decreased when compared to controls (Table 2).

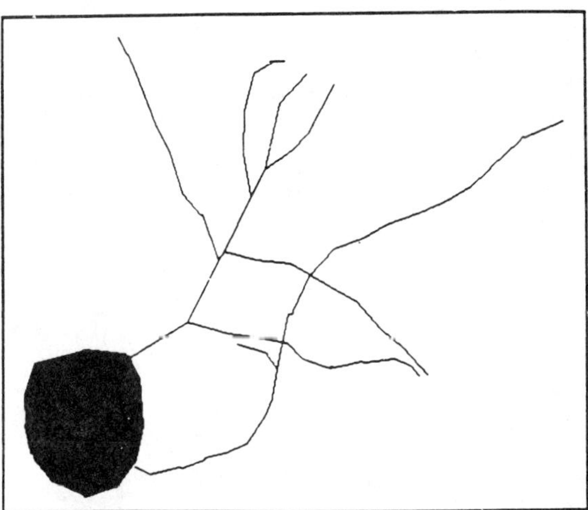

Figure 2. Neurolucida imaging of a pyramidal cell. The solid shape represents the cell body; axons and branches are represented by line drawings. From these images, cell body area and arborization (numbers of branch points going from the cell body) can be computed.

These data suggest a change in neural complexity. The more complex branching pattern observed in cultures of embryos with paternal exposures may be consistent with a failure of pruning, which may be consistent with observations in the visual cortex of postnatally exposed pups (Wilson et al., 2000). Pruning is a specific process by which certain neurons and neuronal processes are removed during early prenatal development in order to facilitate the attainment of the cytoarchitecture of the mature brain. As pointed out by Morrison (1993), interference with this precisely timed stage of neuronal development can compromise for the entire lifetime complex neurobiological functions.

We also conducted preliminary neurophysiological studies in hippocampal cells *in vitro*. For these studies, the hippocampal cells were obtained from GD17 fetuses, to ensure equivalent developmental stage. Single cell neurophysiological measurements were done over 28 days of culture. The results of these studies indicated that paternal lead exposure inhibits the development of mature patterns of NMDA-mediated changes in membrane conductance (Silbergeld et al., 1991). Similar results were reported by Nihei et al (1995) and by Ishihara et al (1995) on hipocampal cultures derived from mice exposed in utero to lead.

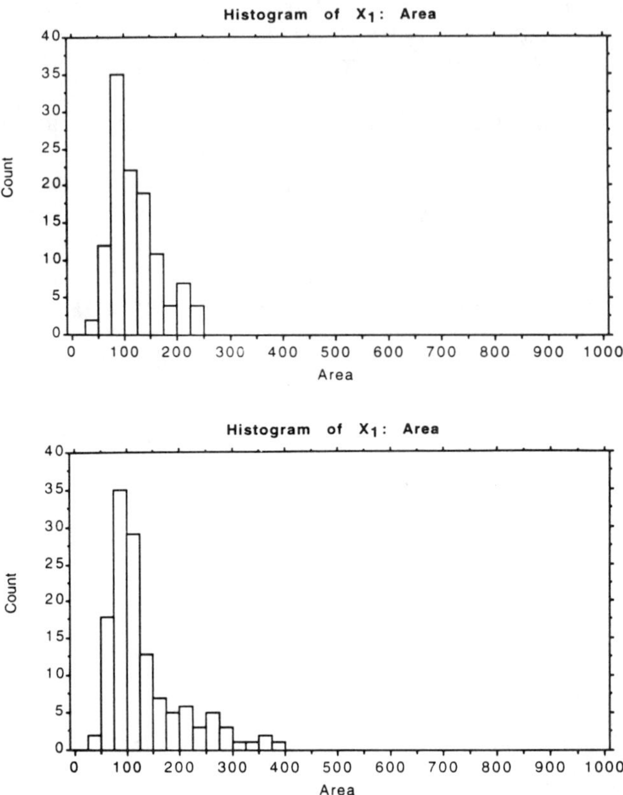

Figure 3. Distribution histogram of pyramidal cell body areas in cultures from controls (3A) and from paternally exposed fetuses (3B) (250-ppm lead). As can be seen in the lead cultures, there are significantly more neurons with cell body areas greater than 150 μm^2. ($p<0.01$, Mann Whitney test).

Table 2. Analysis of Dendrite Number per Cell for all Cells in a Field.

Group	Average Number of Dendrites by Order[a]					
	1st	2nd	3rd	4th	5th	6th
Control						
1	1.8±1.0	2.2±1.6	2.5±1.7	1.8±1.6	1--	
7	2.4±1.1	3.4±1.9	3.6±2.5	1.7±1.0	0.9±0.4	
250-250						
2	2.1±1.0	2.4±1.5	1.9±1.4	1.1±0.6	0.7--	
8	1.7±0.8	2.4±1.7	1.8±1.6	1.0±0.9	0.5--	
Maternal						
3	2.0±0.8	2.3±1.3	1.6±1.2	1.1±1.3	0.6±0.3	0.5--
6	2.3±1.3	3.9±2.1	3.5±2.3	2.5±2.2	1.7±1.7	0.7±.3
Paternal						
5	2.3±1.1	3.1±1.7	3.4±2.5	2.4±2.0	2.9±2.1	4.0--
4	2.8±0.9	4.1±1.4	3.4±1.8	1.8±1.6	0.9±1.3	0.3±.6

[a] The mean number of dendrites per cell ± the standard deviation. The groups were arranged for comparison of treatments as two Control groups, two 250-250 ppm groups. Maternal groups of 25 ppm then 250 ppm, and Paternal groups of 25 ppm then 250 ppm. – as SD indicates only one cell with the order being assessed.

These results confirm earlier reports that paternal lead exposure can induce neurodevelopmental and reproductive changes in animal models. The reproductive effects appear to be functional, in terms of reduced fertility, and to persist into the F1 generation at maturity. The observations of neurotoxicity in our studies did not include full functional characterization of neurobehavioral effects in offspring of lead-exposed males. Nonetheless, these data suggest possible neurobiological correlates for the neurobehavioral toxicity observed by Brady et al (1975) in offspring of lead-exposed male mice.

Mechanisms of Paternally Mediated Lead Toxicity.

We have examined potential mechanisms by which paternal lead exposure may induce developmental effects in offspring. These studies have focused on two aspects: (1) early embryonic events and (2) spermatologic events.

Early embryonic events. We hypothesized that male-mediated effects of lead should be detectable, at the molecular level, at the earliest stages of embryonic development. The male contribution to the zygote can be first observed in pre-implantation embryos when genes of paternal origin are expressed and generate new gene products (Wiekowski et al., 1991). To accomplish this, we modified our earlier study design, using superovulation to obtain a large number of fertilized oocytes. The male rats were exposed, as above, to 0, 25, or 250 ppm lead via drinking water for 5 weeks prior to breeding with unexposed females. The females were superovulated with pregnant mare serum gonadotropic and human chorionic gonadotropin, using an adaptation of murine methods in reproductive biology (Gandley et al., 1999). Approximately two days after mating, 2-cell embryos were recovered from females by flushing excised oviducts. The number of embryos was counted and embryo viability was assessed with methylene blue exclusion. At the higher lead exposure, there was a slight decrease in percentage of ovulated oocytes yielding 2-cell stage embryos, as compared to controls (Gandley et al 1999). No other effects were observed.

Protein synthesis was studied in 2-cell embryos using ^{35}S-methionine labeling and one- and two-dimensional gel electrophoresis (Gandley et al., 1999). Paternal lead exposure increased the amount of both total and newly synthesized proteins (Table 3) and the density of five proteins of similar molecular weight but differing in pI. We have not identified this family of apparent protein isoforms. While we have not ruled out post-expression or post-translational changes, the results indicate that paternal lead exposure may be evident at the level of early embryonic gene expression.

Table 3. 1-Dimensional Autoradiography and Silver Stain Data from 2-Cell Embryos.

Dose	Autoradiography		Silver Stain	
	Volume	% of Control	Volume	% of Control
Control	272.7	100%	1506	100%
25 ppm	386.4	142%	2123	141%
250 ppm	412.8	151%	2673	178%
Control	507.6	100%	1422	100%
25 ppm	659.3	130%	2114	149%
250 ppm	777.9	153%	2671	188%

The table above was generated using the Molecular Dynamics Densitomer. The volumes of bands from both silver strained gels and autoradiographic films were calculated with background substraction. The silver strained gels were loaded with equal embryo numbers for each dose. The autoradiographs were loaded by incorporated activity (cpm).

Effects of lead on sperm chromatin. Lead at subcytotoxic doses is not consistently mutagenic or clastogenic (Silbergeld et al., 2000; Johnson, 1998; Uzych, 1985). However, as reviewed by Johnson (1998), lead may have epigenetic effects through inhibition of DNA repair, inihition of DNA polymerase, or increased error prone repair (Hartwig, 1994). We have explored the hypothesis that lead may increase the potential for genetic damage in germ cells through its interactions with chromatin. The DNA in mammalian spermatozoa is tightly packaged in the nucleus with protamines, small arginine-rich proteins that replace histones during the later stages of spermatogenesis (Balhorn, 1990). The protamine-DNA complex leads to the formation of highly stable and condensed chromatin. The presence of protamines (HP1 and HP2 in humans) and their association with DNA appears to be important in assuring fertility, since some cases of male infertility have been associated with reduced amounts of human protamine 2 (HP2) or reduced binding of HP2 to DNA (Balhorn et al., 1988; deYebra et al., 1998). Moreover, sperm protamine plays an important role in regulating condensation-decondensation events critical to fertilization (Kvist, 1980).

Several studies have suggested that lead may affect male reproduction through incorporation of lead into sperm chromatin. Protamines contain cysteine residues that bind zinc, and HP2 is thought to be a zinc finger protein (Bianchi et al., 1992; Gatewood et al., 1986). Since lead binds tightly to free thiols (Goering, 1993), it may replace or compete with the zinc atoms that are normally bound to cysteine residues in nuclear protamines. This could result in any of the following: (1) prevention of normal disulfide bond formation within and among protamines during the final stages of sperm maturation, (2) alterations in DNA-protamine binding, or (3) prevention or delay in decondensation of sperm chromatin following fertilization. Effects of lead on chromatin structure have been reported by Foster et al (1996), and interference with decondensation has been reported by Johansson and Pellicciari (1988).

We have demonstrated that incubation of lead with purified human protamines alters protein structure and decreases DNA-protamine binding by direct interactions with both protamines and DNA (Quintanilla Vega et al., 2000). Our *in vitro* experiments with HP2 showed that lead interacts with HP2 at more than one binding site, one of which may not contain thiols. As shown in figure 4, HP2 incubated with lead binds less well to DNA. Similar findings have been reported for lead interactions with other DNA and RNA binding

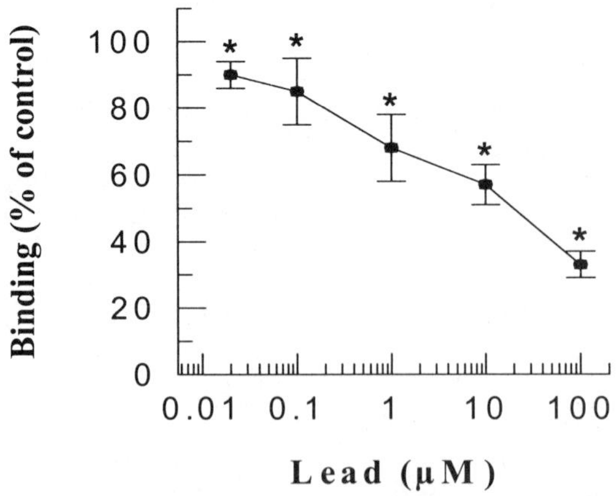

Figure 4. Effects of lead on protamine binding to DNA. Purified HP2 apoprotein was incubated with lead and DNA binding assessed by gel retardation assays, as described in Quintanilla-Vega et al (2001).

proteins with zinc fingers (Hanas et al., 1999; Razmiafshari et al., 2001). It has also been shown that cadmium, nickel and copper can bind to HP2 (Gatewood et al., 1986; Bal et al., 1997).

The timing of mammalian sperm nuclear decondensation and pronucleus formation is related to the S-S bond content of the sperm nucleus (Perreault et al., 1987). *In vivo* exposures have shown that lead is able to increase the stabilization of sperm chromatin, probably by forming more stable disulfide bonds in the sperm chromatin (Johansson and Pellicciari, 1988) or by decreasing the protection of free thiols by zinc (Hernández-Ochoa et al., 2001). This effect of lead on sperm chromatin might prevent or delay the S-S bond reduction that is necessary for sperm nuclear decondensation during fertilization, and consequently compromise male fertility.

CONCLUSIONS

Lead is an appropriate candidate for epidemiological, toxicological, and molecular research on male mediated developmental toxicity. In humans, there is significant but not yet conclusive evidence that lead can affect early development as a consequence of either paternal or maternal exposures. This evidence has stimulated some experimental research, and, in three separate rodent studies, paternal lead exposure has been associated with toxicity in offspring, as reviewed above. In addition, recent studies have reported that lead can induce cancer in offspring after transplacental exposures (reviewed by Silbergeld et al 2000), but no studies have followed paternally exposed offspring to determine if this outcome could also be induced by paternal exposures to lead, an effect that has been reported for chromium (Anderson et al., 1994)

Our studies, reviewed here, found effects of lead on neurological development, as determined by electrophysiological and neuromorphological studies of hippocampal cells cultured from fetuses and neonates sired by lead-exposed males. Whether these effects are associated with the neurobehavioral effects reported by Brady et al. (1975) is not known. Our studies produced effects at blood lead levels lower than the current OSHA standard (US) for protecting lead workers from adverse health effects. This standard (50 $\mu g/dL$) was set in 1978 on the basis of male reproductive toxicity of lead reported by Lancranjan et al (1975). As reported by CDC, the prevalence of elevated blood lead levels (>25 $\mu g/dL$) among adults in the US was 15-20% of those tested (mostly workers) (CDC, 1999).

At the time of this research, we speculated as to potential mechanisms by which paternal lead exposure might affect offspring. Either lead could be transmitted at the time of conception in the sperm or seminal fluid, or lead could alter genetic or nongenetic contributions of the sperm to the zygote. The first hypothesis has not been fully excluded, and concentrations of lead in ejaculate have been reported in animals and humans exposed to lead (Johansson and Wide 1986; Oldereid et al., 1993; Hernández-Ochoa et al., 2001). It is not known if these amounts of lead could induce developmental effects on embryos that persist through the F_1 generation. Hernández-Ochoa et al (2001) have also reported that lead exposures alter concentration of zinc in sperm, which may also affect early events in fertilization and embryogenesis.

The second hypothesis is perhaps more likely, that lead exposure to the F_0 male produces an effect on his germ cells that results in a permanent alternations in his offspring. While lead may induce genotoxic effects, either point mutation or chromosomal damage, the evidence is not consistent (Silbergeld et al., 2000). Moreover, there is no evidence for dominant lethality in studies of lead.

Our mechanistic studies suggest that lead acts on germ cell chromatin in exposed males. Our recent studies have shown that lead can affect the association of protamines with DNA (Quintanilla-Vega et al., 2000). The reduced association between protamines and DNA in the presence of lead may leave sperm chromatin and DNA open to damage from other exposures (Silbergeld et al., 2000). Sperm exposed to lead *in vitro* are also affected functionally in terms of sperm head condensation and decondensation. Whether these or other mechanisms are involved in the functional consequences observed in offspring at the early embryonic, late fetal, and postnatal stages remain to be demonstrated.

At the time we initiated this work in 1993, considerably less was known of the male contribution to offspring development. This conference marks the substantial progress that has been made in developmental biology to demonstrate the role of male factors in mammalian development (Trasler and Doerksen, 1999). Of particular interest is the potential role of paternally imprinted genes in male mediated developmental toxicity (Murphy and Jirtle, 2000; Tycko et al., 1997). Genomic imprinting is an epigenetic process that determines differences in allelic expression depending upon parent of origin (Bartolomei and Tilghman, 1997). While much focus has been on the contribution of paternal and maternal genes to the growth of trophoblast and placenta (Georgiades et al., 2001), other imprinted genes have been associated with neurobiological development. Studies using *C. elegans* suggests that defects in DNA binding proteins may play a role in gene silencing in germ cells (Jedrusik and Schulz, 2001). Thus the chromatin-related effects of lead mediated through lead-protamine interactions may be related to effects on the expression of imprinted genes.

Clearly, it is appropriate to continue investigations of lead, not only as a model compound for research in male-mediated developmental toxicity but also as a continuing occupational hazard to which millions of men continue to be exposed.

REFERENCES

Alexander, B.H., Checkoway, H., van Netten, C., Kaufman, J.D., Vaughan, T.L., Mueller, B.A., Faustman, E.M., 1996, Paternal occupational lead exposure and pregnancy outcome. *Int J Occup Environ Health.* 2:280-285.

Anderson, L.M, Kasprzak, K.S., and Rice, J.M., 1994, Preconception exposure of males and neoplasia in their progeny: effects of metals and consideration of mechanisms, in: Male-mediated developmental toxicity, A.F. Olshan and D.R. Mattison, eds., Plenum Press, New York, 129-140.

Balhorn, R., 1990, Mammalian protamines: Structure and molecular interactions, in: *Molecular Biology of Chromosome Function*, K.W. Adolph, ed., Springer, New York, 366-395.

Balhorn, R., Reed, S., and Tanphaichitr, N., 1988, Aberrant protamine1/protamine2 ratios in sperm of infertile human males. *Experientia.* 44:52-55.

Bal, W., Jezowska-Bojczuk, M., and Kasprzak, K.S., 1997, Binding of nickel(II) and copper (II) to the N-terminal sequence of human protamine HP2. *Chem Res Toxicol.* 10:906-914.

Bartolomei, M.S. and Tilghman, S.M., 1997, Genomic imprinting in mammals. *Annu Rev Genet.* 31:493-525.

Bianchi, F., Rousseaux-Prevost, R., Sautiere, P., and Rosseaux, J., 1992, P2 protamines from human sperm are zinc-finger proteins with one Cys2/His2 motif. *Biochem Biophys Res Commun.* 182:540-547.

Brady, K., Herrera, Y., and Zenick, H., 1975, Influence of paternal lead exposure on subsequent learning ability of offspring. *Pharmacol Biochem Behav.* 3:561-565.

CDC, 1999, Adult blood lead epidemiology and surveillance – United States, second and third quarters, 1998, and annual 1994-1997. *MMWR.* 48:213-223.

De Yebra, L., Ballescá, J.L., Vanrell, J.A., Corzett, M., Balhorn, R., and Oliva, R., 1998, Detection of P2 precursors in the sperm cells of infertile patients who have reduced protamine P2 levels. *Fertil Steril.* 69:755-759.

Faustman, E. and Olshan, A., 1993, Male-mediated developmental toxicity. *Annu Rev Public Health.* 14:159-181.

Foster, W.G., McMahon, A., and Rice, D.C., 1996, Sperm chromatin structure is altered in cynomolgus monkeys with environmentally relevant blood lead levels. *Toxicol Ind Health.* 12:723-735.

Gandley, R.E., Anderson, L.D., and Silbergeld, E.K., 1999, Lead: Male-mediated effects on reproduction and development in the rat. *Environ Res.* 80:355-363.

Gatewood, J.M., Schroth, G.P., Schmid, C.W., and Bradbury, E.M., 1990, Zinc-induced secondary structure transitions in human sperm protamines. *J Biol Chem.* 265:20667-20672.

Georgiades, P., Watkins, M., Burton, G.J., and Ferguson-Smith, A.C., 2001, Roles of genomic imprinting and the zygotic genome in placental development. *Proc Natl Acad Sci USA.* 98:4522-4527.

Goering, P.L., 1993, Lead-protein interactions as a basis for lead toxicity. *Neurotoxicology.* 14:45-60.

Hanas, J.S., Rodgers, J.S., Bantle, J.A., and Cheng, Y.G., 1999, Lead inhibition of DNA-binding mechanism of Cys(2)His(2) zinc finger proteins. *Mol Pharmacol.* 56:982-988.

Hartwig, A., 1994, Role of DNA repair inhibition in lead- and cadmium- induced genotoxicity: a review. *Environ Health Perspect.* 1-2:45-50.

Hernández-Ochoa, I., García-Vargas, G., Morán-Martínez, J., Rubio-Andrade, M., Vera, E., López-Carillo, L., Cebrián, M., and Quintanilla-Vega, B., 2001, Semen quality in environmentally metal-exposed men in northern México. *The Toxicologist.* 60:386.

Irgens, A., Kruger, K., Skorve, A.H., and Irgens, L.M., 2000, Birth defects and paternal occupational exposure. Hypotheses tested in a record linkage based dataset. *Acta Obstet Gynecol Scand.* 79:465-470.

Ishihara, K., Alkondon, M., Montes, J.G., and Albuquerque, E.X., 1995, Ontogenically related properties of N-methyl-D-aspartate receptors in rat hippocampal neurons and the age-specific sensitivity of developing neurons to lead. *J Pharmacol ExpTherap.* 273:1459-1470.

Jedrusik, M.A. and Schulze, E., 2001, A single histone H1 isoform (H1.1) is essential for chromatin silencing and germ line development in *Caenorhabditis elegans. Development.* 128:1069-1080.

Johansson, L. and Pellicciari, C.E., 1988, Lead-induced changes in the stabilization of the mouse sperm chromatin. *Toxicology.* 51:11-24.

Johansson, L., and Wide, M., 1986, Long-term exposure of the male mouse to lead: effects on fertility. *Environ Res.* 41:481-487.

Johnson, F.M., 1998, The genetic effects of environmental lead. *Mutat Res.* 410:123-140.

Kristensen, P., Irgens, L.M., Daltveit, A.K., and Andersen, A., Perinatal outcome among children of men exposed to lead and organic solvents in the printing industry. *Am J Epidemiol.* 137:134-144.

Kuhlmann, A.C., McGlothan, J.L., and Guilarte, T.R., 1997, Developmental lead exposure causes spatial learning deficits in adult rats. *Neurosci Lett.* 233:101-104.

Kvist, U., 1980, Importance of spermatozoal zinc as temporary inhibitor of sperm nuclear chromatin decondensation ability in man. *Acta Physiol Scand.* 109:79-84.

Lancranjan, I., Popescu, H.I., Gavanescu, O., Klepsch, I., and Serbanescu, M., 1975, Reproductive ability of workmen occupationally exposed to lead. *Arch Environ Health* 30:396-401.

McDiarmid, M.A. and Weaver, V., 1993, Fouling one's own nest revisited. *Am J Ind Med.* 24:1-9.

Morrison, J.H., 1993, Differential vulnerability, connectivity, and cell typology. *Neurobiol Aging.* 14:51-54.

Murphy, S.K. and Jirtle, R.L., 2000, Imprinted genes as potential genetic and epigenetic toxicologic targets. *Environ Health Perspect.* 108(Suppl 1):5-11.

Nihei, M.K., Desmond, N.L., McGlothan, J.L., Kuhlmann, A.C., and Guilarte, T.R., 2000, N-methyl-D-aspartate receptor subunit changes are associated with lead-induced deficits of long-term potentiation and spatial learning. *Neuroscience.* 99:233-242.

Oldereid, N.B., Thomassen, Y., Attramadal, A., Olaisen, B., and Purvis, K, 1993, Concentrations of lead, cadmium and zinc in the tissues of reproductive organs of men. *J Reprod Fertil.* 99:421-425.

Olshan, A.F., Teschke, K., and Baird, P.A., 1991, Paternal occupation and congenital anomalies in offspring. *Am J Ind Med.* 20:447-475.

Perreault, S.D., Naish, S.J., and Zirkin, B.R., 1987, The timing of hamster sperm nuclear decondensation and male pronucleus formation is related to sperm nuclear disulfide bond content. *Biol Reprod.* 36:239-244.

Quintanilla-Vega, B., Hoover, D.J., Bal, W., Silbergeld, E.K., Waalkes, M.P., and Anderson, L.D., 2000, Lead interaction with human protamine (HP2) as a mechanism of male reproductive toxicity. *Chem Res Toxicol.* 13:594-600.

Razmiafshari, M., Kao, J., d'Avignon, A., and Zawia, N.H., 2001, NMR identification of heavy metal-binding sites in a synthetic zinc finger peptide: toxicological implications for the interactions of xenobiotic metals with zinc finger proteins. *Toxicol Appl Pharmacol.* 172:1-10.

Silbergeld, E.K., 1992, Neurological perspective on lead toxicity, in: *Human Lead Exposure*, H. Needleman, ed., CRC Press, Boca Raton, FL, 89-103.

Silbergeld, E.K., Akkerman, M., Fowler, B.A., Albuquerque, E.X., and Alkondon, M., 1991, Lead: male-mediated effects on reproduction and neurological development. *Toxicology.* 11:235.

Silbergeld, E.K., Landrigan, P.J., Froines, J.R., and Pfeffer, R.M., 1991, The occupational lead standard: a goal unachieved, a process in need of repair, *New Solutions* Spring:20-30.

Silbergeld, E.K., Waalkes, M., and Rice, J.M., 2000, Lead as a carcinogen: experimental evidence and mechanisms of action. *Am J Ind Med.* 28:316-323.

Stowe, V.M. and Goyer, R., 1971, Reproductive ability and progeny of F1 lead-toxic rats. *Fertil Steril.* 22:755-760.

Trasler, J.M. and Doerksen, T, 1999, Teratogen update: paternal exposures—reproducive risks. *Teratology.* 60:161-172.

Tycko, B., Trasler, J., Bestor, T., 1997, Genomic imprinting: gametic mechanisms and somatic consequences. *J Androl.* 18:480-486.

Uzych, L., 1985, Teratogenesis and mutagenesis associated with the exposure of human males to lead: A review. *Yale J Biol Med.* 58:9-17.

Viskum, S., Rabjerg, L., Jorgensen, P.J., and Grandjean, P., 1999, Improvement in semen quality associated with decreasing occupational lead exposure. *Am J Ind Med.* 35:257-263.

Waalkes, M.P., Diwan, B.A., Ward, J.M., Devor, D.E., Goyer, R.A., 1995, Renal tubular tumors and atypical hyperplasias in B6C3F1 mine exposed to lead acetate during gestation and lactation occur with minimal chronic nephropathy. *Cancer Res.* 55:5265-5271.

WHO, IPCS, 1990, *Inorganic Lead*, World Health Organization: Geneva.

Wiekowski, M., Miranda, M., and DePamphilis, M.L., 1991, Regulation of gene expression in preimplantation mouse embryos: Effects of the zygotic clock and the first mitosis on promoter and enhancer activities. *Dev Biol.* 147:403-414.

Wilson, M.A., Johnston, M.V., Goldstein, G.W., and Blue, M.E., 2000, Neonatal lead exposure impairs development of rodent barrel field cortex. *Proc Natl Acad Sci USA.* 97:5540-5545.

Winder, C., 1989, Reproductive and chromosomal effects of occupational exposure to lead in the male. *Reprod Toxicol.* 3:221-233.

PATERNAL EXPOSURE TO KNOWN MUTAGENS AND HEALTH OF THE OFFSPRING: IONIZING RADIATION AND TOBACCO SMOKE

David A. Savitz, Ph.D.[1]

[1]Department of Epidemiology
School of Public Health
University of North Carolina
Chapel Hill, NC 27599-7435

INTRODUCTION

The potential influence of paternal exposures on the health of pregnancy and the offspring has been of interest to toxicologists and epidemiologists for some time. There are several lines of justification for pursuing the possibility that paternal exposures could affect development during pregnancy and beyond. First, the causes of the vast majority of common reproductive problems, including pregnancy loss, congenital defects, preterm birth, and fetal growth retardation have yet to be identified. In the face of limited success in studies that have focused primarily on fetal exposures through the pregnant woman, a broader conceptualization of potential causes is worthy of exploration. Second, there is extensive laboratory research which demonstrates that isolated exposures to the male before conception can result in a broad spectrum of adverse reproductive health outcomes, including pregnancy loss, congenital anomalies, and growth retardation (Olshan & Faustman, 1993). These studies do not prove that such phenomena occur in humans, but firmly demonstrate that they could. Finally, there is a social and political context for this avenue of investigation involving gender discrimination and equity. The idea that the male partner, not the woman alone, has some accountability for the success of the reproductive process has obvious implications for regulation of workplace exposures (i.e., the notion that developmental concerns extend beyond pregnant women), as well as part of the broader societal theme of increased male responsibility and accountability in parenting. The political context does not make it more or less likely that the father has influence beyond conception, but it does increase enthusiasm for research to address the possibility.

The epidemiologic literature thus far has identified a number of suggestive links primarily between paternal exposure and miscarriage (Savitz et al., 1994) and cancer in the offspring (Colt & Blair, 1998), with a much smaller literature addressing birth defects or fetal growth. Much of the research has focused on paternal occupational exposure, in part because of the hazardous agents that are sometimes found in the workplace but also

Advances in Male Mediated Developmental Toxicity, edited by Bernard Robaire and Barbara F. Hales.
Kluwer Academic/Plenum Publishers, 2003.

because job titles and imputed exposures are readily obtained. One might ask how these particular agents and outcomes have come to the forefront instead of other equally reasonable candidate exposures. The disproportionate emphasis on workplace exposures relative to diet, health behaviors, or medications comes in part from the regulatory concern with occupational hazards as well as the recognition that workplace environments often entail high levels of exposure to hazardous agents in a well-defined population (i.e., employees). Many of the epidemiologic studies of paternal influences on development have been opportunistic, in the form of add-on studies to research in which the primary question involves other outcomes or the other partner. For example, studies focused on potential male fertility may inquire about the health of the offspring or studies of the effects of exposures to pregnant women on development may inquire about the male partner's job. Despite the period of over 20 years that this issue has been of interest to a small number of epidemiologists, there are few if any studies in which the paternal role in development has been the primary focus of major research initiatives. As a result, the quality of information and analytic effort is undoubtedly less thorough.

What would be the ideal set of circumstances for examining the impact of paternal exposures on the health of the offspring? We would want a highly prevalent exposure and common adverse outcome to maximize statistical power. In laboratory experiments, the exposure allocation can be optimized, whereas in epidemiologic studies of occupational exposures and medications, there are relatively few men exposed in the critical interval prior to conception. The choice of outcome should be driven by a well-defined hypothesized biological mechanism. Another optimal feature of a paternal exposure is independence from maternal exposure, that is a lack of correlation based on shared behaviors (e.g., as occurs in tobacco, drug, or alcohol use) and lack of potential for paternal contamination of the maternal environment (e.g., lead exposure, environmental tobacco smoke). Paternal influences on the spouses' behavior are real and potentially quite important, as demonstrated in the partner's assistance or detrimental effect on abstinence from tobacco or illicit drug use, but this behavioral mechanism is not the focus here. Similarly, paternal exposures to the mother through contamination of clothing, sidestream tobacco smoke, or passage of agents through semen are of potential interest, but they really refer to the father merely as another element of the mother's environment. Our interest is in the unique paternal role that is mediated directly through alterations to the sperm resulting in mutation or modification of genetic imprinting to affect the development of the fetus and child.

For these pathways involving genetic damage to the sperm, the most promising exposures to consider would be to known mutagens that are encountered in high doses in sizable segments of the human population and measured with accuracy. Many occupational hazards fall short on the criteria of high exposure prevalence and accurate measurement. Few known mutagens are widespread, given the public concern with such agents and desire to avoid such exposures. This overview focuses on two of the most plausible paternal exposures that might result in developmental toxicity: ionizing radiation and tobacco.

Paternal tobacco use unfortunately remains highly prevalent, with 20-30% of the United States population smoking and a higher proportion elsewhere. The dose levels, relative to those found in the workplace or community environment, are extremely high. Tobacco smoke includes a spectrum of established mutagenic and carcinogenic agents. While the diversity of constituents makes it difficult to attribute its effects to any one chemical, it does increase the chances that one or more such agents will cause developmental toxicity. A major logistical strength is the ease of exposure assessment through self-report. Simply knowing the number of cigarettes smoked per day gives a great deal of information, and the stability of the habit makes it easy to extrapolate across the time periods of interest. There is the potential for environmental tobacco smoke to affect the spouse, as well as the known correlation of smoking among partners, but maternal exposure can be assessed as well and controlled statistically. For many outcomes, maternal smoking is not such an overpowering influence that residual confounding of any paternal effect is a major concern.

External ionizing radiation is rarely encountered in high doses at present, but it has the distinctly favorable attributes of being measured with precision through monitoring with film badges, generating precise information on timing of exposure relative to the time of sperm production resulting in conception. Because ionizing radiation has been extensively studied for many years, it is now a rather well understood mutagen. There is little potential for maternal exposure to external ionizing radiation through tracking exposures at home or correlated couple behaviors.

The endpoints chosen for evaluation include pregnancy loss, congenital defects, and preterm birth/fetal growth. Pregnancy loss is a common event (>10% of recognized conceptions, perhaps 25-30% of all pregnancies), and has diverse causes most of which are currently unknown. There is a clear role for genetics, as demonstrated by the frequency of aneuploidy in early losses (Kline et al., 1989). Logistically, pregnancy loss is increasingly difficult to study as one moves to earlier gestational ages because of the incomplete recognition of pregnancy itself. Known causes include advanced maternal age and probably tobacco use, with prior fetal losses predicting future ones.

Congenital defects are individually rare but collectively rather common, with major defects affecting 3-5% of births. Unfortunately, it seems that the search for etiology requires splitting them into their component groups. The causes are largely unknown with the exception of certain medications (e.g., thalidomide) and low folate intake affecting neural tube defects and perhaps other malformations.

Preterm birth and reduced fetal growth are among the major contributors to infant mortality and quite common depending on the exact measure. Birth prior to the completion of 37 weeks' gestation occurs in approximately 8% of births in the United States, less in other developed countries. The focus of research has been on the intrauterine environment, but the few established predictors of preterm birth are multiple gestations, history of preterm birth, tobacco use (with a modest effect), genital tract infection, and low prepregnancy weight (Savitz & Pastore, 1999).

While new information on the developmental consequences of paternal exposure to tobacco smoke and ionizing radiation will not have regulatory impact, there is substantial scientific value to considering the effects of these known mutagens for understanding mechanisms and especially to address the plausibility of paternal effects on reproduction more generally. If these agents are not associated with development, then how likely is it that solvents, pesticides, and other chemicals will have a discernible effect in epidemiologic studies? This paper summarizes the epidemiologic evidence linking paternal exposure to ionizing radiation and tobacco to spontaneous abortion, birth defects, preterm birth, and intrauterine growth retardation.

METHODS

An effort was made to identify published reports on the association between paternal exposure to tobacco or ionizing radiation in relation to spontaneous abortion, birth defects, preterm birth, and intrauterine growth retardation. Citations were found through computerized literature searches, following up references in published articles, and through review papers. Not every paper with potentially relevant data was included, in that I excluded published reports that inferred ionizing radiation exposure from job title alone. The focus was on studies in which the exposure to ionizing radiation was monitored and quantified for the analysis. While studies of "nuclear workers" or "x-ray technicians" might provide relevant data, one of the reasons for choosing to focus on ionizing radiation was the opportunity to consider a uniquely well-quantified exposure.

For paternal tobacco use, some papers that focused on other aspects of paternal exposure, e.g., occupation, but included information on tobacco use as a potential confounding factor, were not included. Where risk estimates for tobacco were presented, that information was included, but it is likely that the compilation of data is incomplete to at least some extent.

Ionizing radiation would operate purely and directly as an exposure to the sperm, causing mutations or other abnormalities that could be passed to the offspring. In contrast, paternal tobacco use has been studied as often as a source of maternal environmental tobacco smoke as it has been for any sperm-mediated effects on the offspring. Tobacco products are also found in the semen, resulting in another potential exposure pathway. In principle, environmental tobacco smoke exposure has multiple sources other than the father, such as other household members and workplace exposures, so that it would not be possible to isolate the father from others. Conversely, it might be possible to have fathers who smoke but not in the presence of their pregnant spouse. In practice, these distinctions are very difficult to make, so that paternal smoking is an exposure with many possible routes by which the offspring could be affected. Although this poses a challenge for understanding the specific mechanism by which it might operate, it does offer an agent that is commonly encountered, in high doses, with multiple active agents that could operate through any of several biological pathways.

RESULTS

Tobacco

Few studies have addressed paternal tobacco smoke exposure and miscarriage (Table 1). The only studies that reported elevated relative risks found modest elevations (relative risks of under 1.5) (Rupa et al., 1991; Lindbohm et al., 1991). The remaining six epidemiologic studies of tobacco use and miscarriage have yielded relative risks very close to or below the null. Nevertheless, it should be noted that none of the studies combined a detailed assessment of miscarriage with a rigorous evaluation and analysis of paternal tobacco use. That is, one measure or the other was lacking in detail, often both. While the results thus far do not encourage study of this association, they certainly fall far short of putting the question to rest.

The literature on paternal tobacco use and birth defects is more advanced, with several reports building on high quality birth defects research programs (Table 1). The challenge here is to consider the diversity of specific types of birth defects and reconcile findings across the studies. Sporadic positive findings have been reported for all defects, with relative risks of 2.0 (Mau & Netter, 1974) and 1.2 (Zhang et al., 1992). Specific birth defects have generated inconsistent results, with associations of note found for ventricular septal defects (Savitz et al., 1991), foot deformities (Zhang et al., 1992), and limb reduction defects (Wasserman et al., 1996). Two studies have reported increased risk of neural tube defects among offspring of smoking fathers (Zhang et al., 1992; Hearey et al., 1984), but the latter study was based on only 4 exposed cases. None of the associations have been replicated across studies, however, and the magnitudes of association have been modest. At present the findings are a rather scattered set of measures with limited attempt thus far to seek concordant information.

Studies of tobacco use and preterm birth or fetal growth are even scarcer, despite a more substantial body of research on environmental tobacco smoke that includes but does not isolate the paternal smoking contribution. One study suggests a small decrement in birth weight associated with paternal smoking (-34 grams) (Martinez et al., 1994), but it would be difficult to fully isolate any paternal genetic effect from an effect of environmental tobacco smoke.

Table 1. Paternal Tobacco Use and Pregnancy Outcome: Miscarriage and Birth Defects

Reference	No. Exposed Cases	Exposure	Relative risk	95% CI
MISCARRIAGE				
Tokuhata, 1968	290	Father smoking/mother not	1.1	
Spira & Lazar, 1979	Not stated	Any smoking	"no association"	
Beckman & Nordstrom, 1982	66	Any smoking	1.0	
Taskinen et al., 1989	56		0.6	0.4-1.0
Halmesmaki et al., 1989	28	Any smoking	0.9	
Rupa et al., 1991			1.4	1.2-1.5
Lindbohm et al., 1991	110	Any smoking	1.3	0.9-1.9
Windham et al., 1992			1.0	
BIRTH DEFECTS (specific type)				
Mau & Netter, 1974				
All malformations	47	Any smoking	2.0	
Holmberg & Nurminen, 1980				
CNS defects	68	Any smoking	"not significant"	
Beckman & Nordstrom, 1982				
All malformations	24	Any smoking	1.0	
Hearey et al., 1984				
Neural tube defects	4	Any smoking	16.0	
Roeleveld et al., 1992				
Mental retardation	207	Any smoking	1.2	0.8-1.6
	22	Cigar/pipe smoking	2.4	1.2-5.1
Zhang et al., 1992				
All malformations	1012	Any smoking	1.2	1.0-1.5
Anencephalus/spina bifida	15	Any smoking	1.4-3.2	
Cleft lip +/- palate	24	Any smoking	1.0-1.2	
Cleft palate	8	Any smoking	1.4-1.7	
Undescended testicle	25	Any smoking	1.2-2.1	
Polydactyly	20	Any smoking	0.8	
Down's syndrome	9	Any smoking	0.6-1.2	
Varus/vagus foot deformities	14	Any smoking	1.7-1.9	
Savitz et al., 1991				
Refractive errors	147	Any smoking	0.8	0.6-1.1
Cleft lip +/- palate	10	Any smoking	1.7	0.5-6.0
Cleft palate	5	Any smoking	0.9	0.2-3.6
Neural tube defects	5	Any smoking	0.6	0.2-2.5
Ventricular septal defect	31	Any smoking	2.0	0.9-4.3

Ionizing Radiation

Despite the plausibility of a paternally mediated effect, there have been few studies conducted to address the role of ionizing radiation in pregnancy loss (Table 2). None of the studies that use sophisticated methods of exposure assessment have considered miscarriage. Two recent studies have addressed stillbirth among workers exposed to ionizing radiation, and one reported a small increase in risk (Parker et al., 1999) but the other did not (Doyle et al., 2000). Later pregnancy loss can be ascertained reliably from vital records, whereas much more intensive study is required to consider loss before 20 weeks' gestation. Thus, paternal exposure to ionizing radiation and miscarriage remains virtually unexplored.

Table 2. Paternal Exposure to Ionizing Radiation and Pregnancy Outcome: Miscarriage, Birth Defects, and Fetal Growth

Reference	No. Exposed Cases	Exposure	Relative risk	95% CI
MISCARRIAGE				
Boue et al., 1975		Ionizing radiation	1.3	1.1-1.5
Spira & Lazar, 1979		Occupational & medical radiation	1.2	
McDonald et al., 1989		Ionizing radiation	0.9	0.6-1.3
McDonald et al., 1989	18	Probable ionizing radiation	0.9	0.6-1.3
Lindbohm et al., 1991b		X-rays	1.5	0.6-3.6
Parker et al., 1999				
Stillbirth	229 (total)	Preconceptional dose	1.3	1.0-1.5
Stillbirth with congenital anomaly		Preconceptional dose	1.4	0.9-1.9
Stillbirth with neural tube defect		Preconceptional dose	1.7	1.1-2.3
Doyle et al., 2000				
Fetal death	2687	Monitored, >100 mSv	1.1	0.9-1.4
BIRTH DEFECTS				
Green et al., 1997				
Nervous system	5	Dose 6 months before conception	1.2	0.3-4.6
Circulatory system	11	Dose 6 months before conception	0.9	0.4-2.0
Digestive system	10	Dose 6 months before conception	0.8	0.3-1.9
Genitoruinary	12	Dose 6 months before conception	1.0	0.3-2.6
Musculoskeletal	21	Dose 6 months before conception	0.9	0.5-1.6
Doyle et al., 2000				
Congenital malformations	369	Monitored	1.0	0.8-1.2
Neural tube defects	41	Monitored	1.1	0.3-1.5
Heart malformations	49	Monitored	0.7	0.4-1.1
Cleft lip/palate	23	Monitored	1.4	0.6-3.6
Down's syndrome	19	Monitored	0.6	0.3-1.3
FETAL GROWTH				
Shea et al., 1997				
Adjusted birth weight	164	Preconception x-rays	- 53 grams	

Two reports were identified that linked monitoring data on paternal exposure to ionizing radiation with congenital defects in the offspring (Green et al., 1997; Doyle et al., 2000). Relative risks across the multiple malformations considered in the two studies with 5 or more exposed cases ranged from 0.6 to 1.4, providing little indication of an association but also very little data for drawing firm conclusions.

One study examined a possible role of paternal x-ray exposure in relation to birth weight (Shea et al., 1997) and found a small decrease in birth weight.

DISCUSSION

Despite the compelling biologic rationale and relative ease of study, research on paternal tobacco use and pregnancy outcome has been quite limited. A review of the literature on paternal occupation and miscarriage in 1994 identified 39 pertinent studies, and there have undoubtedly been others published after the cutoff for inclusion in that paper. The literature on paternal occupation and childhood cancer is much more extensive than for miscarriage. If the goal is to determine whether paternal exposures in humans are truly capable of affecting the offspring, workplace exposures are certainly of interest but may not be the most promising or at least the only research direction to pursue.

In contrast, tobacco use is among the easiest of behaviors to ascertain, with reasonably complete information obtained from the simple question "how many cigarettes do you smoke each day?" Male habits are very unlikely to change markedly around the time of conception, the prevalence is high enough to study in unselected populations, and the information can be validated readily with biologic markers, at least for concurrent assessment. Perhaps if we are to pursue paternal exposure and health of the offspring, there is a need to develop studies that explicitly take up the challenge of focusing on exposures and circumstances most amenable to definitive study and not simply respond to questions about one environmental agent or another as the issue arises.

Given the limited number and quality of studies to this point, it is far too early to draw the conclusion that paternal tobacco exposure does not affect the outcomes that were considered. The more compelling observation is how limited the research itself has been, with any judgments deferred awaiting expansion of the research.

Ionizing radiation studies are even more limited in quality and quantity than those for tobacco. Despite the construction of numerous cohorts of tens of thousands of men with documented doses of external ionizing radiation, the only application to the outcomes of interest here has come from studies of stillbirth in selected workers in the United Kingdom. Remarkably, not a single study in the United States has reported on the relation between paternal exposure and health of the offspring, representing a missed opportunity to examine the most carefully monitored and best-understood mutagen that could be found. Even the tremendously valuable data generated on atomic bomb survivors has not been analyzed to isolate and address the potential role of paternal exposure in relation to stillbirth and congenital defects, despite the potential for doing so. The more recent analyses considered joint parental exposure to ionizing radiation and reported a very small positive association between exposure and risk of congenital malformations in the aggregate and a very small inverse association with risk of stillbirth (Otake et al., 1990), but these data have little or no value in assessing a potential paternal effect.

What should be done next? Though a case could be made for launching a massive epidemiologic study of tobacco or ionizing radiation or other exogenous agents as potential causes of pregnancy loss or adverse pregnancy outcome, it would be difficult to make the case compelling at this time. The experimental literature supports the plausibility of male-mediated effects in a general way, but it does not make a strong mechanistic case for these exposures affecting specific outcomes. The epidemiologic literature is too weak to provide the support needed, and the evidence that has accrued is largely negative. Nevertheless, the rationale and data would certainly support opportunistic evaluations of these pathways. That is, in a study designed to examine female exposures and spontaneous abortion, for example, some information on the father's age, smoking habits, and occupation would be worthwhile to collect and analyze. Such an approach would reflect investment commensurate with the importance of the topic. A good argument could be made to link the male cohorts with information on ionizing radiation exposure to reproductive health outcomes, if feasible, to address in rather precise, quantitative terms the relation between radiation and health of the offspring. Such studies would be more challenging, but feasible.

Another strategy is to try to fill in some of the gulf between epidemiologic studies and laboratory investigation. To use the analogy of studies of infertility, we could focus exclusively on clinical outcomes, i.e., delayed conception, and try to integrate that information with literature on studies in experimental systems to assemble the evidence. However, the ability to directly study semen parameters among humans has been tremendously valuable to bridge between clinical outcomes and animal studies. Studies with this objective for male-mediated developmental toxicity are just beginning, but the search for biological markers of outcomes in humans that are pertinent both to the hypothesized mechanisms identified in animal studies and to the developmental outcomes of interest would be of tremendous value. Markers in humans that are more sensitive and specific indicators of male exposures and have relevance to the health of the offspring should be enthusiastically supported as a means for moving the entire field forward.

ACKNOWLEDGMENTS

I would like to thank Dr. Andrew Olshan for helpful comments on this chapter.

REFERENCES

Beckman, L., and Nordstrom, S., 1982, Occupational and environmental risks in and around a smelter in northern Sweden. *Hereditas.* 97:1-9.

Boue, J., Boue, A., and Lazar, P., 1975, Retrospective and prospective epidemiological studies of 1500 karyotyped spontaneous human abortions. *Teratology.* 12:11-26.

Colt, J.S., and Blair A., 1998, Parental occupational exposures and risk of childhood cancer. *Environ Health Perspect.* 106:909-926.

Doyle, P., Maconochie, N., Roman, E., Davies, G., Smith, P.G., and Beral V., 2000, Fetal death and congenital malformation in babies born to nuclear industry employees: report from the nuclear industry family study. *Lancet.* 356:1293-1299.

Green, L.M., Dodds, L., Miller, A.B., Tomkins, D.J., Li, J., and Escobar, M., 1997, Risk of congenital anomalies in children of parents occupationally exposed to low level ionizing radiation. *Occup Environ Med.* 54:629-635.

Halmesmaki, E., Valimaki, M., Roine, R., Ylikahri, R., and Ylikorkala, O., 1989, Maternal and paternal alcohol consumption and miscarriage. *Br J Obstet Gynaecol.* 96:188-191.

Hearey, C.D., Harris, J.A., Usatin, M.S., and Epstein, D.M., 1984, Investigation of a cluster of anencephaly and spina bifida. *Am J Epidemiol.* 120:559-564.

Holmberg, P.C., and Nurminen, M. 1980, Congenital defects of the central nervous system and occupational factors during pregnancy. a case-referent study. *Am J Ind Med.* 1:167-176.

Inskip, H., 1999, Stillbirth and paternal preconceptional radiation exposure (commentary). *Lancet.* 354:1400-1401.

Kline, J., Stein, Z., and Susser, M., 1989, *Conception to Birth: Epidemiology of Prenatal Development,* Oxford University Press, New York 83-84.

Lindbohm, M-L., Sallmen, M., and Anttila, A., 1991, Paternal occupational lead exposure and spontaneous abortion. *Scand J Work Environ Health.* 17:95-103.

MacMahon, B., Alpert, M., and Salber, E.J., 1966, Infant weight and parental smoking habits. *Am J Epidemiol.* 82:247-261.

Martinez, F.D., Wright, A.L., Taussig, L.M. and the Group Health Medical Associates, 1994, The effect of paternal smoking on the birthweight of newborns whose mothers did not smoke. *Am J Public Health.* 84:1489-1491.

Mau, G., and Netter, P., 1974, Effects of paternal cigarette smoking on perinatal mortaliy and incidence of malformations. *Dtsch Med Wochenschr.* 99:1113-1118.

McDonald, A.D., McDonald, J.C., Armstrong, B., Cherry, N.M., Nolin, A.D., and Robert, D., 1988, Fathers' occupation and pregnancy outcome. *Br J Ind* Med. 46:329-333.

Olshan, A.F., and Faustman, E., 1993, Male-mediated developmental toxicity. *Annu Rev Public Health.* 14:159-181.

Otake, M., Schull, W.J., and Neel, J.V., 1990, Congenital malformations, stillbirths, early mortality among the children of atomic bomb survivors: a reanalysis. *Radiat Res.* 122:1-11.

Parker, L., Pearce, M.S., Dickinson, H.O., Aitkin, M., and Craft, A.W., 1999, Stillbirths among offspring of male radiation workers at Sellafield nuclear reprocessing plant. *Lancet.* 354:1407-1414.

Roeleveld, N., Vingerhoets, E., Zielhuis, G.A., and Gabreels, F., 1992, Mental retardation associated with parental smoking and alcohol consumption before, during, and after pregnancy. *Prev Med.* 21:110-119.

Rupa, D.S., Reddy, P.P. and Riddi, O.S., 1991, Reproductive performance in population exposed to pesticides in cotton fields in India. *Environ Res.* 55:123-128.

Savitz, D.A., Schwingl, P.J., and Keels, M.A., 1991, Influence of paternal age, smoking, and alcohol consumption on congenital anomalies. *Teratology.* 44:429-440.

Savitz, D.A., Pastore, and L.M., 1999, Causes of prematurity, in: *Prenatal Care Effectiveness and Implementation*, McCormick M.C., Siegel J.E., eds., Cambridge University Press, UK 63-104.

Savitz, D.A., Sonnenfeld, N.L, and Oshan, A.F., 1994, Review of epidemiologic studies of paternal occupational exposure and spontaneous abortion. *Am J Ind Med.* 25:361-383.

Shaw, G.M., Wasserman, C.R., Lammer, E.J., O'Malley, C.D., Murray, J.C., Basart, A.M., and Tolarova, M.M., 1996, Orofacial clefts, parental cigarette smoking, and transforming growth factor-alpha gene variants. *Am J Hum Genet.* 58:551-561.

Shea, K.M., Little, R.E., and the ALSPAC study team, 1997, Is there an association between preconception paternal x-ray exposure and birth outcome?. *Am J Epidemiol.* 145:546-551.

Spira, A., and Lazar, P., 1979, Spontaneous abortions in sibship of children with congenital malformation or malignant disease. *Europ J Obstet Gynec Reprod Biol.* 9:89-95.

Taskinen, H. Anttila, A., Lindbohm, M-L., Sallmen, M., and Hemminki, K., 1989, Spontaneous abortions and congenital malformations among the wives of men occupationally exposed to organic solvents. *Scand J work Environ Health.* 15:345-352.

Tokuhata, G.K., 1968, Smoking in relation to infertility and fetal loss. *Arch Environ Health.* 17:353-359.

Wasserman, C.R., Shaw, G.M., O'Malley, C.D., Tolarova, M.M. and Lammer, E.J., 1996, Parental cigarette smoking and risk for congenital anomalies of the heart, neural tube, or limb. *Teratology.* 53:261-267.

Windham, G.C., Swan, S.H., and Fenster, L., 1992a, Parental cigarette smoking and the risk of spontaneous abortion. *Am J Epidemiol.* 135:1394-1403.

Zhang, J., Savitz, D.A., Schwingl, P.J., and Cai, W-W, 1992, A case-control study of paternal smoking and birth defects. *Int J Epidemiol.* 21:273-278.

FISH (FLUORESCENCE IN SITU HYBRIDIZATION) TO DETECT EFFECTS OF SMOKING, CAFFEINE, AND ALCOHOL ON HUMAN SPERM CHROMOSOMES

Wendie A. Robbins

Center for Occupational and Environmental Health
University of California, Los Angeles
Los Angeles, CA 90095-6919

INTRODUCTION

Sperm chromosomal damage can serve as a marker of adverse effects on the male reproductive system following exposures during sensitive periods of spermatogenesis. A number of laboratory assays have been developed to directly detect and measure chromosome damage in human sperm cells. Fluorescence in situ hybridization (FISH) to detect damage at the level of the chromosome is one example. FISH has been applied in clinical andrology, epidemiology, and reproductive toxicology studies to evaluate effects of workplace, environmental, pharmacologic, and lifestyle exposures on sperm, as well as to assess potential risks for male mediated effects. Laboratory techniques like FISH increase our ability to tease out relationships with increasingly smaller sample sizes, however, are not a panacea for poorly designed or underpowered investigations. Appropriate application and interpretation of laboratory findings within methodologically solid studies is still a basic requirement. The application of sperm-FISH to assess effects of tobacco smoke, caffeine, and alcohol exposures on human sperm chromosomes will be discussed in this chapter.

SPERM-FISH ASSAY IN SEMEN FIELD STUDIES

Sperm-fluorescence in situ hybridization (sperm-FISH) is a relatively simple approach in which DNA probes (fluorescently labeled chromosome-specific sequences) are hybridized to blocks of repetitive DNA on specific chromosomal regions of interest within interphase sperm nuclei. The fluorescent labels are then visualized, scored, and interpreted depending on the labeling scheme. For example, single and multi-color probes of specific sequences from highly repeated human satellite DNA located at the centromere can be used to detect chromosome specific ploidy (Robbins et al. 1993; Downie et al., 1997; Morel et al., 1997) or multicolor probes to chromosomes X, Y plus autosomes to determine meiosis I versus meiosis II errors (Williams et al., 1993; Chevret et al., 1995; Finkelstein et al.,

1998; Martinez-Pasarell et al., 1999). Multicolor probes to adjacent regions of repetitive DNA on the same chromosome can be used to detect chromosome breakage (Rupa et al., 1995; Chen and Eastmond, 1995), and multicolor probes to telemeric plus centromeric regions can be used to describe recombination and meiotic segregation patterns in reciprocal translocation carriers (Van Hummelen et al., 1997). Sperm-FISH has been shown to be reliable for estimation of aneuploidy (Martin and Rademaker, 1995) and diploidy (Rademaker et al., 1997) and has been validated with the human-sperm/hamster-egg sperm karyotyping technique (Robbins et al., 1993; Martin et al., 1996). Due to minimal requirements involved in sample storage and processing for the sperm-FISH assay plus ability to score thousands of sperm cells per sample in relatively short periods of time, sperm-FISH lends itself well to human semen field studies.

Sperm-FISH data have been published on more than 500 men and greater than five million sperm from laboratories around the world (reviewed by Shi and Martin, 2000). Perhaps the most common application of sperm-FISH has been to detect and measure abnormal numbers of chromosomes in ejaculated sperm cells to determine baseline frequencies for the human (Shi and Martin, 2000). Researchers have also incorporated sperm-FISH assays in studies of associations between specific in vivo exposures that might induce aneuploidy and polyploidy in sperm. Exposures studied include pesticides (Rupa et al., 1999; Padungtod et al., 1999; Härkönen et al., 1999; Recio et al., 2001), cancer chemotherapy (Robbins et al., 1997a, Martin et al., 1997; Monteil et al., 1997), air pollution (Robbins et al., 1999), and lifestyle behaviors such as tobacco smoking, alcohol, and caffeine intake. Studies related to tobacco smoking, alcohol and caffeine intake and their association with abnormal numbers of chromosomes in human sperm will be reviewed in this chapter.

TOBACCO SMOKING AND SPERM ANEUPLOIDY

A persuasive body of literature documents adverse effects of cigarette smoking on male reproductive functions including decreased sperm count, poor motility, abnormal morphology, retained cytoplasm, ultrastructural abnormalities, DNA adducts and reduced fertility (Evans et al., 1981; Handelsman et al., 1984; Stillman et al., 1986; Vogt et al., 1986; Saaranen et al., 1987; Vine et al., 1994; Little et al., 1994; Sofikitis et al., 1995; Vine, 1996; Vine et al., 1997; Shen et al., 1997, Curtis et al., 1997; Zavos et al., 1998; Zenzes et al., 1999; Shen and Ong, 2000; Wong et al., 2000; Zinaman et al., 2000; Mak et al., 2000; Wang et al., 2001). It has also been shown that smoking can interact with workplace exposures to exacerbate adverse effects on semen quality (Wang et al., 2001). This synergistic action is of concern. As pointed out by Meistrich (1986) high doses of many agents may be reproductive toxins but the ones of primary concern are those that would adversely affect male fertility at doses to which human populations are likely to be exposed. It may be that smoking potentiates susceptibility to moderate doses of putative reproductive toxicants that are likely to be found in everyday environments and workplaces of the human.

The exact mechanism through which cigarette smoke can affect sperm is still being investigated, however, several pathways seem likely. Cigarette smoke contains a myriad of mutagenic, aneugenic (aneuploidy inducing, causing abnormal numbers of chromosomes in a cell), and clastogenic (chromosome breaking) agents that may act directly on spermatogenic cells undergoing mitosis and meiosis (Hopkin, 1984, Kelsey et al., 1986, Kawano et al., 1987, Lähdetie and Husgafvel-Pursianen, 1987, Izzotti et al., 1991, Rahn et al., 1991, Zenzes et al., 1995, Granella et al., 1996, Mure et al., 1997). Oxidative damage is another mechanism through which semen quality, including integrity of DNA, is compromised (Storey, 1999; Aitken, 1999; Sakkas et al., 1999). Fraga et al. (1996) and

Potts et al. (1999) found that total anti-oxidant activity is significantly lower in the seminal plasma of smokers compared to nonsmokers. In addition, Fraga et al. (1996) found an increase in the number of oxidative DNA adducts in the sperm of smokers. Specifically regarding abnormal numbers of chromosomes in sperm cells (nullisomy, aneuploidy, diploidy), at least four constituents in cigarette smoke have been shown to cause aneuploidy in different test systems (Dellarco et al., 1986). Indirect mechanisms also play a role and may mediate numerical chromosomal errors through constituents in cigarette smoke (example, cadmium) that compromise the biological process of spermatogenesis and consequent quality of sperm.

Three published papers and one published abstract have reported increased chromosome aneuploidy in sperm of men who smoke compared to men who do not smoke (Table 1.). In an abstract by Potts et al. (1999), 20 smokers of more than 15 cigarettes per day were compared to 20 nonsmokers. All men were between the ages of 20- 30 years and were within the WHO (1994) guidelines for normal semen parameters including count, morphology, and motility. The researchers used fluorescence in situ hybridization to determine sperm aneuploidy for chromosomes 1 and Y and reported a significant increase in YY aneuploidy for smokers (0.13% for smokers versus 0.08% nonsmokers, p<0.05) but no statistically significant increase in chromosome 1 aneuploidy. The abstract does not remark on other common lifestyle exposures such as alcohol or caffeine use in the study population, or mention attempts to control for these.

In a 1998 paper by Rubes et al., 10 men who smoked 20 or more cigarettes per day and had been smoking for at least the two years previous to the study were compared to 15 nonsmokers. Smoking status was confirmed by urinary cotinine. All men were eighteen years old and in good health. Fluorescence in situ hybridization was used to determine aneuploidy for chromosomes X, Y and 8. The authors reported increased aneuploidy for all the chromosomes examined. A significant increase in YY aneuploidy for smokers was found (0.05% for smokers versus 0.02% for nonsmokers, p<0.001) although statistical significance was not reached for chromosome X or 8. The authors attempted to control for other exposures in regression modeling such as work, fever, and lifestyle behaviors. Caffeine consumption was not associated with sperm aneuploidy in this study group. Regarding alcohol use, all the smokers in the Rubes study drank alcohol so the role of alcohol in the aneuploidy estimates could not be separated from smoking.

Table 1. Summary Table of Smoking and Sperm Aneuploidy Studies

Study by Authors and Date	Number of Smokers*	Number of Nonsmokers	Increase in Sperm Chromosome Aneuploidy	P Value
Härkönen et al., 1999	9	19	51% for 1-1	<0.02
Potts et al., 1999	20	20	63% for YY	<0.05
Rubes et al., 1998	10	15	105% for YY	<0.001
Robbins et al. 1997	10	28	88% for XX	<0.05

*Härkönen defined smokers as men smoking currently and the sample ranged from 1-30 cigarettes per day, Potts defined smokers as men smoking 15 cigarettes/day or more, Rubes & Robbins defined smokers as men smoking 20 cigarettes/day or more.

Härkönen et al. (1999) conducted an investigation of the effects of pesticide exposure on aneuploidy in sperm of 28 Danish farmers in which they also looked at smoking and alcohol intake. Men ranged in age from 29 to 49 years, self-reported the number of alcohol drinks per week, and number of cigarettes smoked per day (range 1 to 30). In this study, exposure to fungicides was not associated with sperm aneuploidy however smoking was significantly associated with sperm aneuploid for chromosome 1. Using Poisson regression modeling and controlling for effects of age, sperm density, and alcohol intake, smoking increased the frequency of sperm cells aneuploid for chromosome 1 prior to exposure to pesticides (mean 0.16% for smokers versus 0.11% nonsmokers, p=0.02) as well as after exposure (mean 0.14% for smokers versus 0.11% nonsmokers) although the second comparison did not reach statistical significance. Alcohol did not have a stable relationship with sperm aneuploidy and the authors suggest it may be due to only moderate consumption of alcohol in this study group (farmers reported a mean alcohol intake of six drinks per week).

Robbins and collaborators suggested an effect of cigarette smoking on sex chromosome aneuploidy comparing a group of 10 men who smoked 20 or more cigarettes per day to 28 nonsmokers (Robbins et al., 1997b). Men in the study were a subset of 88 healthy volunteers recruited from the Chapel Hill, North Carolina community for a larger study of the effects of smoking on semen quality. The age ranged from 19-35 years and all men evaluated for aneuploidy were within the WHO (1994) guidelines for normal semen parameters with an added restriction that density be greater than 30 million per mL. A group of men who smoked between 1-19 cigarettes per day were also included in the study design, however, the wide range of exposure to tobacco smoke within this middle group proved uninformative regarding smoking. A significant increase in XX aneuploidy was noted for the smokers of 20+ cigarettes per day compared to the nonsmokers (0.030% compared to 0.016%, p=0.02) after controlling for donor age. Aneuploidy for YY increased from 0.011% to 0.017% although this was not statistically significant. After controlling for alcohol and caffeine use in addition to donor age, XX aneuploidy remained elevated with an associated p value of 0.07 when using categorical values for smoking (20+ cigarettes per day versus none), and covariates alcohol and caffeine consumption. A p value equal to 0.05 was obtained when logged urine cotinine was used for smoking exposure and covariates (alcohol, caffeine, age) were treated as continuous variables. A good portion of the effect of smoking on sex chromosome aneuploidy in this study group could be explained by concurrent alcohol and/or caffeine use. However, limited numbers of men remained in each strata when controlling for smoking, alcohol, and caffeine together and might have been insufficient to detect a stronger effect size for the smoking variable. It is clear, however, that all three lifestyle behaviors (smoking, alcohol, and caffeine) need to be considered as potential confounders in future epidemiological investigations of sperm aneuploidy.

All of these studies were cross sectional in design and involved healthy men without known reproductive problems. All studies used fluorescence in situ hybridization (FISH) to detect and measure aneuploidy for sex chromosomes and autosomes in sperm. All studies found an increase in chromosomal aneuploidy in smokers compared to nonsmokers for the chromosomes studied and this was significant for chromosome Y in two studies, X, and chromosome 1 in the other studies. Although the aneuploidy estimates were always increased over the estimates for nonsmoking controls, no statistically significant associations were noted with the autosomes 7, 8, and 18.

CAFFEINE INTAKE AND SPERM ANEUPLOIDY

Caffeine is the most frequently consumed behaviorally active substance in the world (Boyer, 1993) and is found in coffee, tea, chocolate, cola drinks, and more than 2,000 nonprescription drugs (Robson, 1992). Caffeine has a relatively short half-life in men ~ 3.5 to 5 hours (Balogh et al., 1992) and serum levels show wide fluctuations in response to intake (Lelo et al., 1986a). Interperson variability in metabolism and elimination has been demonstrated (Lelo et al., 1986a; Balogh et al., 1992) and can be affected by other exposures. For example, Brown et al. (1988) showed that smokers metabolize caffeine more rapidly than nonsmokers. Caffeine consumption has been shown to vary considerably in individuals over time, for example related to seasonal weather changes. Caffeine circulates rapidly through all tissues of the body, including testes, and there appears to be no physiological barrier (Boyer, 1993).

The majority of published studies looking at caffeine effects on in vivo human reproduction have used self report to calculate total milligrams of consumption, for example, one cup of coffee contains 100 mg caffeine, tea 50 mg, and cola drinks 40 mg, etc. However, it has been shown that even if individuals accurately report consumption using this scheme, additional sources of error come into play such as brewing related variation in caffeine content or concentration of instant brands (Lelo et al., 1986b). This variability plus the large inter-person variability in metabolism and elimination has been suggested as a cause for conflicting findings regarding effects of caffeine on reproductive outcomes and thus some investigators have used paraxanthine, the main caffeine metabolite in the human, as a biomarker of exposure (James et al., 1989; Klebanoff et al., 1999).

Studies examining reproductive effects of caffeine in human field studies have centered on women, particularly related to adverse pregnancy outcomes. Very few published studies are available on male reproductive effects although a number of researchers report collecting data on caffeine intake as a potential confounder or moderator of other main effects. In the few published reports looking directly at caffeine intake related to semen parameters, semen quality has been shown to be better in moderate caffeine consumers compared to non-caffeine users (Marshburn et al., 1989; Sidhu et al., 1997) or not significantly associated with intake in linear regression modeling using healthy volunteers (Vine et al., 1996; Fenster et al., 1997). Decreased fecundity was shown by Florack et al. (1994) with caffeine intake greater than 700 mg per day but not at moderate or low doses. Lack of effects at moderate to low doses is consistent with work by Oldereid et al. (1992), as well. Increased time to pregnancy has been associated with drinking more than three cups of tea a day by Curtis et al. (1997) in a study of Canadian farmers, although similar associations with coffee were not demonstrated in this same study group.

A sperm-FISH study by Robbins et al. (1997) specifically addressing caffeine intake associated with sperm aneuploidy found significant associations between self-reported caffeine intake and categories of sex chromosome disomy and sperm diploidy. Total caffeine intake was converted to coffee cup equivalents per day and analyzed as a categorical value (0, 1 cup equivalent per day, 2+ equivalents per day) and as a continuous variable. Both analyses showed a statistically significant association between caffeine consumption and XX disomy, XY disomy and diploid sperm. These effects remained after adjusting for age, tobacco smoking, and alcohol intake.

ALCOHOL INTAKE AND SPERM ANEUPLOIDY

Alcohol is the second most popular drug of choice after caffeine in the United States with 70-80% of adults reporting alcohol intake (Braun, 1996). With the exception of water, most popular adult beverages contain either caffeine or alcohol (Braun, 1996). Alcohol is rapidly absorbed through the digestive tract, mixes with blood, and diffuses through cell membranes.

Chronic alcohol abuse has been associated with adverse effects on the male reproductive system including erectile inhibition, atrophied seminiferous tubules, abnormal sperm morphology, reduced sperm count, reduced testicular weight, and infertility (reviewed by Boyer, 1993). Alcohol has also been shown to affect testosterone levels in healthy men without liver disease (Ruusa et al., 1997; Rasmussen et al., 1999; Frias et al.,

2000) and has been shown to exert this effect after only a few months of daily intake (Gordon et al., 1976). However, low to moderate alcohol intake has not been found to be significantly correlated with male fertility in human field studies (Dunphy et al., 1991; Florack et al., 1994; Olsen et al., 1997; Jensen et al., 1998; Zinaman et al., 2000) or semen quality (Dunphy et al., 1991; Oldereid et al., 1992; Goverde et al., 1995; Vine et al., 1996) except by Parazzini et al. (1993) who reported increased risk of dyspermia with increased number of alcohol drinks.

A standard drink is generally considered to be one 12 ounce can of beer, a 5-ounce glass of wine, or a 1.5 ounce shot of liquor, each containing the equivalent of one half-ounce of pure ethanol (Braun, 1996). Field studies evaluating the effects of alcohol often categorize intake as drinks per week, for example 0, 1-5, 6-10, etc., as well as evaluate intake as a continuous variable. Looking at specific drink categories such as beer, wine or spirits has not changed interpretation of findings in studies reporting this. As with caffeine, very few studies look specifically at the effects of alcohol on human male reproduction but a number of researchers collect data on alcohol intake as a potential confounder of other main effects.

Robbins et al. (1997) investigated sperm aneuploidy related to alcohol intake in healthy young men as part of a study examining the effects of caffeine, alcohol, and tobacco smoking. Alcohol was defined in two ways: continuous as drinks per week and then grouped as fewer than six drinks per week, six to 14 drinks per week, and 14 or more drinks per week based on the distribution in the data. A significant association was found between both continuous and categorized alcohol intake with XX aneuploidy, diploidy, and the category of duplicated XX-18-18. These associations remained after controlling for donor age, smoking status, and caffeine intake. However, Härkönen et al. (1999) in a study of workers exposed to pesticides did not find an association between increasing alcohol intake and disomy or diploidy for autosomes 1 and 7 and even noted a significant negative association before exposure to pesticides. It may be that the sex chromosomes and autosomes are different with respect to induction of meiotic errors following alcohol exposure. The inconsistent results may also reflect lower consumption of alcohol in the occupational study of Danish farmers, mean 6.2 drinks per week, compared to community volunteers in the Robbins et al. (1997) study, mean 8.9 drinks per week. Increased aneuploidy in lymphocytes of alcoholics has been demonstrated using micronuclei and FISH (Maffei et al., 1997). Additionally, in vitro work has shown that alcohol can induce chromosome damage and interfere with chromosome segregation (Obe and Ristow, 1979; Obe et al., 1977, 1986; Obe and Anderson, 1987; IARC Working Group, 1988). This is an area that may lend itself to further research using sperm-FISH techniques.

SPERM ANEUPLOIDY SUMMARY

Humans seem to be especially susceptible to meiotic chromosomal errors, in great excess compared to other mammals including rodents (Martin, 1984). The factors responsible for this high frequency of abnormalities in human germ cells are currently being studied and include things that interfere with pre-recombination alignment, recombination, separation, and segregation of chromosomes. Factors may be intrinsic to the chromosomes themselves such as the large blocks of heterochromatin on chromosome 1, the nucleolar organizer on chromosome 15 (Spriggs et al., 1996) or the pseudoautosomal region and the difficulties encountered during X:Y pairing and recombination (Rappold, 1993). Factors may be related to the recombination and segregation process itself, for example, incorrect release of distal chiasmata (Hawley et al., 1994; Lamb et al., 1996), or anaphase lag. Factors could also be external, for example agents that interfere with microtubule assembly, attachment of chromosomes to the spindle, disruption of centromeres, cell cycle check points, or cytokinesis (Allen et al., 1986; Bond, 1987; Onfelt, 1987).

Using the review publication of Shi and Martin (2000) as guide, the average disomy frequency reported for human sperm for autosomes across laboratories is 0.15% and for sex chromosomes is 0.26%. High interindividual variability across men has been noted in numerous studies (Shi and Martin, 2000) although intra-man variability is limited (that is, men seem stable in their sperm aneuploidy estimates) across time (Benet et al., 1992; Robbins et al., 1993). Diploidy frequencies also demonstrate wide interindividual variability with a range from 0.05-0.51% (Rademaker et al., 1997).

With such a high load of sperm aneuploidy reported for the human, it is important to determine if sperm aneuploidy may be induced by exogenous exposures. Animal studies have shown that DNA damaging exposures to the father can result in impaired embryonic development and/or birth defects in offspring (reviewed by Hales and Robaire, 1996). Chromosomally unbalanced sperm cells have been shown to be able to transmit the defect to offspring when the sperm fertilizes an ovum (Hassold, 1998). If exposures that lead to chromosomal imbalances in germ cells could be identified and mediated, the potential for transmission of the damage to offspring could subsequently be reduced.

Tobacco smoking, caffeine intake, and alcohol consumption are common human exposures. Caffeine is probably the most popular licit drug on earth and the majority of adults report some alcohol use. There is currently too little data available on caffeine and alcohol effects on human sperm to suggest either might induce aneuploidy. In contrast, the weight of published evidence on tobacco smoking suggests there is an association with increased risk for aneuploid sperm. This may reflect a publication bias and more study needs to be done, however, there is enough evidence to warrant inclusion of smoking status information in all clinical, toxicological, and epidemiologic studies evaluating the relationship between environmental or workplace exposures and sperm aneuploidy. Above an independent contribution of smoking to sperm chromosome damage, it may be found that smoking moderates effects of other workplace or environmental insults, similar to the findings for semen quality measures (Wang et al. 2001). Confounding factors affect responses in all cell types (Anderson, 1999) including those of the male reproductive system and supporting tissues. Thus, it is clear, that all three lifestyle behaviors (smoking, alcohol, and caffeine) need to be considered as potential confounders or moderators of effects in future epidemiologic investigations of exposures related to induction of sperm aneuploidy.

Design Issues in Sperm-FISH Studies

There are a number of study design issues to consider when incorporating sperm-FISH assays into human studies. The difficulties seen in interpreting the findings of investigations of the effects of tobacco smoking, caffeine intake, and alcohol consumption exemplify this.

First, participation rates can be small in semen studies, ranging from 10% to over 80% with control group participation often being a particular problem (Ratcliffe et al., 1989; Lähdetie, 1995, Bonde et al., 1996). Small study groups combined with the rarity of cytogenetic outcomes of interest plus the expense of conducting the cytogenetic assays result in summarizing and testing disomy/diploidy frequencies across groups. Generally, group-based strategies will result in larger standard errors compared to individual-based strategies (Tiellemans et al., 1998) if measures are stable for individuals. For conventional semen parameters, both men with good semen quality and men with poor semen quality show tremendous interindividual variability as well as intraindividual changes over time (Sherins, 1995; Meschede et al., 1995). With sperm chromosomal damage, there is also significant interindividual variability (Rademaker et al., 1997) but not intraindividual variability over time (Benet et al., 1992; Robbins et al., 1993). Thus, ineffective groupings (groupings with a large ratio of within-group variance) will yield larger standard errors and obscure differences across comparison groups. For example, in the study of the effects of smoking on sperm aneuploidy by Robbins et al. (1997), for one analysis men were categorized into three groups: smokers of >20 cigarettes per day, smokers of 1-19 cigarettes per day, and nonsmokers. Clearly in terms of tobacco exposures the men who smoked one cigarette per day were more similar to non-smokers and men who smoked 19 cigarettes per day more similar to the smokers of 20 or more cigarettes per day. When the data were summarized and compared for the three groups, the middle group behaved erratically, yielded unstable estimates and was not informative. Thus, grouping schemes must be optimized and the range of exposure is key.

Related to this, precision for the individual aneuploidy estimates must be weighed against study costs. With the advent of commercially available DNA probes for use in sperm-FISH, coupled with stringent scoring protocols, the variability for sperm aneuploidy frequencies reported across laboratories has attenuated considerably. However, commercially available DNA probes are expensive on a per test basis. In terms of precision for aneuploidy estimates, preliminary work in our laboratory suggests diminishing returns for increasing samples sizes above 20 men whereas increasing the number of cells scored per man (much less expensive in terms of DNA probe and recruitment costs) improves dramatically over 7,500 cells scored per man (Matthews et al., 1999).

Because sperm disomy and diploidy frequencies seem stable for individual men over time, a number of researchers are using longitudinal designs with repeated measures where the exposure is varied over time. Longitudinal studies generally cover a three-month period to capture effects on a complete cycle of spermatogenesis. Attrition and loss to follow-up can be a problem for longitudinal designs. In a recent community-based longitudinal study done in our lab to follow healthy men over a 16-week period as they quit smoking, only 50% followed through with three repeated semen samples over the 4-month study period. In contrast, in a clinic-based study of essentially healthy HIV infected men taking antiretroviral therapy, 100% of the men originally enrolled continued through the entire four month sampling period although fewer than 30% donated all semen specimens at the correct sampling time (Robbins et al., 2001).

A third study design issue for sperm-FISH relates to poor semen quality as a moderator of sperm aneuploidy frequencies. Men showing poor semen quality comprise a high-risk group for meiotic nondisjunction according to many researchers (Moosani et al., 1995; Bonaccorsi et al., 1997; Bernardini et al., 1998; Finkelstein et al., 1998; Li and Hoshiai, 1998; Storeng et al., 1998; Pfeffer et al., 1999; Pang et al., 1995,1999) but not all (Miharu et al., 1994; Guttenbach et al., 1997). The diversity in study groups may explain the varying findings. For example, Finkelstein et al. (1998) enrolled only varicocele patients and found a clear association between poor semen quality and aneuploidy whereas Miharu et al. (1994) included men with sperm antibodies, unexplained infertility, and oligozoospermia in their study group. However, even in investigations where a statistically significant association between sperm aneuploidy and poor semen quality is not found, individual men with unusually high frequencies of abnormalities have been found in the poor semen quality group (Guttenbach et al., 1997). A proportion of infertile males are known to have a genetic or chromosomal disorder underlying their infertility and it has been estimated that 5 to 7% of men with oligozoospermia will have an abnormal karyotype (Retief, 1986; De Braekeller and Dao, 1991). There are a number of reports of more XY disomy than XX or YY in men with poor semen quality suggesting segregation errors for these men are generated at first meiotic division (similar to the age dependent errors in oogenesis) rather than second (Moosani et al., 1995; Finkelstein et al., 1998).

If semen quality is not predetermined, then men with poor semen quality will be enrolled in population-based epidemiologic field studies with regularity. The expected proportion is difficult to estimate since the bulk of previous work describing semen properties has been conducted with special populations presenting to fertility clinics, semen donors, or men chosen for other specific research reasons. However, in a population based field study in the U.S. looking at fertility related to semen quality, Zinamen et al. (2000), found 3% of the men on entry to be azoospermic. Bonde et al. (1998) in their population based study conducted in Denmark found 19% of men (80/418) to have sperm concentrations on entry of <20 x 10^6/mL. If men with poor semen quality have higher disomy and diploidy frequencies, as suggested by a number of researchers (Moosani et al., 1995; Pang et al., 1999; Finkelstein et al., 1998; Storeng et al., 1998; Pfeffer et al., 1999; Columbero et al., 1999) and men sort into study groups without regard to the exposure of interest, then the overall variation in the data set will increase and may make it difficult to detect differences in aneuploidy between exposure groups. If the men with poor semen quality sort into study groups in a nonrandom manner related to the exposure of interest, findings will be biased. Thus it appears that if you recruit without regard to semen quality on study entry, then larger numbers of men will be needed. If you restrict according to healthy semen parameters, then fewer men will be needed but you may also be missing a population subset especially susceptible to meiotic errors that may be of interest in risk assessment determinations related to exposures of interest.

Finally, the implications of even moderate increases in sperm aneuploidy frequencies in terms of male mediated effects on offspring are poorly understood for the human. Due to the rarity of events, it is unlikely that prospective field studies such as those investigating conventional semen quality measures and fertility conducted by Zinamen et al. (2000) and Bonde et al. (1998) would be practical in this case. Male-mediated effects occur when chromosomally abnormal sperm contribute genetic abnormalities to the conceptus and result in fetal wastage or affected offspring (Hassold, 1986, Sakkas et al., 1998). Nearly one-third of pregnancies detectable by hCG elevation in a study by Wilcox et al. (1988) did not survive to delivery and about two-thirds of these losses occurred before the pregnancy was clinically recognized. Yet, most of these losses were followed by successful pregnancies. Could a portion of these adverse events have been associated with sperm chromosomal abnormalities and given the rarity of disomic and diploid sperm likely be followed by successful pregnancies, even in males who routinely carry a high disomy frequency in sperm? What about transient increases in abnormal sperm complements induced by independent effects of lifestyle behaviors such as tobacco smoking, caffeine, and alcohol or their moderation of workplace and environmental effects? Would our current study designs be adequate to detect these? It is clear that the simplicity and efficiency of the sperm-FISH assay has outpaced our ability to adequately interpret findings in terms of human risk. As the body of knowledge generated from carefully designed studies increases, our abilities to interpret risk will improve.

ACKNOWLEDGEMENTS

The work of the author reported here and the writing of this review were supported by the UCLA Center for Occupational and Environmental Health and the UCLA School of Nursing. Thank you to Dr. Lin Xun, Thuan Ong, Karen Young for critical comment on the manuscript.

REFERENCES

Aitken, R.J., 1999, The human spermatozoon- a cell in crisis? *J Reprod Fertil.* 115:1-7.
Anderson, D., 1999, Factors contributing to biomarker responses in exposed workers. *Mutat Res.* 428:197-202.
Allen, J.W., Liang, J.C., Carrano, A.V., Preston, R.J., 1986, Review of literature on chemical-induced aneuploidy in mammalian male germ cells. *Mutat Res.* 167:123-127.
Balogh, S., Harder, R., Vollandt and Staib, A.H., 1992, Intra-individual variability of caffeine elimination in healthy subjects. *Int J Clin Pharmacol Ther Toxicol.* 30:383-387.
Benet, J., Genesca, A., Navarro, J., Egozcue, J., and C. Templado, 1992, Cytogenetic studies in motile sperm from normal men. *Hum Genet.* 889:176-180.
Bernardini, L., Borini, A., Preti, S., Conte, N., Flamigni, C., Capitanio, G.L. and Venturini, P.L., 1998, Study of aneuploidy in normal and abnormal germ cells from semen of fertile and infertile men. *Hum Reprod.* 13:3406-3413.
Bonaccorsi, A.C., Franco, J.G., Martins, R.R.S., Botler, J., Vargas, F.R., 1997, Genetic disorders in normally androgenized infertile men and the use of intracytoplasmic sperm injection as a way of treatment. *Fertil Steril.* 67:928-931.
Bond, D.J., 1987, Mechanisms of aneuploid induction. *Mutat Res.* 181:257-266.
Bonde, J.P., Giwercman, A., Ernst, E., 1996, Identifying environmental risk to male reproductive function by occupational sperm studies: logistics and design options. *Occ Environ Med.* 53:511-519.
Bonde, J.P.E., Ernst, E., Jensen T.K., Hjollund, N.H.I., Kolstad, H., Henriksen, T.B., Scheike, T., Giwercman, A., Olsen, J., Skakkebaek, N.E., 1998, Relation between semen quality and fertility: a population-based study of 430 first-pregnancy planners. *Lancet.* 352:1172-1177.
Boyer, H., 1993, Effects of licit and illicit drugs, alcohol, caffeine, and nicotine on infertility, in understanding infertility: risk factors affecting fertility, Research Studies. *Royal Commission on New Reproductive Technologies.* 7:173-208.
Braun, S., 1996, Buzz: The Science and Lore of Alcohol and Caffeine, Oxford University Press, New York, pp.6
Brown, C.R., Jacob, P., Wilson, M. and Benowitz, N.L., 1988, Changes in rate and pattern of caffeine metabolism after cigarette abstinence. *Clin Pharmacol Ther.* 43:488-491.
Chen, H., Eastmond, D.A., 1995, Synergistic increase in chromosomal breakage within the euchromatin induced by an interaction of the benzene metabolites phenol and hydroquinone in mice. *Carcinogenesis.* 16:1963-1969.
Chevret, E., Rousseaux, M., Monteil, R., Pelletier, R., Cozzi, J., and Sele, B., 1995, Meiotic segregation of the X and Y chromosomes and chromosome 1 analyzed by three-color FISH in human interphase spermatozoa. *Cytogent Cell Genet.* 71:126-130.
Colombero, L.T., Hariprashad, J.J., Tsai, M.C., Rosenwaks, Z., Palermo, G.D., 1999, Incidence of sperm aneuploidy in relation to semen characteristics and assisted reproductive outcome. *Fertil Steril.* 72:90-96.
Curtis, K.M., Savitz, D.A., Arbuckle, T.E., 1997, Effects of cigarette smoking, caffeine, and alcohol intake on fecundability. *Am J Epidemiol.* 146:32-41.
De Braekeleer, M. and Dao, T.N., 1991, Cytogenetic studies in male infertility: a review. *Hum Reprod.* 6:245-250.
Dellarco, V.L., Mavournin, K.H., Waters, M.D., 1986, Aneuploidy data review committee: Summary compilation of chemical data base and evaluation of test methodology. *Mutat Res.* 167:149-169.
Downie, S.E., Flaherty, S.P., Swann N.J., and Matthews C., 1997, Estimation of aneuploidy for chromosomes 3, 7, 16, X and Y in spermatozoa from 10 normospermic men using fluorescence in-situ hybridization. *Mol Hum Reprod.* 18:815-819.
Dunphy, B.C., Barratt, C.L., Cooke, I.D., 1991, Male alcohol consumption and fecundity in couples attending an infertility clinic. *Andrologia.* 23:219-221.
Estop, A.M., Munne, S., Cieply, K.M., Vandermark, K.K., Lamb, A.N., Fisch, H., 1998, Meiotic products of a Klinefelter, 47, XXY male as determined by sperm fluorescence in-situ hybridization analysis. *Hum Reprod.* 13:124-127.
Evans, H.J., Fletcher, J., Torrance, M., Hargreave, T.B., 1981, Sperm abnormalities and cigarette smoking. *Lancet.* 1:627-629.
Fenster, L., Katz, D.F., Wyrobek, A.J., Pieper, C., Rempel, D.M., Oman, D., Swan, S.H., 1997, Effects of psychological stress on human semen quality. *J Androl.* 18:194-202.
Florak, E.I., Zielhous, G.A., Rolland, R., 1994, Cigarette smoking, alcohol consumption and caffeine intake and fecundability. *Prev Med.* 23:175-180.
Finkelstein, S., Mukamel, E., Yavetz, H., Paz, G., Avivi, L., 1998, Increased rate of nondisjunction in sex cells derived from low-quality semen. *Hum Genet.* 102:129-137.

Fraga, C.G., Motchnik, P.A., Wyrobek, A.J., Rempel, D.M., Ames, B.N., 1996, Smoking and low antioxidant levels increase oxidative damage to sperm DNA. *Mutat Res.* 351:199-203.

Frias, J., Rodriguez, R., Torres, J.M., Ruiz, E., Ortega, E., 2000, Effects of acute alcohol intoxication on pituitary-gonadal axis hormones, pituitary-adrenal axis hormones, beta-endorphin and prolactin in human adolescents of both sexes. *Life Sci.* 67:1081-1086.

Gordon, G.G., Altman, K., Southern, A.L., Rubin, E., Lieber, C.S., 1976, Effects of alcohol (ethanol) administration on sex-hormone metabolism in normal men. *N Engl J Med.* 295:793-797.

Goverde, H.J., Dekker, H.S., Janssen, H.J., Bastiaans,B.A., Rolland, R., Zielhuis, G.A., 1995, Semen quality and frequency of smoking and alcohol consumption---an explorative study. *Int J Fertil Menopausal Stud.* 40:135-138.

Granella, M., Priante, E., Nardini, B., Bono, R., Clonfero, E., 1996, Excretion of mutagens, nicotine and its metabolites in urine of cigarette smokers. *Mutagenesis.* 11:207-211.

Guttenbach, M., Martinez-Exposito, M.J., Michelmann, H.W., Engel, W., Schmid, M., 1997, Incidence of diploid and disomic sperm nuclei in 45 infertile men. *Hum Reprod.* 12:468-473.

Hales, B.F., Robaire, B., 1996, Paternally mediated effects on development, In: Handbook of Developmental Toxicology, Hood, R..D., ed., CRC Press, Boca Raton, FL, pp91-107.

Handelsman, D.J., Conway, A.J., Boylan, L.M., Turtle, J.R., 1984, Testicular function in sperm donors: normal ranges and the effects of smoking and vaicocele. *Int J Androl.* 7:369-382.

Härkönen, K., Viitanen, T., Larsen, S., Bonde, J., Lähdetie, J., 1999, Aneuploidy in sperm and exposure to fungicides and lifestyle factors. ASCLEPIOS. A European concerted action on occupational hazards to male reproductive capability. *Environ Mol Mutagen.* 34:39-46.

Hassold, T.J., 1986, Chromosome abnormalities in human reproductive wastage, T.I.G., 105-10.

Hassold, T.J., Pettay, D., Robinson, A., Uchida, I., 1992, Molecular studies of parental origin and mosaicism in 45, X conceptuses. *Hum Genet.* 89:647-652.

Hassold, T.J., 1998, Nondisjunction in the human male. *Curr Top Dev Biol.* 37:383-406.

Hawley, R.S., Frazier, J.A., Rasooly, R., 1994, Separation anxiety: the etiology of nondisjunction in flies and people. *Hum Mol Genet.* 3:1521-1528.

Hopkin, J.M., 1984, Sister chromatid exchange induction by cigarette smoke. *Basic Life Sci.* 29 Pt B:927-37.

IARC Working Group, 1988, Alcohol drinking: Monographs on the Evaluation of Carcinogenic Risks to Humans. 44. Lyon: IARC.

Izzotti, A., Rossi, G.A., Bagnasco, M., De Flora, S., 1991, Benzo[a]pyrene diolepoxide-DNA adducts in alveolar macrophages of smokers. *Carcinogenesis.* 12:1281-1285.

James, J.E., Bruce, M.S., Lader, M.H. and Scott, N.R., 1989, Self-report reliability and symptomatology of habitual caffeine consumption. *Br J Clin Pharmacol.* 27:507-514.

Jensen, T.K., Hjollund, N.H.I., Henriksen, T.B., Scheike, T., Kolstad, H., Giwercman, A., Ernst, E., Bonde, J.P., Skakkebaek, N.E., Olsen, J., 1998, Does moderate alcohol consumption affect fertility? Follow up study among couples planning first pregnancy. *BMJ.* 317:505-510.

Kawano, H., Inamasu, T., Ishizawa, M., Ishinishi, N., Kumazawa, J., 1987, Mutagenicity of urine from young male smokers and nonsmokers. *Int Arch Occup Environ Health.* 59:1-9.

Kelsey, K.T., Christiani, D.C., Little, J.B., 1986, Enhancement of benzo[a]pyrene-induced sister chromatid exchanges in lymphocytes from cigarette smokers occupationally exposed to asbestos. *J Natl Cancer Inst.* 77:321-327.

Klebanoff, M.A., Levine, R.J., Dersimonian, R., Clemens, J.D., Wilkins, D.G., 1998, Serum caffeine and paraxanthine as markers for reported caffeine intake in pregnancy. *Ann Epidemiol.* 8:107-111.

Lähdetie, J., Husgafvel-Pursiainen, K., 1987, Cigarette smoking--a possible genotoxic hazard to male germ cells?. *Int J Androl.* 10:431-434.

Lähdetie, J., 1995, Occupation- and exposure-related studies on human sperm. *JOEM.* 37:922-930.

Lähdetie, J. Ajosenpaa-Saari, M., Mykkanen, J., 1996, Detection of aneuploidy in human spermatozoa of normal semen donors by fluorescence in situ hybridization. *Environ Health Perspec.* 104, supp 3:629-632.

Lamb, N.E., Freeman, S.E., Savage-Austin, A., Pettay, D., Taft L., Hersey, J., Gu, Y., Shen, J., Saker, D., May, K.M., Avramopoulos, D., Petersen, M.B., Hallberg, A., Mikkelsen, M., Hassold, T.J., Sherman, S.L., 1996, Susceptible chiasmata configurations of chromosome 21 predispose to non-disjunction in both maternal meiosis I and meiosis II. *Nat Genet.* 14:400-405.

Lelo, A., Birkett, D.J., Robson, R.A., Miners, J.O. 1986a, Quantitative assessment of caffeine partial clearances in man. *Br J Clin Pharmacol.* 22:183-186.

Lelo, A., Miners, J.O., Robson, R. and Birkett, D.J., 1986b, Assessment of caffeine exposure: caffeine content of beverages, caffeine intake, and plasma concentrations of methylxanthines. *Clin Pharmacol Ther.* 39:54-59.

Li, P., and Hoshiai, H., 1998, Detection of numerical chromosome abnormalities in human spermatozoa by three-color fluorescence in situ hybridization. *J Obstet Gynaecol Res.* 24:385-392.

Little, J., Vainio, H., 1994, Mutagenic lifestyles? A review of evidence of associations between germ-cell mutations in humans and smoking, alcohol consumption and use of 'recreational' drugs. *Mutat Res.* 313:131-151.

Mak, V., Jarvi, K., Buckspan, M., Freeman, M., Hechter, S., Zini, A., 2000, Smoking is associated with the retention of cytoplasm by human spermatozoa. *Urology.* 56:463-466.

Martin, R.H., 1984, Comparison of chromosomal abnormalities in hamster egg and human sperm pronuclei. *Biol Reprod.* 31:819-825.

Martin, R.H., Rademaker, A., 1995, Reliability of aneuploidy estimates in human sperm: results of fluorescence in situ hybridization studies using two different scoring criteria. *Mol Reprod Dev.* 42:89-93.

Martin, R.H., Spriggs, E., Rademaker, A.W., 1996, Multicolor fluorescence in situ hybridization analysis of aneuploidy and diploidy frequencies in 225, 846 sperm from 10 normal men. *Biol Reprod.* 54:394-398.

Martin, R.H., Ernst, S., Radmaker, A., Barclay, L., Ko, E., Summers, N., 1997, Chromosomal abnormalities in sperm from testicular cancer patients before and after chemotherapy. *Hum Genet.* 99:214-218.

Martin, R.H., Green, J., Ko, E., Barclay, L., Rademaker, A., 2000, Analysis of aneuploidy frequencies in sperm from patients with hereditary nonpolyposis colon cancer and an hMSH2 mutation. *Am J Hum Genet.* 66:1149-1152.

Martinez-Pasarell, O., Nogues, C., Bosch, M., Egozcue, J. Templado, C., 1999, Analysis of sex chromosome aneuploidy in sperm from fathers of Turner syndrome patients. *Hum Genet.* 104:345-349.

Mashburn, P.B., Sloan, C.S., Hammond, M.G., 1989, Semen quality and association with coffee drinking, cigarette smoking, and ethanol consumption. *Fertil Steril.* 52:162-165.

Matthews, S., Blakey, D.H., Arbuckle, T., Robbins, W.A., 1999, Aneuploidy estimates in small sample size sperm-FISH studies. *Environ Mol Mutagen.* 33 (Suppl 30):43.

Meistrich, M.L, 1986, Critical components of testicular function and sensitivity to disruption. *Biol Reprod.* 34:17-28.

Meschede, D., Nieschlag, E., Horst, J., 1995, The importance of clinical documentation in genetic studies of male infertility. *Hum Genet.* 96:500-501.

Miharu, N., Best, R.G., Young, S.R., 1994, Numerical chromosome abnormalities in spermatozoa of fertile and infertile men detected by fluorescence in situ hybridization. *Hum Genet.* 93:502-506.

Monteil, M., Rousseaux, S., Chevret, E., Pelletier, R., Cozzi, J., Sele, B., 1997 Increased aneuploid frequency in spermatozoa from a Hodgkin's disease patient after chemotherapy and radiotherapy. *Cytogenet Cell Genet.* 76:134-138.

Moosani,N., Pattinson, H., Carter, M., Cox, D., Rademaker, A., Martin, R., 1995, Chromosomal analysis of sperm from men with idiopathic infertility using sperm karyotyping and fluorescence in situ hybridization. *Fertil Steril.* 64:811-817.

Morel, F., Mercier, S., Roux, C., Clavequin M.C., Bresson J., 1997, Estimation of aneuploidy levels for 8, 15, 18, X and Y chromosomes in 97 human sperm samples using fluorescence in situ hybridization. *Fertil Steril.* 67:1134-1139.

Mure, K., Hayatsu, H., Takeuchi, T., Takeshita, T., Morimoto, K., 1997, Heavy cigarette smokers show higher mutagenicity in urine. *Mutat Res.* 373:107-111.

Obe, G., Anderson, D., 1987, Genetic effects of ethanol. *Mutat Res.* 186:177-200.

Obe, G., Jones, R., Schmidt, S., 1986, Metabolism of ethanol in vitro produces a compound which induces sister chromatid exchanges in human peripheral lymphocytes in vitro: acetaldehyde not ethanol is mutagenic. *Mutat Res.* 174: 47-51.

Obe, G., Ristow, H., 1979, Mutagenic, carcinogenic, and teratogenic effects of alcohol. *Mutat Res.* 65:229-259.

Obe, G., Ristow, H.J., Herha, J., 1977, Chromosomal damage by alcohol in vitro and in vivo. *Advances in Experimental Medicine.* 85A:47-70.

Oldereid, N.B., Rui, H., Purvis, K., 1992, Life styles of men in barren couples and their relationship to sperm quality. *Int J Fertil.* 37:343-349.

Olsen, J., Bolumar, F., Boldsen, J., Bisanti, L., 1997, Does moderate alcohol intake reduce fecundability? A European multicenter study on infertility and subfecundity. European study group on infertility and subfecundity. *Alcohol Clin Exp Res.* 21:206-212.

Onfelt, A., 1987, Spindle disturbances in mammalian cells. III. Toxicity, c-mitosis and aneuploidy with 2 different compounds. Specific and unspecific mechanisms. *Mutat Res.* 182:135-154.

Padungtod, C., Hassold, T.J., Millie, E., Ryan, L.M., Savitz, D.A., Christiani, D.C., Xu, X.., 1999, Sperm aneuploidy among Chinese pesticide factory workers: scoring by the FISH method. *Am J Ind Med.* 36:230-238.

Pang, M.G., Zackowski, J.L., Hoegerman, S.F., Moon S.Y., Cuticchia, A.J., Acosta, A.A., et al., 1995, Detection by fluorescence in situ hybridization of chromosome 7, 11, 12, 18, X and Y abnormalities in sperm from oligoasthenoteratozoospermic patients of an in vitro fertilization program. *J Assist Reprod Genet.* 12:supp. 53S.

Pang, M.G., Hoegerman, S.F., Cuticchia, A..J., Moon, S.Y., Doncel, G.F., Acosta, A.A. and Kearns, W.G., 1999, Detection of aneuploidy for chromosomes 4, 6, 7, 8, 9, 10, 11, 12, 13, 17, 18, 21, X and Y by fluorescence in-situ hybridization in spermatozoa from nine patients with oligoasthenoteratozoospermia undergoing intracytoplasmic sperm injection. *Hum Reprod.* 14:1266-1273.

Parazzini, F., Marchini, M., Tozzi, L., Mezzopane, R., Fedele,L., 1993, Risk factors for unexplained dyspermia in infertile men. A case-control study. *Arch Androl.* 31:105-113.

Pfeffer, J., Pang, M.G., Hoegerman, S.F., Osgood, C.J., Stacey, M.W., Mayer, J., Oehninger, S., Kearns, W.G., 1999, Aneuploidy frequencies in semen fractions from ten oligoasthenoteratozoospermic patients donating sperm for intracytoplasmic sperm injection. *Fertil Steril.* 72:472-478.

Potts, R.J., Newbury, C.J., Notarianni, L.J., Jefferies, T.M., 1999, Smoking is associated with sperm aneuploidy and lowered seminal plasma anti-oxidant activity. *Toxicologist.* 48:383.

Rademaker, A., Spriggs, E. Ko, E., and Martin, R., 1997, Reliability of estimates of diploid human spermatozoa using multicolor fluorescence in-situ hybridization. *Hum Reprod.* 12:77-79.

Rahn, C.A., Howard, G., Riccio, E., Doolittle, D.J., 1991, Correlations between urinary nicotine or cotinine and urinary mutagenicity in smokers on controlled diets. *Environ Mol Mutagen.* 17:244-252.

Rappold, G.A., 1993, The pseudoautosomal regions of the human sex chromosomes. *Hum Genet.* 92:315-324.

Ratcliffe, J.M., Schrader, S.M., Clapp, D.E., Halperin, W.E., Turner, T.W., Hornung, R.W., 1989, Semen quality in workers exposed to 2-ethoxyethanol. *Br J Ind Med.* 46:399-406.

Recio, R, Robbins W.A, Ocampo-Gomez G, Borja-Aburto V, Moran-Martinez J, Froines J.R, Garcia Hernandez, R.M, Cebrian ME, 2001 Organophosphorous pesticide exposure increases the frequency of sperm sex null aneuploidy. *Environ Health Perspec.* :in press.

Retief, A.E., 1986, Cytogenetics and recombinant technology in male infertility. *Arch Androl.* 17:119-123.

Robbins, W.A., Segraves, R., Pinkel, D., Wyrobek, A.J., 1993, Detection of aneuploid human sperm by fluorescence in situ hybridization: Evidence for a donor difference in frequency of sperm disomic for chromosomes 1 and Y. *Am J Hum Genet.* 52: 79-807.

Robbins, W.A., Meistrich, M.L., Cassel,M.J., Weier,H U, Hagemeister, F.B., Wilson, G.,Eskenazi, B , Wyrobek, A.J., 1997a, Chemotherapy induces transient sex chromosomal and autosomal aneuploidy in human sperm. *Nat Genet.* 16:74-78.

Robbins, W.A., Vine, M.F., Truong, K.Y., Everson, R.B., 1997b, Use of FISH (fluorescence in situ hybridization) to assess effects of smoking, caffeine, and alcohol on aneuploidy load in sperm of healthy men. *Environ Mol Mutagen.* 30:175-183.

Robbins, W.A., Rubes, J., Seleven, S.G., Perreault, S.D., 1999, Air pollution and sperm aneuploidy in healthy young men. *Environ Epi Tox.* 1:125-131.

Robbins, WAY., Witt, KILL., Haseman, O.K., Dunson, DAB., Troiano, L., Cohen, M.S., Hamilton, C.D., Perreault, S.D., Bishop, JOB., Beyler, SEA., Raburn, D., Libbus, B., Sanders, L., Tedder, S.T., Morris, R., Johndrow, D., Jameson, O.K., Shelby, M.D., 2001, Antiretroviral therapy effects on genetic and morphologic endpoints in lymphocytes and sperm of men with Human Immunodeficiency Virus infection. *J Infectious Dis.* 184:127-135.

Robson, R.A., 1992, The effects of quinolines on xanthine pharmacokinetics. *Am J Med.* 92:22.

Rubes, J., Lowe, X.., Moore, D., Perreault, S., Slott, V., Evenson, D., Selevan, S.G., Wyrobek, A.J., 1998, Smoking cigarettes is associated with increased sperm disomy in teenage men. *Fertil Steril.* 70:715-724.

Rupa, D.S., Hasegawa, L., Eastmond, D.A., 1995, Detection of chromosomal breakage in the 1cen-1q12 region of interphase human lymphocytes using multicolor fluorescence in situ hybridization with tandem DNA probes. *Cancer Res.* 55:640-645.

Rupa, D.S., Eastmond, D.A., Robbins, W.A., Reddy, P.P., 1999, Comparison of chromosome alterations affecting the 1cen-q12 region in G0-lymphocytes, cultured lymphocytes and sperm of Indian pesticide workers using FISH with tandem labeling. *Environ Mol Mutag.* (Suppl): 33: 54.

Ruusa, J., Bergman,B., Sundell,M.L., 1997, Sex hormones during alcohol withdrawal: a longitudinal study of 29 male alcoholics during detoxification. *Alcohol.* 32:591-597.

Saaranen M., Suonio S., Kauhanen O., Saarikoski S., 1987, Cigarette smoking and semen quality in men of reproductive age. *Andrologia.* 19:670-676.

Sakkas D., Urner F., Bizzaro D., Manicardi G., Bianchi P.G., Shoukir Y., Campana A., 1998, Sperm nuclear DNA damage and altered chromatin structure: effect on fertilization and embryo development. *Hum Reprod.* 4 (Suppl):11-19.

Sakkas D, Mariethoz E, Manicardi G, Bizzaro D, Bianchi PG, Bianchi U, 1999, Origin of DNA damage in ejaculated human spermatozoa. *Rev Reprod.* 4:31-7.

Shen H.M., Chia S.E., Ni Z.Y., New A.L., Lee B.L., Ong C.N., 1997, Detection of oxidative DNA damage in human sperm and the association with cigarette smoking. *Reprod Toxicol.* 11:675-80.

Shen, H.M. and Ong, C.N., 2000, Detection of oxidative DNA damage in human sperm and its association with sperm function and male infertility. *Free Radical Biol Med.* 28:529-536.

Sherins, R.J., 1995, Are semen quality and male fertility changing? *N Engl J Med.* 332:327-328.

Shi,Q., Martin, R.H., 2000, Aneuploidy in human sperm: a review of the frequency and distribution of aneuploidy, effects of donor age and lifestyle factors. *Cytogenet Cell Genet.* 90:219-226.

Sidhu, R.S., Sharma, R.K., Kachoria, S., Curtis, C., Agarwal, A., 1997, Reasons for rejecting potential donors from a sperm bank program. *J Assist Reprod Genet.* 14:354-360

Sofikitis, N., Miyagawa, I., Dimitriadis, D., Zavos, P., Sikka, S., Hellstrom, W., 1995, Effects of smoking on testicular function, semen quality and sperm fertilizing capacity. *J Urol.* 154:1030-1034.

Spriggs, E., Rademaker, A., Martin, R., 1996, Aneuploidy in human sperm: The use of multicolor FISH to test various theories of nondisjunction. *Am J Hum Genet.* 58:356-362.

Stillman, R.J., Rosenberg, M.J., Sachs, B.P., 1986, Smoking and reproduction. *Fertil Steril.* 46:545-566.

Storeng, R.T., Plachot, M., Theophile, D., Mandelbaum, J., Belaisch-Allart, J., Vekemans, M., 1998, Incidence of sex chromosome abnormalities in spermatozoa from patients entering an IVF or ICSI protocol. *Acta Obstet Gynecol Scand.* 77:191-197.

Storey, B.T., 1999, Biochemistry of the induction and prevention of lipoperoxidative damage in human spermatozoa. *Mol Hum Reropd.* 3:203-213.

Tiellemans, E., Kupper, L.L., Kromhout, H., Heederik, D., Houba, R., 1998, Individual-based occupational exposure assessment: Some equations to evaluate different strategies. *Ann Occup Hyg.* 42:115-119.

Van Hummelen, P., Lowe, X.R., Wyrobek, A.J., 1996, Simultaneous detection of structural and numerical chromosome abnormalities in sperm of healthy men by multicolor fluorescence in situ hybridization. *Hum Genet.* 98:608-615.

Van Hummelen, P., Manchester, D., Lowe, X., Wyrobek, A.J., 1997, Meiotic segregation, recombination, and gamete aneuploidy assessed in a t(1;10)(p22.1;q22.3) reciprocal translocation carrier by three- and four-probe multicolor FISH in sperm. *Am J Hum Genet.* 61:651-659.

Vine, M.F., Margolin, B.H., Morrison, H.I., Hulka, B.S., 1994, Cigarette smoking and sperm density: a meta analysis. *Fertil Steril.* 61:35-43.

Vine, M.F., Tse, C.K.J., Hu, P.C., Truong, K.Y., 1996, Cigarette smoking and semen quality. *Fertil Steril.* 65:835-842.

Vine, M.F., Setzer, R.W., Everson, R.B., Wyrobek, A.J., 1997, Human sperm morphometry and smoking, caffeine, and alcohol consumption. *Reprod Toxicol.* 11:179-184.

Vogt, H.J., Heller, W.D., Borelli, S., 1986, Sperm quality of healthy smokers, ex-smokers, and never-smokers. *Fertil Steril.* 45:106-110.

Wang, S.L., Wang, X.R., Chia, S.E., Shen, H.M., Song, L., Xing, H.X., Chen, H.Y., Ong, C.N., 2001, A study on occupational exposure to petrochemicals and smoking on seminal quality. *J Androl.* 22:73-78.

Wilcox, A.J., Weinberg, C.R., O'Connor, J.F., Baird, D.D., Schlatterer, J.P., Canfield, R.E., Armstrong, E.G., Nisula, B.C., 1988, Incidence of early loss of pregnancy. *N Engl J Med.* 319:189-194.

Williams, B.J., Ballenger, C., Malter, H., Bishop, F., Tucker, M., Zwingman, T.A., Hassold, T.J., 1993, Nondisjunction in human sperm: Results of fluorescence in situ hybridization studies using two and three probes. *Hum Mol Genet.* 2:1929-1936.

Wong, W.Y., Thomas, C.M.G., Merkus, H.M.W.M., Zielhuis, G.A, Doesburg, W.H., Steegers-Theunissen, R.P.M., 2000, Cigarette smoking and the risk of male factor subfertility: minor association between cotinine in seminal plasma and semen morphology. *Fertil Steril.* 74:930-935.

World Health Organization, 1992, Laboratory manual for the examination of human semen and semen-cervical mucus interaction. 3rd ed. New York: Cambridge University Press, 1-107.

Yurov, Y.B., Saias, M..J., Vorsanova, S.G., et al., 1996, Rapid chromosomal analysis of germ-line cells by FISH: an investigation of an infertile male with large-headed spermatozoa. *Mol Hum Reprod.* 2:665-668.

Zavos, P.M., Correa, J.R., Karagounis, C.S., Ahparaki, A., Phoroglou, C., Hicks, C.L., Zarmakoupis-Zavos, P.N., 1998, An electron microscope study of the axonemal ultrastructure in human spermatozoa from male smokers and nonsmokers. *Fertil Steril.* 69:430-434.

Zenzes M.T., Wang P., Casper R.F., 1995, Cigarette smoking may affect meiotic maturation of human oocytes. *Hum Reprod.* 10:3213-3217.

Zenzes, M.T., Bielecki, R., Reed, T.E., 1999, Detection of benzo(a)pyrene diol epoxide-DNA adducts in sperm of men exposed to cigarette smoke. *Fertil Steril.* 72:330-335.

Zinaman, M.J., Brown, C.C., Selevan, S.G., Clegg, E.D., 2000, Semen quality and human fertility: A prospective study with healthy couples. *J Androl.* 21:145-153.

SPERM NUCLEAR DNA DAMAGE IN THE HUMAN

Denny Sakkas[1], Gian Carlo Manicardi[2], Davide Bizzaro[3]

[1]Department of Obstetrics and Gynaecology, Yale University School of Medicine, Connecticut, USA
[2]Department of Animal Biology, University of Modena and Reggio Emilia, Italy
[3]Institute of Biology and Genetics, University of Ancona, Italy

INTRODUCTION

The male partner is now recognized as the single most common contributor to a couple's reason for infertility, with sperm defects accounting for around 30-50% of cases presenting to infertility clinics (Hull, 1992). A standard semen analysis, in particular the assessment of sperm concentration and motility, should therefore always be viewed as a mandatory test for inclusion in the investigation of the infertile couple (WHO, 1992; WHO, 1999). A semen analysis, however, only helps to provide broad diagnostic categories.

With the increased use of assisted reproduction to treat infertile couples, we ultimately need to identify specific sperm defects in order to allocate patients more appropriately to treatments, such as intrauterine insemination (IUI), in-vitro fertilization (IVF) or intracytoplasmic sperm injection (ICSI). Semen analysis, therefore, needs to develop to the point where we can not only provide patients with some idea of prognosis in terms of natural conception but also assess whether assisted reproduction will have a successful outcome, both in terms of fertilization and of producing a healthy conceptus. An increased understanding of sperm function means that new methods of diagnosis will lead us to investigate spermatozoa at different levels, so as to ascertain if a male's sperm is competent. This assessment will include investigations at the nuclear, organelle and cytoskeletal levels.

One form of analysis which may have widespread use for assessing spermatozoa is the analysis of the integrity of the sperm nuclear chromatin/DNA. Links between sperm chromatin defects and decreased fertilization have now been reported widely in the context of assisted reproduction techniques (Bianchi et al., 1996a; Bianchi et al., 1996b; Esterhuizen et al., 2000a; Lopes et al., 1998; Sakkas et al., 1996; Sun et al., 1997). In addition, numerous studies have shown that spermatozoa with abnormal nuclear chromatin organization are more frequent in sub-fertile or infertile men (Bianchi et al., 1996a; Bianchi et al., 1996b; Esterhuizen et al., 2000b; Evenson et al., 1980; Irvine et al., 2000; Sun et al., 1997). Interestingly, spermatozoa with a morphologically normal aspect can show

Advances in Male Mediated Developmental Toxicity, edited by Bernard Robaire and Barbara F. Hales.
Kluwer Academic/Plenum Publishers, 2003.

differences in chromatin packaging when comparing sperm from normozoospermic men with those possessing abnormal sperm parameters (Bianchi et al., 1996b). It is clear that men with poor semen parameters are more likely to show a higher percentage of sperm nuclear DNA damage than men with normal semen parameters (Figure 1).

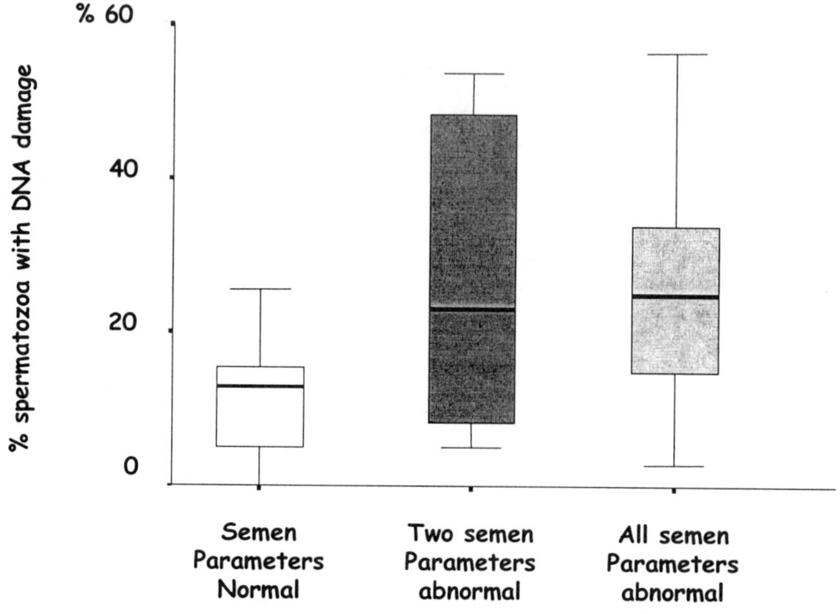

Figure 1. The percentage of ejaculated human spermatozoa per sample (shown as box plots) containing DNA damaged nuclei in relation to the assessment of the patient's semen parameters. Patients with two or all parameters showing abnormalities have combinations of oligozoospermia, asthenozoospermia and teratozoospermia.

The question that arises, however, is what information are we obtaining when we assess nuclear DNA damage in ejaculated human spermatozoa? To answer this we must first investigate how the nuclear DNA damage may arise.

THE ORIGIN OF DNA DAMAGE IN EJACULATED HUMAN SPERMATOZOA

Two theories have been proposed to explain the phenomenon of why ejaculated human spermatozoa possess anomalies in their nuclear DNA. The first theory arises from studies performed in animal models and is linked to the unique manner in which mammalian sperm chromatin is packaged. Endogenous nicks in DNA have been shown to be normally present at specific stages of spermiogenesis in rats and mice (McPherson and Longo, 1993a; McPherson and Longo, 1992; McPherson and Longo, 1993b; Sakkas et al., 1995). In the rodent species, these endogenous nicks are evident during spermiogenesis but are not observed once chromatin packaging is completed (rat: McPherson and Longo, 1993a; mouse: Sakkas et al., 1995; Smith and Haaf, 1998). The presence of nicks is maximal during the transition from round to elongated spermatids in the testis and occurs before the completion of protamination in maturing rat and mouse spermatozoa. It has been proposed that the endogenous nuclease, topoisomerase II, may play a role in both creating and ligating nicks during spermiogenesis, that these nicks are thought to provide relief of torsional stress and that they aid chromatin rearrangement during the displacement of histones by the protamines (Chen and Longo, 1996; McPherson and Longo, 1992). Recently, we have also found the presence of topoisomerase II in the seminiferous tubules of human testes (unpublished results).

When examining sections of seminiferous tubules from human testes we have found that a similar appearance of DNA nicks is apparent, when compared to rodent species, in that there appears to be an increase in DNA strand breaks in spermatids (Figure 2). We have observed that in cross sections of seminiferous tubules from men with normal spermatogenesis there is an increased presence of DNA strand breaks in early to mid spermatid stage cells, as detected by TUNEL immunohistochemistry. This is in agreement with that seen in other mammals. In contrast, a similar study (Oldereid et al., 2001) in the human has reported that the frequency of TUNEL positive cells was 17% and 26% in spermatids and primary spermatocytes, respectively.

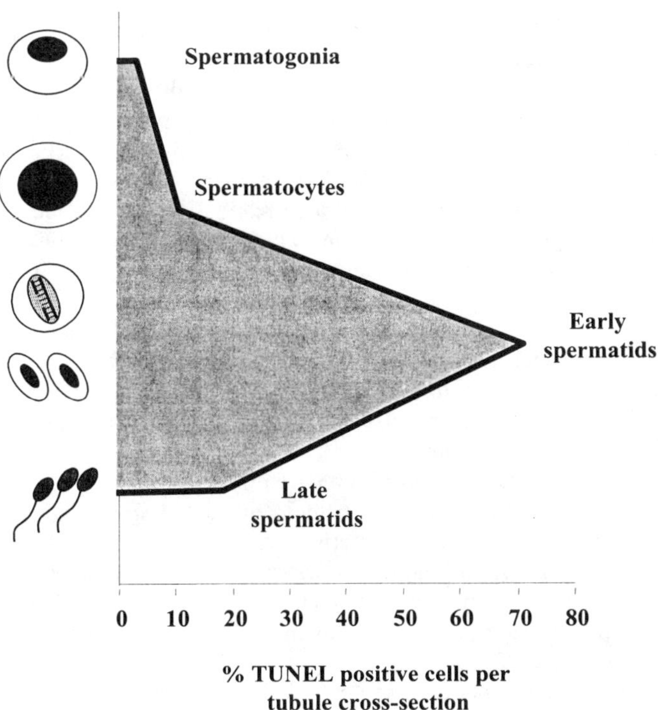

Figure 2. The percentage of TUNEL positive cells per human seminiferous tubule cross-section in relation to the stage of spermatogenesis.

The second theory of why sperm nuclear DNA damage is present in ejaculated human spermatozoa arises from the use of the TUNEL assay as a marker of apoptosis. Following from these results, a number of studies have stated that the presence of TUNEL positive sperm indicates that these sperm are indeed apoptotic (Gorczyca et al., 1993; Lopes et al., 1998; Sun et al., 1997; Tesarik et al., 1998).

THE PRESENCE OF APOPTOTIC MARKERS IN EJACULATED SPERMATOZOA

In a number of animal models, overproliferation of early germ cells is tempered by selective apoptosis of their progeny (Allan et al., 1992; Bartke, 1995; Billig et al., 1995; Furuchi et al., 1996; Rodriguez et al., 1997; Sinha Hikim et al., 1997; Tapanainen et al., 1993). Testicular germ cell apoptosis occurs normally and continuously throughout life. One factor postulated to be implicated in sperm apoptosis is the cell surface protein, Fas (Lee et al., 1997). Fas is a type I membrane protein that belongs to the tumour necrosis factor / nerve growth factor receptor family and mediates apoptosis (Krammer et al., 1994; Schulze-Osthoff et al., 1994; Suda et al., 1993). Binding of Fas ligand (FasL) or agonistic anti-Fas antibody to Fas kills cells by apoptosis (Suda et al., 1993). In mice and rats, it has been shown that, in the normal state, Sertoli cells express FasL and signal the killing of Fas positive germ cells, thus limiting the size of the germ cells population to numbers they can support (Lee et al., 1999; Lee et al., 1997; Rodriguez et al., 1997). In addition, after injury, Sertoli cells increase FasL expression to reach a new equilibrium state that matches the reduced capacity of the dysfunctional Sertoli cells with fewer germ cells. Thus, upregulation of Fas in germ cells is seen as a self- elimination process for cells that are destined to die because of inadequate support. As in other mammalian systems, it appears that the Fas mediated system and its related pathways are also responsible during the germ cell stages for controlling spermatogenesis in the human (Pentikainen et al., 1999). The study of Pentikainen et al. (1999) would also implicate the caspase pathway in this initial control of programmed cell death.

Studies on sections of seminiferous tubule from human testes also indicate that apoptosis plays a role in spermatogenesis. As evidenced by Pentikainen et al. (1999), the Fas mediated pathway plays a role, however we and others have also found that members of the Bcl-2 protein family, in particular Bcl-x, are more prevalent in spermatogonia (Oldereid et al., 2001; Sakkas, unpublished results).

In a previous study we have found that men with abnormal sperm parameters display higher levels of the apoptotic protein Fas on their ejaculated spermatozoa (Sakkas et al., 1999). The presence of Fas on ejaculated spermatozoa correlates strongly with a decreased sperm concentration and sperm with abnormal morphology. More recently, we and others have also found that other apoptotic markers, such as p53 and annexin V are also present on ejaculated human spermatozoa and show distinct relationships with abnormal semen parameters (Barroso et al., 2000).

The relationship of these apoptotic proteins with poor semen parameters may indicate that indeed the sperm that have DNA damage are also the ones that are positive for certain apoptotic proteins. When we examined whether TUNEL expression was evident in the same samples that had high Fas, p53 or Bcl-x expression, we found that there was no strict relationship. When double labeling was performed to see if the same population of sperm that were TUNEL positive would also show either Bcl-x or p53 expression, we found that although the majority of p53 positive sperm also showed TUNEL positivity only about half of the Bcl-x positive sperm showed TUNEL positivity (Figure 3). It appears from these results that ejaculated sperm exhibiting DNA damage do not necessarily show distinct apoptotic markers.

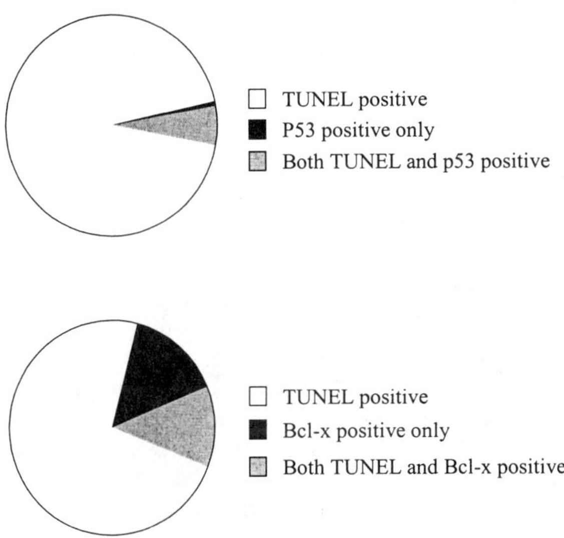

Figure 3. Double labeling of ejaculated human spermatozoa to examine whether p53 or Bcl-x are expressed in the same sperm that are TUNEL positive. Spermatozoa were assessed after labeling with fluorescent markers and examined using flow cytometry.

We have previously hypothesized that the presence of Fas on ejaculated human spermatozoa may be indicative of sperm cells that are 'earmarked' for apoptosis, but they escape apoptosis because there are too many for the available Fas-L to induce apoptosis or the signaling through Fas is not functional. We have called this "abortive apoptosis" (Sakkas et al., 1999). How these Fas labelled sperm escape apoptosis is unknown, however ,the persistence of increased expression levels of some apoptotic markers on mature spermatozoa may involve a similar mechanism to that reported in the rat (Blanco-Rodriguez and Martinez-Garcia, 1999). Blanco-Rodriguez and Martinez-Garcia (1999) have shown that apoptotic signaling molecules can be restricted to a specific cytoplasmic compartment during rat spermatogenesis and form residual bodies similar to apoptotic bodies. These residual bodies have increased expression levels of caspase-1, c-jun, p53 and p21. Although human spermatozoa do not normally possess residual bodies they do sometimes retain cytoplasm. Therefore, a defect in the cytoplasmic remodeling may be responsible for some of the mature spermatozoa showing high levels of apoptotic markers. This would explain the relationship with poor morphology. Cytoplasmic retention has already been shown to vary in human spermatozoa that exhibit different biochemical markers such as creatine-kinase (Gergely et al., 1999; Huszar et al., 1998; Huszar et al., 2000).

The relationship between the presence of certain apoptotic proteins and nuclear DNA damage may therefore be explained by the nature of spermatogenesis.

APOPTOSIS MEETS CELLULAR AND NUCLEAR REMODELLING DURING SPERMATOGENESIS

We propose the following preliminary model to explain the presence of nuclear DNA damage and apparent anomalies in apoptosis during human spermatogenesis (Figure 4). As in other mammalian systems, it appears that the Fas mediated system and its related pathways are responsible during the germ cell stages for controlling spermatogenesis (Pentikainen et al., 1999). The study of Pentikainen et al. (1999) would also implicate the caspase pathway in this initial control of programmed cell death. In addition, the Bcl-2 family of proteins, in particular Bcl-x, would also be candidates in the selection process of which cells proceed to maturity. These systems would have their greatest influence pre-spermiogenesis, prior to the remodeling of the nucleus and the extrusion of the cytoplasm. During spermiogenesis DNA breaks appear normally in the early spermatids, and we would predict that the ejaculated spermatozoa possessing nuclear DNA breaks are those in which the DNA breaks have not been repaired. Furthermore, the nuclear remodeling that takes place during spermiogenesis could also change or derail the classic programmed cell death pathway/s which were activated in the earlier stages. The extensive cytoplasmic remodeling that occurs during the later stages of spermatogenesis may also disrupt the apoptotic pathways that are functional prior to spermiogenesis. Keeping this model in mind, the apparition of defective spermatozoa in the ejaculate could appear to be related to anomalies in the repair of the DNA breaks that appear in spermatids and / or an increased apoptosis during early spermatogenesis. The two processes may be operating independently, but they also may obstruct the function of each other.

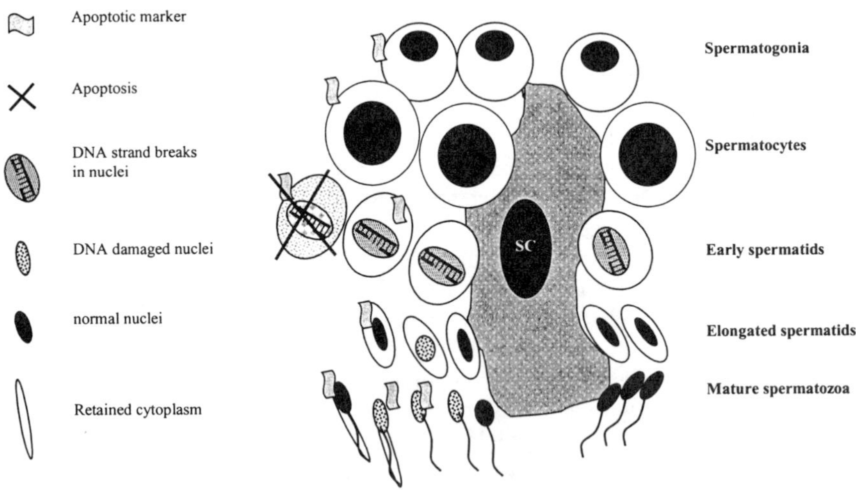

Figure 4. A hypothetical model of the relationship between apoptosis and cellular remodeling. Normal spermatogenesis, arising in normal mature spermatozoa, is indicated on the right side of the Sertoli Cell (SC). On the left of the SC are a number of scenarios giving rise to abnormal spermatozoa; these include failed apoptosis and/or a failure to repair DNA strand breaks appearing during the early spermatid stage.

The final outcome would be the production of spermatozoa that possess a range of anomalies. This would include spermatozoa that have single or combinations of abnormallities such as abnormal chromatin packaging [indicated by low levels of protamine], DNA strand breaks, presence of abnormal levels of apoptotic proteins and/or cytoplasmic retention. The relative proportion of these sperm in relation to those that have normal chromatin packaging, no DNA strand breaks, normal levels of apoptotic proteins, and no cytoplasmic retention could be reflective of a male's likelihood to be infertile.

What occurs once sperm enter into the epididymis is not known. A large body of evidence exists indicating that sperm from infertile men are more susceptible to reactive oxygen species exposure within the male reproductive tract (Aitken et al., 1989; Aitken et al., 1998) and that this increases levels of nuclear DNA damage (Hughes et al., 1998; Twigg et al., 1998a; Twigg et al., 1998b). The compact nature of the sperm nucleus should protect it from DNA damage post-ejaculation. Indeed, even incubation of mature mouse and human spermatozoa with DNAse is unable to induce DNA damage (Sakkas et al., 1995; Weil et al., 1998; unpublished results). However, abnormalities in the chromatin structure will make some sperm more susceptible to attack from external factors such as reactive oxygen species.

The mechanisms responsible for producing abnormal spermatozoa in the human ejaculate have been poorly understood. The presence of spermatozoa with DNA damage in the ejaculate may be related to a number of factors in combination or acting separately. The nuclear DNA damage may arise due to anomalies in nuclear remodeling and could be a direct result of faulty ligation of the nuclear DNA during spermiogenesis. A different population of sperm could exist that has escaped programmed cell death and expresses various apoptotic markers, a procedure we have previously called "abortive apoptosis" (Sakkas et al., 1999). This may not be classical apoptosis and could be due to defects in the remodeling of the cytoplasm that takes place during spermatogenesis. In the literature, many of these ejaculated sperm are being called apoptotic but the above evidence indicates that we may have to amend our terminology.

Furthermore, when we searched for the presence of nuclear DNA damage [using in situ nick translation] and for the integrity of chromatin packaging [using the Chromomycin A_3 (CMA_3) fluorochrome, which is a surrogate indicator of the presence of protamine (Bizzaro et al., 1998)] on the same sperm population, we found the existence of different classes of spermatozoa in the human ejaculate. Subsequently, we proposed that in most ejaculates the majority of healthy spermatozoa contain compact chromatin (highly protaminated CMA_3 negative sperm) and unbroken DNA. The other two main classes, however, represent sub-groups of spermatozoa that have anomalies, and contain either loosely packaged chromatin (probably under-protaminated CMA_3 positive sperm) but unbroken DNA or contain both loosely packaged chromatin (probably under-protaminated) and nicked DNA (Manicardi et al., 1995).

The end product of faulty nuclear remodeling and/or discrepancies in the cytoplasmic alterations leads to a heterogeneous population of ejaculated sperm, and it appears that this heterogeneity becomes greater as the quality of the semen parameters (i.e., concentration, motility and morphology) decrease. In using different techniques to assess sperm nuclear DNA damage, we may be assessing various levels of this heterogeneity.

WHAT ARE WE ASSESSING WHEN WE MEASURE NUCLEAR DNA DAMAGE?

A variety of techniques have been used to measure sperm nuclear DNA damage. All these techniques provide valuable information but also differ in that they assess DNA damage at various levels. As stated above, the relative heterogeneity of the population of ejaculated spermatozoa is represented differently when using the various assessment techniques. Techniques such as the sperm chromatin structure assay (SCSA) (Evenson and Jost, 1994) measure the susceptibility of sperm nuclear DNA to heat or acid induced denaturation in situ for a period of 30 seconds followed by staining with acridine orange; quantitation is by the metachromatic shift of acridine orange fluorescence from green (native double stranded DNA) to red (denatured single stranded DNA). This technique is likely to measure the population of spermatozoa that also have poorly packaged chromatin in addition to those with DNA damage. The single cell gel electrophoresis (COMET) assay also involves performing a lysis step that could lead to assessing spermatozoa that are not solely showing nuclear DNA damage. This technique, however, is more sensitive and is also able to distinguish between the presence of double and single stranded DNA breaks in the sperm nucleus. The COMET assay also enables the analysis of single cells (Figure 5).

(A)

(B)

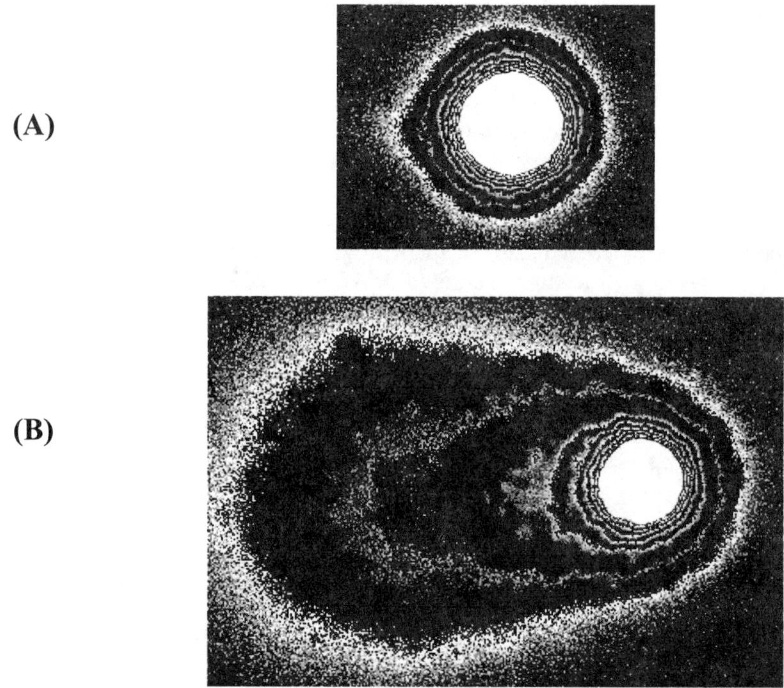

Figure 5. The single cell gel electrophoresis (COMET) assay of a normal (a) human spermatozoa showing no COMET tail and an abnormal (b) human spermatozoa showing a tail of DNA strand breaks.

The other commonly used techniques, such as the TUNEL assay or in situ nick translation, are more likely to directly measure only nuclear DNA damage if used without any prior preparation of the sperm. The use of flow cytometry, in conjunction with techniques such as the SCSA and TUNEL assays, allows a greater number of sperm to be assessed and, in the future, may be more readily accessible if these techniques are to be used as a routine analytical tool

The assessment of sperm nuclear DNA damage, apart from providing basic information about nuclear packaging in human sperm, may provide us with clinical information in treating men who have reduced fertility.

THE USE OF DNA ASSESSMENT TO PREDICT THE FERTILITY OF MEN AND SUCCESS AFTER ASSISTED REPRODUCTION TREATMENT

A relationship between sperm nuclear DNA integrity and fertility has been reported in two studies which have shown that the SCSA could be used as a prognostic factor for human fertility (Evenson et al., 1999; Spano et al., 2000). Both of these studies have examined couples attempting to achieve pregnancy without the aid of assisted reproduction techniques. As stated above, the SCSA measures the susceptibility of sperm nuclear DNA to heat or acid induced denaturation in situ for a period of 30 seconds followed by staining with acridine orange; it is quantitated by the metachromatic shift of acridine orange fluorescence from green (native double stranded DNA) to red (denatured single stranded DNA). Hence patients with a greater shift from the normal native double stranded DNA pattern towards denatured single stranded DNA are believed to possess less competent spermatozoa. Human SCSA data is also believed to be more constant over time than the classical measures of spermatozoa, as shown in a study of 45 men who provided monthly semen samples over a 9 month period (Evenson et al., 1991). In a study by the group of Evenson and colleagues (1999), 200 presumably fertile couples wishing to achieve pregnancy and who were discontinuing contraception were evaluated. The authors calculated the number of sperm with denatured DNA by calculating the percentage of cells falling out of the normal sperm area and found that, among these couples, men who had a value of greater or equal to 30% had difficulties in achieving pregnancy. The authors stated that "of great importance to the andrology clinic are those cases where the classical criteria (concentration, motility and morphology) are within the normal ranges, but the SCSA values are poor, and probably not compatible with good fertility". In addition, using selected cut-off values, the SCSA data predicted seven of 18 miscarriages (39%).

Similarly, another study by Spano et al. (2000) examined 250 first pregnancy planners who had no previous knowledge of their fertility capability. They found that the probability of fathering a child sharply declined when the fraction of cells with abnormal chromatin was >20% and was negligible when a value of 40% was obtained. These studies showed that the SCSA, which is an indicator of abnormal nuclear chromatin organization, was highly indicative of male subfertility, regardless of the concentration, motility and morphology of spermatozoa from an individual patient.

In addition to a relationship between sperm nuclear DNA damage and fertility, the assessment of sperm DNA damage has been found to correlate with various reproductive outcomes after assisted reproduction treatment. In a study that investigated the relationship between the outcome of intracytoplasmic sperm injection (ICSI) and the percentage of spermatozoa possessing nuclear DNA damage, as assessed by the TUNEL assay, Lopes *et al.* (1998) showed that DNA damage in sperm might contribute to fertilisation failure after ICSI. They found a significant negative association between the percentage of sperm with DNA fragmentation and the ICSI fertilization rate. Indeed, they found that men who had more than 25% of their sperm possessing DNA damage were more likely to experience fertilisation rates of less than 20% after ICSI. We also found that failed fertilized oocytes, examined after ICSI, contained significantly more condensed spermatozoa that failed to initiate decondensation when they were selected from patients with a higher percentage of sperm nuclear DNA damage (Sakkas et al., 1996).

In a study conducted by Twigg et al. (1998b), it was found that although sperm exposed to hydrogen peroxide, NADPH or activated leucocytes showed increased amounts of DNA strand breakage, their rate of decondensation and pronuclear formation, when microinjected into hamster oocytes, was the same as untreated control sperm. They went on to state that the ability of genetically damaged spermatozoa to achieve normal fertilization had implications for assisted conception therapy. This result is of interest as it suggests that sperm nuclear DNA damage may not preclude sperm from fertilizing an oocyte and their effects may be seen at a later stage. Interestingly, it is evident that embryos created by ICSI, in which the chance of selecting a spermatozoa possessing DNA damage is greatly increased, have a lower capacity to reach the blastocyst stage when compared to embryos created after routine in vitro fertilization (Dumoulin et al., 2000; Shoukir et al., 1998). A relationship between pregnancy outcome after in vitro fertilization and ICSI has also been found when using SCSA (Larson et al., 2000) and in situ nick translation (Tomlinson et al., in press) to assess sperm being used in the treatment of patients.

CONCLUSION

The presence of nuclear DNA strand breaks in ejaculated human spermatozoa has now been evidenced in a multitude of studies and using a range of different techniques. The reason why human spermatozoa, in particular from men with abnormal semen parameters, possess these abnormalities in their nuclear DNA is still not clear. Understanding the mechanisms responsible will improve our knowledge about certain causes of male infertility. More importantly, the impact of such sperm, if selected to perform ICSI, needs to be better understood so that any detrimental paternal effects can be avoided.

REFERENCES

Aitken, R.J., Clarkson, J.S., Hargreave, T.B., Irvine, D.S. and Wu, F.C., 1989, Analysis of the relationship between defective sperm function and the generation of reactive oxygen species in cases of oligozoospermia. *J Androl.* 10:214-220.

Aitken, R.J., Gordon, E., Harkiss, D., Twigg, J.P., Milne, P., Jennings, Z. and Irvine, D.S., 1998, Relative impact of oxidative stress on the functional competence and genomic integrity of human spermatozoa. *Biol Reprod.* 59:1037-1046.

Allan, D., Harmon, B. and Roberts, S., 1992, Spermatogonial apoptosis has three morphologically recognizable phases and shows no circadian rhythm during normal spermatogenesis in the rat. *Cell Proliferation.* 25:241-250.

Barroso, G., Morshedi, M. and Oehninger, S., 2000, Analysis of DNA fragmentation, plasma membrane translocation of phosphatidylserine and oxidative stress in human spermatozoa. *Hum Reprod.* 15:1338-1344.

Bartke, A., 1995, Apoptosis of male germ cells, a generalized or a cell type-specific phenomenon? *Endocrinology.* 136:3-4.

Bianchi, P.G., Manicardi, G., Bizzaro, D., Campana, A., Bianchi, U. and Sakkas, D., 1996a, Use of the guanine-cytosine (GC) specific fluorochrome, chromomycin A3, as an indicator of poor sperm morphology. *J Assist Reprod Genet.* 13:246-250.

Bianchi, P.G., Manicardi, G.C., Urner, F., Campana, A. and Sakkas, D., 1996b, Chromatin packaging and morphology in ejaculated human spermatozoa: evidence of hidden anomalies in normal spermatozoa. *Mol Hum Reprod.* 2:139-144.

Billig, H., Furuta, I., Rivier, C., Tapanainen, J., Parvinen, M. and Hsueh, A.J., 1995, Apoptosis in testis germ cells: developmental changes in gonadotropin dependence and localization to selective tubule stages. *Endocrinology.* 136:5-12.

Bizzaro, D., Manicardi G.C., Bianchi, P.G., Bianchi, U., Mariethoz, E. and Sakkas D., 1998, In-situ competition between protamine and fluorochromes for sperm DNA. *Mol Hum Reprod.* 4:127-132.

Blanco-Rodriguez, J. and Martinez-Garcia, C., 1999, Apoptosis is physiologically restricted to a specialized cytoplasmic compartment in rat spermatids. *Biol Reprod.* 61:1541-1547.

Chen, J.L. and Longo, F.J., 1996, Expression and localization of DNA topoisomerase II during rat spermatogenesis. *Mol Reprod Dev.* 45:61-71.

Dumoulin, J.C., Coonen, E., Bras, M., van Wissen, L.C., Ignoul-Vanvuchelen, R., Bergers-Jansen, J.M., Derhaag, J.G., Geraedts, J.P. and Evers, J.L., 2000, Comparison of in-vitro development of embryos originating from either conventional in-vitro fertilization or intracytoplasmic sperm injection. *Hum Reprod.* 15:402-409.

Esterhuizen, A.D., Franken, D.R., Lourens, J.G., Prinsloo, E. and van Rooyen, L.H., 2000a, Sperm chromatin packaging as an indicator of in-vitro fertilization rates. *Hum Reprod.* 15:657-661.

Esterhuizen, A.D., Franken, D.R., Lourens, J.G., Van Zyl, C., Muller, II and Van Rooyen, L.H., 2000b, Chromatin packaging as an indicator of human sperm dysfunction. *J Assist Reprod Genet.* 17:508-514.

Evenson, D. and Jost, L., 1994, Sperm chromatin structure assay: DNA denaturability. *Methods Cell Biol.* 42:159-176.

Evenson, D.P., Darzynkiewicz, Z. and Melamed, M.R., 1980, Relation of mammalian sperm chromatin heterogeneity to fertility. *Science.* 210:1131-1133.

Evenson, D.P., Jost, L.K., Baer, R.K., Turner, T.W. and Schrader, S.M., 1991, Individuality of DNA denaturation patterns in human sperm as measured by the sperm chromatin structure assay. *Reprod Toxicol.* 5:115-125.

Evenson, D.P., Jost, L.K., Marshall, D., Zinaman, M.J., Clegg, E., Purvis, K., de Angelis, P. and Claussen, O.P., 1999, Utility of the sperm chromatin structure assay as a diagnostic and prognostic tool in the human fertility clinic. *Hum Reprod.* 14:1039-1049.

Furuchi, T., Masuko, K., Nishimune, Y., Obinata, M. and Matsui, Y., 1996, Inhibition of testicular germ cell apoptosis and differentiation in mice misexpressing Bcl-2 in spermatogonia. *J Cell Biol.* 122:1703-1709.

Gergely, A., Kovanci, E., Senturk, L., Cosmi, E., Vigue, L. and Huszar, G., 1999, Morphometric assessment of mature and diminished-maturity human spermatozoa: sperm regions that reflect differences in maturity. *Hum Reprod.* 14:2007-2014.

Gorczyca, W., Traganos, F., Jesionowska, H. and Darzynkiewicz, Z., 1993, Presence of DNA strand breaks and increased sensitivity of DNA in situ to denaturation in abnormal human sperm cells: analogy to apoptosis of somatic cells. *Exp Cell Res.* 207:202-205.

Hughes, C.M., Lewis, S.E., McKelvey-Martin, V.J. and Thompson, W., 1998, The effects of antioxidant supplementation during Percoll preparation on human sperm DNA integrity. *Hum Reprod.* 13:1240-1247.

Hull, M.G., 1992, Infertility treatment: relative effectiveness of conventional and assisted conception methods. *Hum Reprod.* 7:785-796.

Huszar, G., Patrizio, P., Vigue, L., Willets, M., Wilker, C., Adhoot, D. and Johnson, L., 1998, Cytoplasmic extrusion and the switch from creatine kinase B to M isoform are completed by the commencement of epididymal transport in human and stallion spermatozoa. *J Androl.* 19:11-20.

Huszar, G., Stone, K., Dix, D. and Vigue, L., 2000, Putative creatine kinase M-isoform in human sperm is identified as the 70-kilodalton heat shock protein HspA2. *Biol Reprod.* 63:925-932.

Irvine, D.S., Twigg, J.P., Gordon, E.L., Fulton, N., Milne, P.A. and Aitken, R.J., 2000, DNA integrity in human spermatozoa: relationships with semen quality. *J Androl.* 21:33-44.

Krammer, P., Behrmann, I., Daniel, P., Dhein, J. and Debatin, K., 1994, Regulation of apoptosis in the immune system. *Current Opinions in Immunology.* 6:279-289.

Larson, K.L., DeJonge, C.J., Barnes, A.M., Jost, L.K. and Evenson, D.P., 2000, Sperm chromatin structure assay parameters as predictors of failed pregnancy following assisted reproductive techniques. *Hum Reprod.* 15:1717-1722.

Lee, J., Richburg, J.H., Shipp, E.B., Meistrich, M.L. and Boekelheide, K., 1999, The Fas system, a regulator of testicular germ cell apoptosis, is differentially up-regulated in Sertoli cell versus germ cell injury of the testis. *Endocrinol.* 140:852-858.

Lee, J., Richburg, J.H., Younkin, S.C. and Boekelheide, K., 1997, The Fas system is a key regulator of germ cell apoptosis in the testis. *Endocrinol.* 138:2081-2088.

Lopes, S., Sun, J.G., Jurisicova, A., Meriano, J. and Casper, R.F., 1998, Sperm deoxyribonucleic acid fragmentation is increased in poor-quality semen samples and correlates with failed fertilization in intracytoplasmic sperm injection. *Fertil Steril.* 69:528-532.

Manicardi, G.C., Bianchi, P.G., Pantano, S., Azzoni, P., Bizzaro, D., Bianchi, U. and Sakkas, D., 1995, Presence of endogenous nicks in DNA of ejaculated human spermatozoa and its relationship to chromomycin A3 accessibility. *Biol Reprod.* 52:864-867.

McPherson, S. and Longo, F.J., 1993a, Chromatin structure-function alterations during mammalian spermatogenesis: DNA nicking and repair in elongating spermatids. *Eur J Histochem.* 37:109-128.

McPherson, S.M. and Longo, F.J., 1992, Localization of DNase I-hypersensitive regions during rat spermatogenesis: stage-dependent patterns and unique sensitivity of elongating spermatids. *Mol Reprod Dev.* 31:268-279.

McPherson, S.M. and Longo, F.J., 1993b, Nicking of rat spermatid and spermatozoa DNA: possible involvement of DNA topoisomerase II. *Dev Biol.* 158:122-130.

Oldereid, N.B., Angelis, P.D., Wiger, R. and Clausen, O.P., 2001, Expression of Bcl-2 family proteins and spontaneous apoptosis in normal human testis. *Mol Hum Reprod.* 7:403-408.

Pentikainen, V., Erkkila, K. and Dunkel, L., 1999, Fas regulates germ cell apoptosis in the human testis in vitro. *Am J Physiol.* 276:E310-316.

Rodriguez, I., Ody, C., Araki, K., Garcia, I. and Vassalli, P., 1997, An early and massive wave of germinal cell apoptosis is required for the development of functional spermatogenesis. *EMBO J.* 16:2262-2270.

Sakkas, D., Manicardi, G., Bianchi, P.G., Bizzaro, D. and Bianchi, U., 1995, Relationship between the presence of endogenous nicks and sperm chromatin packaging in maturing and fertilizing mouse spermatozoa. *Biol Reprod.* 52:1149-1155.

Sakkas, D., Mariethoz, E. and St John, J.C., 1999a, Abnormal sperm parameters in humans are indicative of an abortive apoptotic mechanism linked to the Fas-mediated pathway. *Exp Cell Res.* 251:350-355.

Sakkas, D., Urner, F., Bianchi, P.G., Bizzaro, D., Wagner, I., Jaquenoud, N., Manicardi, G. and Campana, A., 1996, Sperm chromatin anomalies can influence decondensation after intracytoplasmic sperm injection. *Hum Reprod.* 11:837-843.

Schulze-Osthoff, K., Walczak, H., Droge, W. and Krammer, P., 1994, Cell nucleus and DNA fragmentation are not required for apoptosis. *J Cell Biol.* 127:15-20.

Shoukir, Y., Chardonnens, D., Campana, A. and Sakkas, D., 1998, Blastocyst development from supernumerary embryos after intracytoplasmic sperm injection: a paternal influence?. *Hum Reprod.* 13:1632-1637.

Sinha Hikim, A., Rajavashisth, T., Sinha Hikim, I., Lue, Y., Bonavera, J., Leung, A., Wang, C. and Swerdloff, R., 1997, Significance of apoptosis in the temporal and stage-specific loss of germ cells in the adult rat after gonadotropin deprivation. *Biol Reprod.* 57:1193-1201.

Smith, A. and Haaf, T., 1998, DNA nicks and increased sensitivity of DNA to fluorescence in situ end labeling during functional spermiogenesis. *Biotechniques.* 25:496-502.

Spano, M., Bonde, J.P., Hjollund, H.I., Kolstad, H.A., Cordelli, E. and Leter, G., 2000, Sperm chromatin damage impairs human fertility. The Danish First Pregnancy Planner Study Team. *Fertil Steril.* 73:43-50.

Suda, T., Takahashi, T., Golstein, P. and Nagata, S., 1993, Molecular cloning and expression of the Fas ligand a novel member of the tumor necrosis factor family. *Cell.* 75:1169-1178.

Sun, J.G., Jurisicova, A. and Casper, R.F., 1997, Detection of deoxyribonucleic acid fragmentation in human sperm: correlation with fertilization in vitro. *Biol Reprod.* 56:602-607.

Tapanainen, J., Tilly, J., Vihko, K. and Hsueh, A., 1993, Hormonal control of apoptotic cell death in the testis: gonadotropins and androgens as testicular cell survival factors. *Mol Endocrinol.* 7:643-650.

Tesarik, J., Greco, E., Cohen-Bacrie, P. and Mendoza, C., 1998, Germ cell apoptosis in men with complete and incomplete spermiogenesis failure. *Mol Hum Reprod.* 4:757-762.

Tomlinson, M.J., Moffatt, O., Manicardi, G.C., Bizzaro, D., Afnan, M. and Sakkas, D, Interrelationships between seminal parameters and sperm nuclear DNA damage before and after density gradient centrifugation: Implications for assisted conception. *Hum Reprod.* (in press)

Twigg, J., Fulton, N., Gomez, E., Irvine, D.S. and Aitken, R.J., 1998a, Analysis of the impact of intracellular reactive oxygen species generation on the structural and functional integrity of human spermatozoa: lipid peroxidation, DNA fragmentation and effectiveness of antioxidants. *Hum Reprod.* 13:1429-1436.

Twigg, J.P., Irvine, D.S. and Aitken, R.J., 1998b, Oxidative damage to DNA in human spermatozoa does not preclude pronucleus formation at intracytoplasmic sperm injection. *Hum Reprod.* 13:1864-1871.

Weil, M., Jacobson, M.D. and Raff, M.C., 1998, Are caspases involved in the death of cells with a transcriptionally inactive nucleus? Sperm and chicken erythrocytes. *J Cell Sci.* 111:2707-2715.

WHO., 1992, *World Health Organization Laboratory Manual for Examination of Human Semen*, Cambridge University Press, Cambridge.

WHO., 1999, *World Health Organization Laboratory Manual for Examination of Human Semen*, Cambridge University Press, Cambridge.

THE HUMAN SPERMATOZOON – NOT WAVING BUT DROWNING

R. John Aitken and Dennis Sawyer

Discipline of Biological Sciences
Centre for Life Sciences
University of Newcastle
Callaghan, NSW 2308

INTRODUCTION

The human male produces the poorest quality semen of all mammalian species studied to date. This alarming fact is illustrated by the World Health Organization's recent reference text on human seminology (World Health Organization, 1999) which suggests that up to 85% of spermatozoa may be morphologically abnormal, even in the fertile male population. The abnormally shaped spermatozoa that appear endemic to our species also seem to possess a diminished capacity for fertilization, given that defective sperm function is the commonest, defined cause of human infertility (Hull et al., 1985). The morphological abnormalities that dominate the human semen profile may take many forms, but retention of excess residual cytoplasm in the midpiece of the spermatozoa and defects in the acrosomal region appear to be particularly important (Keating et al., 1997; Garrett et al., 1997). Impaired sperm motility, asthenozoospermia, is another common cause of defective human sperm function. This is a very dynamic aspect of sperm biology that is particularly susceptible to environmental interference, such as workplace exposure to organic solvents (Wang et al., 2001; Xiao et al., 2001). In the wake of such defects in semen quality, the human species is distinguished by a higher rate of spontaneous abortions and birth defects than any other mammal. In light of such data, it is difficult to avoid the conclusion that something is seriously wrong with spermatogenesis in the human male.

Not only is human semen quality poor, but also claims have been made that it is getting progressively poorer. Several studies have reported a reduction in sperm counts since the Second World War, in concert with the sudden growth of the chemical industry. The suggestion that sperm counts might be declining originated from a meta-analysis performed by Carlsen et al. (1992) on data sets covering a fifty-year period (1938-1990). Subsequent studies in various European countries have provided some supporting evidence for the falling sperm count hypothesis in Italy (Bilotta et al., 1999), France (Auger *et al.,* 1995), Scotland (Irvine et al., 1996), and Greece (Multigner and Spira, 1997). A careful analysis of semen donors to a sperm bank in Paris, for example, revealed a fall of sperm counts from 89 to 60×10^6/ml between 1973 to 1992 (Auger *et al.,* 1995). After taking into

Advances in Male Mediated Developmental Toxicity, edited by Bernard Robaire and Barbara F. Hales.
Kluwer Academic/Plenum Publishers, 2003.

account all potential covariates (year of birth, age and abstinence time), it was concluded that within this donor population sperm concentration was declining by 2.6% per annum. Similarly, analysis of semen donors in Edinburgh revealed a dramatic decline in semen quality in birth cohorts sampled over an 11 year period commencing in 1959 (Irvine *et al.*, 1996). Reanalysis of existing data sets (Swan et al., 1998) has supported the notion of a temporal decline in sperm counts in the United States (studies from 1938-1988; slope = -1.50; 95% confidence interval (CI), -1.90 to -1.10) and Europe (1971-1990; slope = -3.13; CI, -4.96 to -1.30), but not in non-Western countries (1978-1989; slope = 1.56; CI, -1.00 to 4.12). In contradiction to Swann et al (1998), other studies in modern industrialized countries such as Japan and Spain have failed to find any evidence for a decline in semen quality or male fertility over the past twenty years (Andolz et al., 1999; Itoh et al., 2001).

It is as difficult to reconcile these disparate results, as it is to design a study where selection bias does not seriously confound the analysis (Handelsman, 1997). By their very nature, all of these studies were serendipitous and retrospective and not the outcome of a carefully designed prospective analysis. The existing data clearly indicate that there are significant, as yet unexplained, inter-regional variations in semen quality, as recently confirmed in a multi-centered European study (Jorgensen et al., 2001). However they are no more than suggestive of a time-dependent decline in semen quality in certain geographical locations. They clearly cannot be taken as hard evidence for such a decline, because there are too many weaknesses in the study designs and too much intrinsic variation in the criteria assessed. If time dependent changes in male reproductive function are occurring, we have to look beyond the conventional semen profile.

TESTICULAR CANCER

Additional evidence for a time-dependent decline in male reproductive function has come from analyses of reproductive tract pathologies that are developmental in origin. Thus, the incidence of reproductive tract pathologies such as testicular cancer, hypospadias, and cryptorchidism has risen alarmingly in many highly industrialized areas, while cancers of the female reproductive tract (cervix, uterus, and ovary) have remained relatively constant or declined (Matlai and Berala, 1985; Brown et al., 1986; Osterlind, 1986; Giwercman et al., 1993). A recent analysis of testicular cancer trends in Denmark for example, revealed that the rate of increase of testicular cancer was about 2.6% per annum (Moller, 2001). The incidence was more strongly dependent on the man's birth cohort than on the calendar period. In New South Wales the incidence of testicular cancer reflects this global trend, increasing in a progressive manner, in complete contrast to cancers of the female reproductive tract that appear to be static or, in the case of cervical cancer, decreasing (Figure 1). The increased incidence of testicular cancer cannot be related to the increasing longevity of the population since it is a disease of relatively young men, peaking in the early 30's. It is also unlikely to be due to better methods of detection since testicular cancer is relatively easily diagnosed. In contrast the dramatic increase in prostate cancer observed in New South Wales and elsewhere (Figure. 1; Moller, 2001) could easily involve both of these factors.

This global increase in testicular cancer, when taken together with the increasing incidence of hypospadias and cryptorchidism and the controversial decline in sperm concentration, have been taken to indicate the existence of a single underlying pathological condition, 'the testicular dysgenesis syndrome' (TDS). At the population level, the anomalous elements that make up TDS appear to be linked. The male population in Denmark, for example, exhibits the poorest semen quality and a high incidence of testicular cancer in Europe. The mechanisms by which these conditions might be associated are currently unknown. It is possible that the origins of TDS lie in the fetal or neonatal development of the testes. For example, testicular cancer is associated with a failure of the primordial germ cell to spermatogonial stem cell transition during the early development of the testes. Poor semen quality might theoretically be linked to a neonatal decline in Sertoli cell numbers, since the latter determines how many germ cells the testes can support. A key determinant of Sertoli cell replication is the circulating concentration of FSH that, in turn, is exquisitely sensitive to exogenously administered estrogens. Estrogen administration to animals in fetal or early neonatal life leads to reduced sperm counts and small testes in the offspring. It has been suggested that xenobiotics in the environment that behave as weak estrogens might affect the human fetus by suppressing FSH-induced Sertoli cell replication, thereby limiting sperm production in the adult, post puberty (Sharpe and Skakkebaek, 1993). These environmental estrogens include a large number of phenols and organohalogens that are extremely resistant to environmental degradation. At present, it is difficult to understand how these weak estrogens can compete with the high levels of potent natural estrogens produced by the human placenta to influence the normality of male development. It is also not clear why a decline of sperm number via this mechanism would be accompanied by a decline in sperm function, or how this mechanism might explain the rising prevalence of hypospadias, cryptorchidism or testicular cancer. Nevertheless 'the estrogen hypothesis' is an important contribution to the field that merits critical examination.

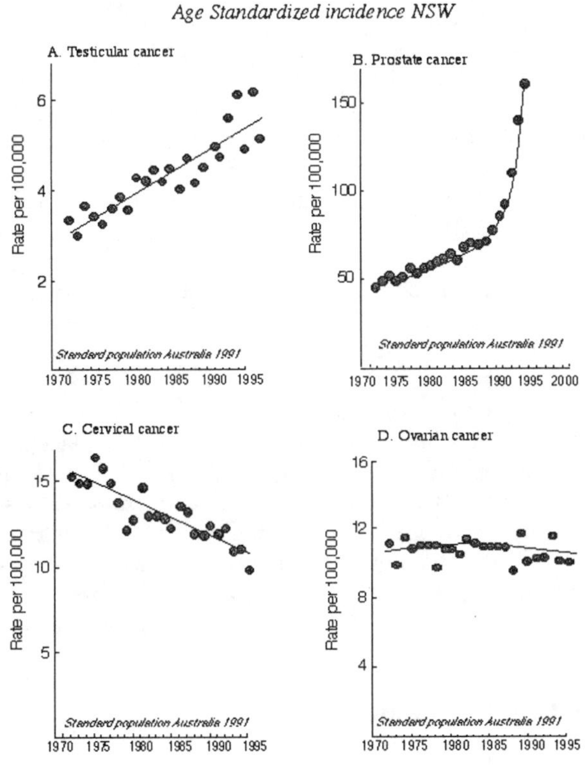

Figure 1. Changes in the rates of reproductive tract cancers over the past 30 years in New South Wales.

GENETIC FACTORS AND SEMEN QUALITY

Genetic factors are also an undeniable feature of male infertility. Specifically patients exhibiting azoospermia or severe oligozoospermia may exhibit DNA deletions on the long arm of the Y chromosome that remove genes crucial for the progress of normal spermatogenesis. The incidence of such Y chromosome microdeletions among infertile men varies between studies, from 1% to 55% (Krausz and McElreavey, 1999), the major factor influencing this parameter being the composition of the study population. Approximately 15% of azoospermic men have Y chromosome deletions compared with 5-10% in the severely oligozoospermic population. (Aitken and Krausz, 2001). Yq microdeletions are specific for spermatogenic failure since no such deletions have been detected in normozoospermic men (Krausz and McElreavey, 1999; Aitken and Krausz, 2001). Vogt and colleagues (1996) observed that Y chromosome microdeletions follow a certain deletion pattern, with three recurrently deleted non-overlapping subregions in proximal, middle, and distal Yq11, designated "AZFa", "AZFb," and "AZFc", respectively. In each region a number of candidate genes have been proposed (McElreavey and Krausz, 1999). The fact that no single gene deletions have been found in more than 200 oligo/azoospermic men screened for 6 AZF genes (Krausz and McElreavey, 1999; Krausz et al. 1999) suggests that gene-specific deletions are probably rare events and that only large Y deletions removing several genes, are normally associated with infertility.

The Y chromosome appears to be particularly susceptible to genetic damage that will affect male fertility. This chromosome is derived largely from a single autosomal region that was added to the sex chromosomes 80-130 million years ago and contained many of the key genes involved in the regulation of spermatogenesis (Waters et al., 2001). Most of the genes from the original sex chromosome have decayed on the Y with the result that the latter now contains just a small number of genes (largely performing house-keeping functions) with homologues on the X. Most of the remaining functional genes on the human Y chromosome appear to have evolved from transferred autosomal homologues, or in one case (CRY) retroposition of the corresponding mRNA (Chai et al., 1997; Lahn and Page 1997; Saxena et al., 1996; Lahn and Page, 1999). Once located on the Y, there has been a strong tendency for these genes to undergo amplification presumably because the presence of multiple gene copies creates a buffer against attrition (Burgoyne, 1998). These Y chromosome-specific genes encode molecules that are essential for sex determination (SRY) or male fertility (RBM, DAZ) (Delbridge and Graves, 1999; Marshall Graves, 2000). The ultimate fate of these fertility genes on the Y chromosome will presumably be to erode in the same way as the original X genes have done. The degeneration of genes on the Y is a reflection of the inability of this chromosome to replace lost genetic information by homologous recombination; particularly in the haploid state when sister chromatid exchange is not possible. Given that this is the case, then the progressive deterioration of spermatogenesis genes on the Y chromosome, and with it a parallel reduction of male fertility, would seem inevitable in the long term. Of course, the rates at which these genes deteriorate and male infertility is observed will be exacerbated by environmental factors that stimulate the occurrence of DNA fragmentation in the male germ line.

DNA DAMAGE IN THE GERM LINE

Most aspects of semen quality are independent of age, including fertility (Paulson et al., 2001). What does decline with paternal age is the genetic normality of the genome present in the sperm nucleus (Crow, 1997). Thus the incidence of dominant genetic mutations leading to conditions such as Aperts syndrome or achondroplasia, increases dramatically with paternal age presumably as a consequence of replication errors; the older the male, the more rounds of replication his germ cells have been through in generating the fertilizing spermatozoon. In contrast to these dominant genetic conditions, mutations in the male germ line associated with other diseases in the offspring, such as cancer or infertility, bear no apparent relationship with paternal age. Other, possibly environmental, factors are involved in their etiology.

Clearly environmental factors have an important effect on human semen quality. It is well documented that long-term exposure to pesticides has a negative impact on sperm count and morphology (Abell et al., 2000). Smoking can also induce a time-dependent decline in semen quality as can exposure to petrochemicals; moreover, the latter may aggravate the negative effects of smoking (Wang et al., 2001). Air pollution has also been shown to induce acute changes in semen quality, as reflected in sperm movement and morphology, without significantly affecting sperm number (Selevan et al., 2000).

The mutagenic effects of radiation and chemicals on the male germ line have been studied extensively in the mouse (Dubrova et al., 1993, 1998; Nomura, 1982,1988). Many different types of mutations arise in these cells after irradiation or chemical insult including base substitutions, frameshifts, deletions, and chromosomal abnormalities. The conesquences of these changes range from infertility, through pregnancy loss to birth defects, depending on the level of DNA damage induced. Because the timing of mouse spermatogenesis is known precisely, all stages of germ cell development can be tested for mutagenic sensitivity. An important finding from these studies is that in all but a few cases, post-meiotic germ cells are more sensitive to mutation induction than spermatogonial stem cells. In particular, late spermatids and spermatozoa are very susceptible to chemical mutagenesis. At least two mechanisms are responsible for this; the lack of DNA repair in germ cells beyond the late spermatid stage, and a selection process whereby spermatogonia harbouring mutations are selectively removed prior to meiosis. Thus DNA damage in spermatogonia is either efficiently repaired or, at higher levels, induces apoptosis. At the other end of spermatogenesis, late spermatids have lost their capacity for DNA repair and apoptosis, so DNA damage induced at these stages persists in the spermatozoa (Sawyer and Aitken, 2000). Such damage comprises a pre-mutational change that must be repaired by the oocyte following fertilization, before initiation of the first cleavage division (Figure 2). If the oocyte makes an error when repairing the DNA damage brought into the zygote by the fertilizing spermatozoon, a mutation will be fixed in the embryo that, if it is not repaired, will be passed onto the next generation through the germ line. The damage induced via this mechanism would be expected to include both genetic and chromosomal mutations (Figure 2). If the oocyte is not able to repair the sperm-derived DNA damage before the first mitotic division, further attempts could be made by the embryo to effect this repair during cleavage. Indeed an important objective of the embryo during cleavage might be to screen individual blastomeres for genetic damage before cells are finally committed to the embryonic lineage. If excessive damage is detected, the responsible blastomeres will be deleted through the induction of apoptosis. In the case of badly affected embryos, the net result of this strategy will be to terminate embryonic development; the greater the genetic damage, the earlier the termination of pregnancy (Figure 2). Partial repair of DNA damage during cleavage could be one mechanism by which germ line mosaicisms for a given mutation might be established (Figure 2).

The scheme proposed in Figure 2 for the causes and consequences of DNA damage in the male germ line includes a large number of suppositions, unknowns and questions. Foremost amongst the latter is the precise nature of the mechanisms by which DNA damage is induced in male germ cells and the environmental factors, if any, that trigger this change.

Mechanisms of DNA damage in the germ line

As previously indicated, one clear cut mechanism for the introduction of mutations into the male germ line involves replication errors that are a direct function of paternal age. The mutations that result in dominant genetic diseases such as achondroplasia or Apert's syndrome arise via this mechanism (Crow, 1997). Of course replication errors in the paternal germ line could be responsible for many different kinds of disease in the offspring; it is just that the mutations that cause the genetic conditions mentioned above are easily identifiable because of the dominant nature of the mutation and the clarity of the phenotype. Such replication error mutations are presumably introduced into the germ line at the spermatogonial stage of development and are sufficiently subtle to escape the cell cycle checkpoints that normally result in the deletion of mutated cells by apoptosis.

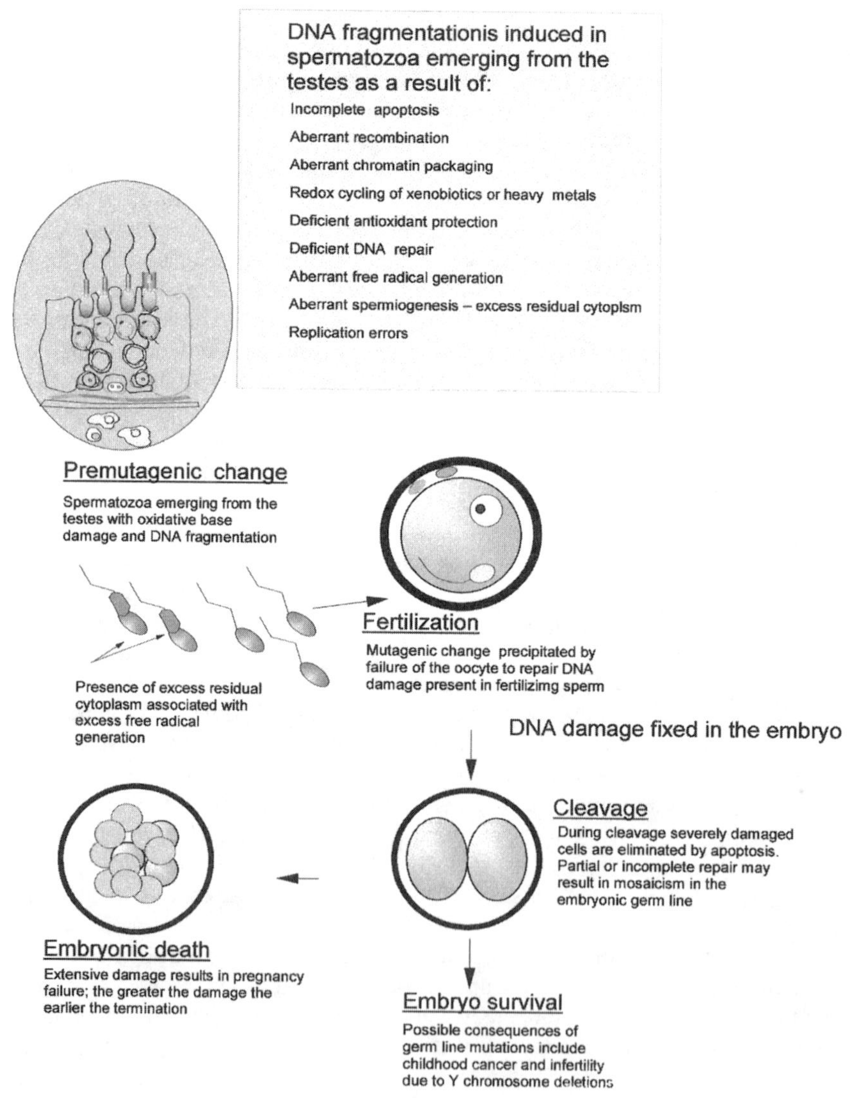

Figure 2. A proposed scheme for causes and consequences of DNA damage in the male germ line.

Additional possible sources of mutation in the male germ line include a variety of mechanisms for inducing DNA fragmentation in these cells. Causes of DNA fragmentation include abortive apoptosis (Sakkas et al., 1999ab), aberrant recombination, defective chromatin packaging (Bianchi et al., 1993), and oxidative stress (Aitken and Krausz, 2001). At present there is no consensus as to the major mechanism responsible for the induction of DNA fragmentation in the male germ line. Heavy smoking, for example, is associated with DNA fragmentation in germ cells that could be due to either oxidative stress and/or abortive apoptosis. Men who smoke heavily are clearly under oxidative stress as reflected by the diminished levels of antioxidant vitamins in their serum and seminal plasma (Fraga et al., 1996). This depletion of antioxidant vitamins induces a pro-oxidant state within the male reproductive tract and the appearance of significant levels of DNA fragmentation in the spermatozoa in association with oxidative base damage (Fraga et al., 1996). The DNA damage associated with heavy smoking is also associated with the retention of excess residual cytoplasm by the spermatozoa (Mak et al., 2000), and this too appears to be involved in the etiology of DNA fragmentation in the germ line due to oxidative stress.

Thus during spermiogenesis, spermatozoa undergo a unique morphological change, i.e., the removal of most of their cytoplasm, a cellular constituent that most cells depend on for their survival. In man this change is effected with varying degrees of efficiency such that the retention of excess residual cytoplasm is a common feature of defective human spermatozoa. The suggestion that defective sperm function might be associated with cytoplasmic retention stems from a series of independent studies revealing that poor semen quality was associated with the high cellular contents of key cytoplasmic enzymes such as creatine kinase (Huszar et al., 1988), lactic acid dehydrogenase (Casano et al., 1991), superoxide dismutase (Aitken et al., 1996) and glucose-6-phosphate dehydrogenase (Aitken et al., 1994; Gomez et al., 1996). The feature that all these enzymes share in common is that they are cytosolic, their level of cellular activity correlating with the amount of residual cytoplasm retained in the midpiece of human spermatozoa (Gomez et al., 1996). The loss of sperm motility and fertilizing potential associated with varicocoeles and idiopathic male infertility is associated with the retention of excess residual cytoplasm by the spermatozoa (Zini et al., 1998, 1999, 2000) as is the loss of fertility and DNA damage associated with heavy smoking, as indicated above (Mak et al., 2000). Recent studies of patients undergoing IVF therapy have also demonstrated a strong negative correlation between fertilization rate and the presence of residual cytoplasm in the sperm midpiece (Keating et al., 1997).

The mechanism by which the presence of excess residual cytoplasm disrupts human sperm function is thought to involve the induction of oxidative stress (Gomez et al, 1996). Thus the presence of excess glucose-6-phophate dehydrogenase is held to enhance the cellular generation of NADPH that, in turn, fuels the production of free radicals by a proposed sperm NADPH oxidase as indicated in Figure 3 (Aitken et al., 1994). There is a very clear relationship between free radical generation by mammalian spermatozoa and the cellular content of glucose-6-phosphate dehydrogenase (Figure 3). Moreover, NADPH is a powerful inducer of reactive oxygen species (ROS) generation by these cells (Aitken et al., 1997; Figure 3). Furthermore the ability of spermatozoa to generate ROS is negatively correlated with semen quality and the fertilizing capacity of human spermatozoa in vitro (Aitken, 1999; Gomez et al., 1998). Why the spermatozoa of sub-fertile patients should retain excess residual cytoplasm is unknown. One possibility is that this condition reflects an underlying defect of Sertoli cell function, the latter failing in their task to properly remove the cytoplasm from spermatozoa before spermiation.

Thus the effects of smoking on DNA fragmentation in human spermatozoa may involve oxidative base damage subsequent to antioxidant depletion and the possible generation of excess free radicals as a consequence of cytoplasmic retention. It has also been claimed that the sperm DNA damage observed in heavy smokers reflects an abortive apoptotic process initiated at an earlier stage of spermatogenesis. While dividing spermatogonia are able to complete apoptosis and effectively delete damaged cells, this

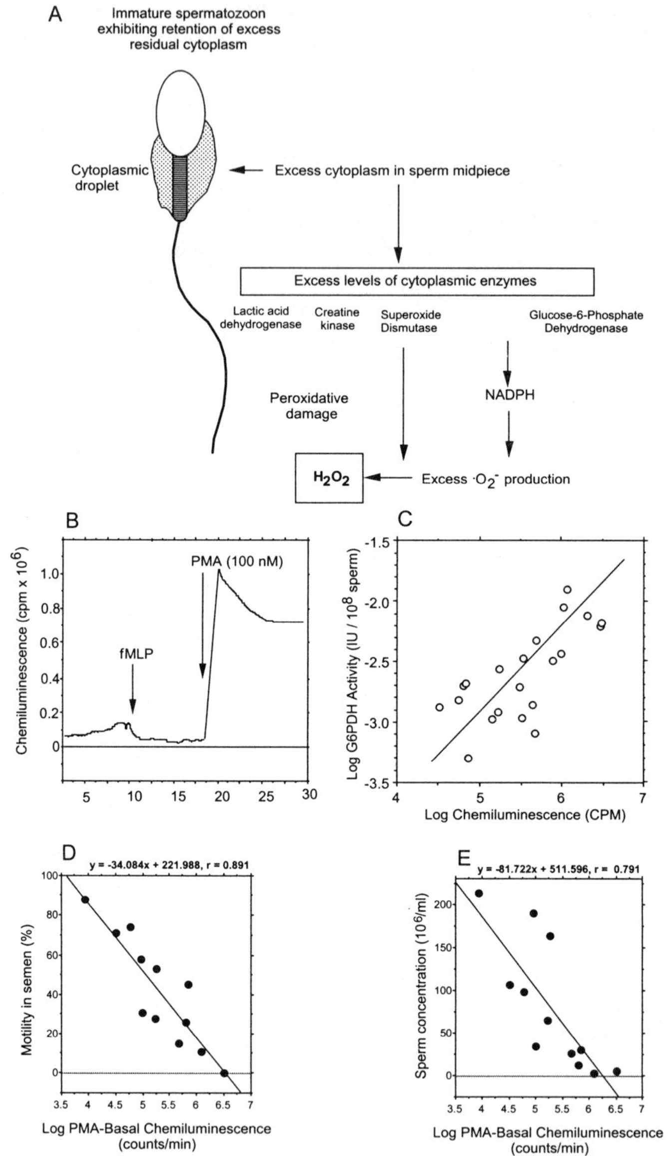

Figure 3. Generation of reactive oxygen species by human spermatozoa. A, retention of excess residual cytoplasm is associated with high cellular contents of cytoplasmic enzymes such as lactic acid dehydrogenase, creatine kinase and glucose-6-phosphate dehydrogenase. B, in leukocyte free sperm suspensions 12-myristate, 13-acetate phorbol ester can induce a burst of ROS. C, the intensity of this burst is highly correlated with the glucose-6-phosphate dehydrogenase concentration and reflects the underlying quality of spermatogenesis as reflected by the powerful negative correlations observed with percentage motility (D) and sperm concentration (E) in the original semen sample.

may become more difficult for germ cells at advanced stages of meiosis, as the mechanisms responsible for inducing cell death are closed down. There are many stimuli capable of inducing apoptosis in the testes, including gonadotrophin depletion, heat and DNA adduct formation. For example, the aromatic hydrocarbons present in cigarette smoke are known to form DNA adducts and may initiate germ cell apoptosis following activation of an aryl hydrocarbon receptor-dependent transcription factor (Benedict et al., 2000). The testicular apoptotic cascade is generally thought to involve the Fas-ligand mediated pathway. The induction of apoptosis via this pathway is clearly an important mechanism by which Sertoli cells regulate germ cell numbers, particularly in times of stress (Boekelheide et al., 2000). Accordingly the testes of men exhibiting deficiencies in the semen profile, particularly oligozoospermia, possess a high number of spermatozoa expressing the Fas receptor, prompting the suggestion that these dysfunctional cells are the product of an incomplete apoptotic cascade (Sakkas et al., 1999a,b). Whether defective apoptosis accounts for a significant proportion of the DNA damage seen in the spermatozoa of infertile men is still an open question. A recent analysis of DNA damage in the germ line could not find any ultrastructural evidence for apoptosis in association with DNA damage (Barroso et al., 2000), while another study found no correlation between DNA damage and Fas expression (Muratori et al., 2000). Of course, Fas binding and ROS generation are not mutually exclusive phenomena; ROS can induce Fas-mediated signal transduction in some cell types (Huang et al., 2000) while in others, Fas induced apoptosis appears to be mediated by ROS (Sayers et al., 2000).

Environmental Factors and DNA damage in germ cells

Radiation. Although clear-cut effects of radiation on germ line mutagenesis have been reported in mice, few effects have been found following exposure of humans to radiation. The first and largest epidemiological study of human radiation exposure was performed on a cohort of atomic bomb survivors in Hiroshima, Japan (Neel and Schull, 1956). This and subsequent studies of this population found no significant increase in the frequencies of stillbirths, congenital malformations, or childhood cancer, nor was there a change in sex ratio (Schull et al., 1966; Neel et al., 1990). Furthermore, no excess of chromosomal abnormalities was found, including structural rearrangements, sex chromosome abnormalities, or polyploidy. Later, more subtle mutations in the children of atomic bomb survivors were sought by examining electrophoretic patterns of proteins, and minisatelite DNA mutations. Again, no significant changes were observed (Neel et al., 1988; Kodaira et al., 1995).

Another massive human exposure to radiation occurred after the Chernobyl nuclear plant disaster in 1986. Increased rates of thyroid cancer in children were seen, most likely due to direct exposure of the children to radiation, rather than an effect on the paternal germ line (Lomat et al., 1997; Schwenn and Brill, 1997). Similar to the Japanese cohort discussed above, no clear associations have been found between exposure to Chernobyl radiation and increased rates of spontaneous abortion or birth defects, although recent data suggest that Down's syndrome rates were increased (Little, 1993; Ericson and Kallen, 1994; Rytomaa 1996). The only indication that germ-line mutation rates were affected by Chernobyl is found in studies of minisatellite DNA, although these results have recently been challenged (Dubrova et al, 1997; Neel, 1999). Overall, these results suggest that the male germ line is relatively resistant to the radical species generated in response to radiation. However, this resistance does not extend to xenobiotics.

Occupational Exposures. Despite the breadth of knowledge concerning chemical mutagenesis in the germ cells of laboratory animals, this process is poorly understood in humans. Very little information is available regarding genetic effects of specific chemicals to which humans are exposed. Exposure data are difficult to obtain and epidemiological studies frequently suffer from significant problems including small sample sizes and confounding variables including unreported or unknown exposures to other toxicants (Joffe, 1992). Thus, rather than assessing the effects of exposure to specific compounds, attempts have been made to link specific abnormal reproductive outcomes with broad occupational categories. Despite these limitations, significant associations have been made between environmental or occupational exposures, and abnormal reproductive outcomes (Table 1).

Table 1. Associations between paternal exposures and adverse reproductive outcomes

Agent	Embryo/fetal loss*	Birth defects*	Childhood cancer*
Radiation	0.9-1.5	1.4-5.6	N/T[§]
Solvents	0.9-2.3	N/T	1.7-7
Anesthetic gases	1.5-1.8	N/T	N/T
Heavy metals	0.9-2.3	1.5-249	3.5-7
Smoking	0.6-1.4	1.3	1.2-3.9
Herbicides/ pesticides	N/T	5.7-405	2.4-7.1
Cancer drugs	N/T	4.1	N/T

*Values represent the range of OR/RR (Odds Ratios/Relative Risk) found by different studies
[§] N/T = not tested

Thus, in contrast to radiation exposure, a large number of male occupations have been associated with increased risk of spontaneous abortion, birth defects, and childhood cancer in the offspring (Table 1; Sawyer and Aitken, 2001). In all of the occupations studied, the men are routinely exposed to various chemicals that might have an adverse affect on the male germ line. Occupations that have been studied include printers, painters, mechanics, farm, chemical, and metal workers, electricians, and firemen. Compounds to which these workers may be exposed include heavy metals, solvents, polycyclic aromatic hydrocarbons, pesticides, and radiation.

Epidemiological studies of paternal exposures and birth defects are still in the hypothesis-generating phase. In many instances, the risk ratio is in the hundreds, and thus, cannot be truly indicative of the actual risk (Table 1). Most likely, these studies suffer from the shortcomings mentioned above, in particular, small sample numbers, misclassify-cation of exposure, and/or confounding variables. Despite these limitations it is undeniable that significantly increased risks have been found, and further studies should be undertaken.

More realistic risks have been found for spontaneous abortion (Table 1). The first analyses done in this area were of dentists and dental technicians exposed to anesthetic gases (Cohen et al., 1980). In these studies, an increased risk for miscarriage was found with RR values ranging from 1.5-1.8. Exposure of males to mercury has been shown to increase the risk of fetal loss (Alcser et al., 1989; Cordier et al., 1991), and these studies were some of the very few which actually analyzed a specific biomarker of exposure (urine mercury levels). Again, most of these types of studies simply use occupation as the sole variable, rather than specific exposures. Other agents that have been associated with paternally mediated miscarriage include solvents, hydrocarbons, and pesticides (Savitz, 1992).

In regard to childhood cancer, similar types of paternal exposures and occupations as mentioned above have been found to increase the risk of childhood cancer. The most extensively studied agents include smoking, solvents, heavy metals, and pesticides (Sawyer and Aitken, 2001). In all cases, an increased risk of childhood cancer was found when fathers were exposed to these agents. Specific cancers include brain tumors, neuroblastoma, hepatoblastoma, leukemia/lymphoma, and Wilm's tumors. To date little research has been conducted establishing that the toxicants examined in these epidemiological studies are actually capable of generating DNA damage in the male germ line. This, together with the chemistry underlying any effects observed, remains an important objective for the future.

SUMMARY

The poor quality of the human ejaculate sets man apart from all other mammalian species. Even in normal fertile men the ejaculate may contain up to 85% abnormal forms according to the World Heath Organization (1999). In the wake of this poor semen quality comes extremely poor fertility (Hull et al, 1985) and the highest rates of aneuploidy, pregnancy loss and birth defects in viviparous vertebrates. Thus, the poor quality of human spermatozoa is reflected in both their capacity for fertilization and their genetic integrity. The ultimate cause of defective sperm function is unknown. In certain patients a genetic basis for male infertility has been identified involving DNA deletions on the long arm of the Y chromosome. Such deletions might explain the impoverished semen quality seen in about 10-14% of men with severely impaired spermatogenesis, but fail to explain the infertility seen in most (>85%) cases of male infertility. One of the key attributes and probable causes of defective sperm function is oxidative stress created by excessive ROS generation by the spermatozoa and/or the disruption of antioxidant defence systems in the male reproductive tract. Excess free radical generation frequently involves an error in spermiogenesis resulting in the release of spermatozoa from the germinal epithelium exhibiting abnormally high levels of cytoplasmic retention. The excess cytoplasm contains enzymes that fuel the generation of ROS by the spermatozoa's plasma membrane redox systems. The consequences of such oxidative stress include a loss of motility and fertilizing potential and the induction of DNA damage in the sperm nucleus. The loss of sperm function is due to the peroxidation of unsaturated fatty acids in the sperm plasma membrane as a consequence of which the latter loses its fluidity and the cells lose their function. The causes and consequences of oxidative damage to the DNA in the sperm nucleus are still not known with certainty. The available evidence suggests that early pregnancy loss and morbidity in the offspring, including childhood cancer, are associated with such damage. Developing animal models with which to establish the validity of these relationships and identifying the environmental factors associated with the proposed 'testicular dysgenesis syndrome' will clearly be important tasks for the future.

REFERENCES

Abell, A., Ernst, E. and Bonde, J.P., 2000, Semen quality and sexual hormones in greenhouse workers. *Scand J Work Environ Health.* 26:492-500.

Aitken, R.J., 1999, The Amoroso Lecture. The human spermatozoon--a cell in crisis? *J Reprod Fertil.* 115:1-7.

Aitken, R.J. and Krausz, C., 2001, Oxidative stress, DNA damage and the Y chromosome. *Reproduction.* 122:497-506.

Aitken, R.J., Buckingham, D.W., Carreras, A. and Irvine D.S., 1996, Superoxide dismutase in human sperm suspensions: relationships with cellular composition, oxidative stress and sperm function. *Free Rad Biol Med.* 21:495-504.

Aitken, R.J., Fisher, H., N. Fulton, Knox, W. and Lewis, B., 1997, Reactive oxygen species generation by human spermatozoa is induced by exogenous NADPH and inhibited by the flavoprotein inhibitors diphenylene iodonium and quinacrine. *Molec Reprod Dev.* 47:468-482.

Aitken, R.J., Krausz, C. and Buckingham, D.W., 1994, Relationships between biochemical markers for residual sperm cytoplasm, reactive oxygen species generation and the presence of leucocytes and precursor germ cells in human sperm suspensions. *Molec Reprod Dev.* 39:268-279.

Alcser, K.H., Brix, K.A., Fine, L.J., Kallenbach, L.R. and Wolfe, R.A., 1989, Occupational mercury exposure and male reproductive health. *Am J Ind Med.* 15: 517-529.

Andolz, P., Bielsa, M.A. and Vila, J., 1999, Evolution of semen quality in North-eastern Spain: a study in 22,759 infertile men over a 36 year period. *Hum Reprod.* 14:731-735.

Auger, J., Kunstmann, J.M., Czyglik, F. and Jouannet ,P., 1995, Decline in semen quality among fertile men in Paris during the past 20 years. *N Engl J Med.* 332:281-285.

Barroso, G., Morshedi, M. and Oehninger, S., 2000, Analysis of DNA fragmentation, plasma membrane translocation of phosphatidylserine and oxidative stress in human spermatozoa. *Human Reprod.* 15:1338-1344.

Benedict,J.C., Lin, T.M., Loeffler, I.K., Peterson, R.E. and Flaws, J.A. (2000) Physiological role of the aryl hydrocarbon receptor in mouse ovary development. *Toxicol Sci.* 56:382-388.

Bianchi, P., Manicardi, G.C., Bizzaro, D., Bianchi, U. and Sakkas, D., 1993, Effect of DNA protamination on fluorochrome staining and in situ nick-translation of murine and human mature spermatozoa. *Biol Reprod.* 49:1038-1043.

Bilotta, P., Guglielmo, R. and Steffe, M., 1999, Analysis of decline in seminal fluid in the Italian population during the past 15 years. *Minerva Ginecol.* 51:223-231.

Boekelheide, K., Fleming, S.L., Johnson, K.J., Patel, S.R. and Schoenfeld, H.A., 2000, Role of Sertoli cells in injury-associated testicular germ cell apoptosis. *Proc Soc Exp Biol Med.* 225:105-115.

Brown, L.M., Pottern, L.M., Hoover, R.N., Devesa, S.S., Aselton, P. and Flannery, J.T., 1986, Testicular cancer in the United States: trends in incidence and mortality. *Int J Epidemiol.* 15:164-170.

Burgoyne, P.S., 1998, The mammalian Y chromosome: a new perspective. *Bioessays.* 20:363-366.

Carlsen, E., Giwercman, A., Keiding, N. and Skakkebaek, N.E., 1992, Evidence for decreasing quality of semen during past 50 years. *Brit Med J.* 305:609-613

Casano, R., Orlando, C., Serio, M. and Forti, G., 1991, LDH and LDH-X activity in sperm from normospermic and oligozoospermic men. *Int J Androl.* 14:257-263.

Chai, N.N., Salido, E.C. and Yen, P.H., 1997, Multiple functional copies of RBM gene family, a spermatogenesis candidate on the human Y chromosome. *Genomics.* 45:355-361.

Cohen, E.N., Brown, B.W., Wu, M.L., Whitcher, C.E., Brodsky, J.B., Gift, H.C., Greenfield, W., Jones, T.W. and Driscoll, E.J., 1980, Occupational disease in dentistry and chronic exposure to trace anesthetic gases. *JADA.* 101:21-31.

Cordier, S., Deplan, F., Mandereau, L. and Hemon, D., 1991, Paternal exposure to mercury and spontaneous abortions. *Br J Ind Med.* 48:375-381.

Crow,J.F., 1997, The high spontaneous mutation rate: is it a health risk? *Proc Nat Acad Sci.* 94:8380-8386.

Delbridge, M.L. and Graves, J.A., 1999, Mammalian Y chromosome evolution and the male-specific functions of Y chromosome-borne genes. *Rev Reprod.* 4:101-109.

Dubrova, Y.E., Jeffereys, A.J., Malashenko, A.M., 1993, Mouse minisatellite mutations induced by ionizing radiation. *Nat Genetics.* 5:92-94.

Dubrova, Y.E., Nesterov, V.N., Krouchinsky, N.G., Ostapenko, V.A., Vergnaud, G., Giraudeau, F., Buard, J., Jeffreys, A.J., 1997, Further evidence for elevated human minisatellite mutation rate in Belarus eight years after the Chernobyl accident. *Mut Res.* 381:267-278.

Dubrova, Y.E., Plum, M., Brown, J. and Jeffereys, A.J., 1998, Radiation-induced germline instability at minisatellite loci. *Int J Radiat Biol.* 74:698-696.

Ericson, A. and Kallen, B., 1994, Pregnancy outcome in Sweden after the Chernobyl accident. *Environ Res.* 67:149-159.

Fraga, C.G., Motchnik, P.A., Wyrobek, A.J., Rempel, D.M., Ames, B.N., 1996, Smoking and low antioxidant levels increase oxidative damage to sperm DNA. *Mut Res.* 351:199-203.

Garrett, C., Liu, D.Y. and Baker, H.W., 1997, Selectivity of the human sperm-zona pellucida binding process to sperm head morphometry. *Fertil Steril.* 67:362-371.

Giwercman, A., Carlsen, E., Keiding, N. and Skakkebaek, N.E., 1993, Evidence for increasing incidence of abnormalities of the human testis: a review. *Environ Health Perspect.* 101:65-71.

Gomez, E., Buckingham, D.W., Brindle, J., Lanzafame, F., Irvine, D.S. and Aitken, R.J., 1996, Development of an image analysis system to monitor the retention of residual cytoplasm by human spermatozoa: correlation with biochemical markers of the cytoplasmic space, oxidative stress and sperm function. *J Androl.* 17:276-287.

Gomez, E., Irvine, D.S. and Aitken, R.J., 1998, Evaluation of a spectrophotometric assay for the measurement of malondialdehyde and 4-hydroxyalkenals in human spermatozoa: relationships with semen quality and sperm function. *Int J Androl.* 21:81-94.

Handelsman, D.J., 1997, Sperm output of healthy men in Australia: magnitude of bias due to self-selected volunteers. *Hum Reprod.* 12:2701-2705.

Huang, C., Li, J., Zheng, R. and Cui, K., 2000, Hydrogen peroxide-induced apoptosis in human hepatoma cells is mediated by CD95(APO-1/Fas) receptor/ligand system and may involve activation of wild-type p53. *Molec Biol Rep.* 27:1-11.

Hull, M.G.R., Glazener, C.M.A., Kelly, N.J., Conway, D.I., Foster, P.A., Hunton, R.A., Coulson, C., Lambert, P.A., Watt, E.M. and Desai, K.M., 1985, Population study of causes, treatment and outcome of infertility. *Brit Med J.* 291:1693-1697.

Huszar, G., Vigue, L. and Corrales, M., 1988, Sperm creatine phosphokinase quality in normospermic, variablespermic and oligospermic men. *Biol Reprod.* 38:1061-1066.

Irvine, S., Cawood, E., Richardson, D., MacDonald, E. and Aitken, R.J., 1996, Evidence of deteriorating semen quality in the United Kingdom: birth cohort study in 577 men in Scotland over 11 years. *Brit Med J.* 312:467-471.

Itoh, N., Kayama, F., Tatsuki, T.J. and Tsukamoto, T., 2001, Have sperm counts deteriorated over the past 20 years in healthy, young Japanese men? Results from the Sapporo area. *J Androl.* 22:40-44.

Joffe M., 1992, Detection of agents causing genetic or reproductive damage. *Br J Indust Med.* 49:1-4.

Jorgensen, N., Andersen, A.G., Eustache, F., Irvine, D.S., Suominen, J., Petersen, J.H., Andersen, A.N., Auger, J., Cawood, E.H., Horte, A., Jensen, T.K., Jouannet ,P., Keiding ,N., Vierula, M., Toppari, J. and Skakkebaek, N.E., 2001, Regional differences in semen quality in Europe. *Hum Reprod.* 16:1012-1019.

Keating, J., Grundy, C.E., Fivey, P.S., Elliott, M. and Robinson, J., 1997, Investigation of the association between the presence of cytoplasmic residues on the human sperm midpiece and defective sperm function. *J Reprod Fertil.* 110:71-77.

Krausz, C. and McElreavey, K., 1999, Y chromosome and male infertility. *Front Biosci.* 4:1-8.

Krausz, C., Bussani-Mastellone, C., Granchi, S., McElreavey, K., Scarsell, G. and Forti, G., 1999, Screening for microdeletions of Y chromosome genes in patients undergoing intracytoplasmic sperm injection. *Human Reprod.* 14:1717-1721.

Kodaira, M., Satoh, C., Hiyama, K. and Toyama, K. 1995, Lack of effects of atomic bomb radiation on genetic instability of tandem-repetitive elements in human germ cells. *Am J Hum Genet.* 57:1275-1283.

Krausz, C. and McElreavey, K., 1999, Y chromosome and male infertility. *Front Biosci.* 4:1-8.

Krausz, C., Bussani-Mastellone, C., Granchi, S., McElreavey, K., Scarsell, G. and Forti, G., 1999, Screening for microdeletions of Y chromosome genes in patients undergoing intracytoplasmic sperm injection. *Human Reprod.* 14:1717-1721.

Lahn, B.T. and Page, D.C., 1997, Functional coherence of the human Y chromosome. *Science.* 278:675-680.

Lahn, B.T. and Page, D.C., 1999, Retroposition of autosomal mRNA yielded testis-specific gene family on human Y chromosome. *Nature Genetics.* 21:429-433.

Little, J., 1993, The Chernobyl accident, congenital anomalies, and other reproductive outcomes. *Paediatr Perinat Epidemiol.* 7:121-151.

Lomat, L., Galburt, G., Quastel, M.R., Polyakov, S., Okeanov, A. and Rozin, S., 1997, Incidence of childhood disease in Belarus associated with the Chernobyl accident. *Environ Health Perspec.* 105:1529-1532.

Mak, V., Jarvi, K., Buckspan, M., Freeman., M., Hechter, S. and Zini, A., 2000, Smoking is associated with the retention of cytoplasm by human spermatozoa. *Urology.* 56:463-466.

Marshall Graves, J., 2000, Human Y chromosome, sex determination, and spermatogenesis- a feminist view. *Biol Reprod.* 63:667-676.

Matlai, P. and Beral, V., 1985, Trends in congenital malformations of external genitalia. *Lancet.* 1:108.

McElreavey, K. and Krausz, C., 1999, Male infertility and the Y chromosome. *Am J Hum Genet.* 64:928-933

Moller, H., 2001, Trends in incidence of testicular cancer and prostate cancer in Denmark. *Hum Reprod.* 16:1007-1011.

Multigner, L. and Spira, A., 1997, The epidemiology of male reproduction, in *Genetics of Human Male Fertility* pp 43-65 C. Barratt, C. de Jong, D. Mortimer and J. Parinaud, eds., EDK Press, Paris.

Muratori, M., Piomboni, P., Baldi, E., Filimberti, E., Pecchioli, P., Moretti, E., Gambera, L., Baccetti, B., Biagiotti, R., Forti, G., and Maggi, M., 2000, Functional and ultrastructural features of DNA-fragmented human sperm. *J Androl.* 21:903-912.

Neel, J.V., 1999, Two recent radiation-related genetic false alarms: leukemia in West Cumbria, England, and minisatellite mutations in Belarus. *Teratology* 59:302-306.

Neel, J.V. and Schull, W.J., 1956, The effect of exposure to the atomic bombs on pregnancy termination in Hiroshima and Nagasaki. *Nat Acad Science* Washington. 461:241.

Neel, J.V., Satoh, C., Goriki, K., Asakawa, H., Fujita, M., Tadahashi, N., Dageoda, T. and Fazama, R., 1988, Search for mutations altering protein charge and/or function in children of atomic bomb survivors: final report. *Am J Hum Genet.* 42:663-676.

Neel, J.V., Schull, W.J., Awa, A.A., Satoh, C., Kato, H., Otake, M. and Yoshimoto, Y., 1990, The children of parents exposed to atomic bombs: estimate of the genetic doubling dose of radiation for humans. *Am J Hum Genet.* 46:1053-1072.

Nomura, T., 1982, Parental exposure to X-rays and chemicals induces heritable tumors and anomalies in mice. *Nature.* 296:575-577.

Nomura, T., 1988, X-ray and chemically induced germline mutation causing phenotypical anomalies in mice. *Mut Res.* 198:301-320

Osterlind A., 1986, Diverging trends in incidence and mortality of testicular cancer in Denmark, 1943-1982. *Brit J Cancer.* 53:501-505.

Paulson, R.J., Milligan, R.C. and Sokol, R.Z., 2001, The lack of influence of age on male fertility. *Am J Obstet Gynecol.* 184:818-822.

Rytomaa T., 1996, Ten years after Chernobyl. *Ann Med.* 28:83-87.

Sakkas, D., Mariethoz, E., Manicardi, G., Bizzaro, D., Bianchi, P.G. and Bianchi, U., 1999a, Origin of DNA damage in ejaculated human spermatozoa. *Rev Reprod.* 4:31-37.

Sakkas, D., Mariethoz, E. and St John, J.C., 1999b, Abnormal sperm parameters in humans are indicative of an abortive apoptotic mechanism linked to the Fas-mediated pathway. *Exp Cell Res.* 251:350-355.

Savitz, D.A., 1994, Paternal exposures and pregnancy outcome: miscarriage, stillbirth, low birth weight, preterm delivery, in: *Male-Mediated Developmental Toxicity* pp. 177-184, A.F. Olshan and D.R. Mattison, eds., Plenum Press, New York, 1994.

Sawyer, D.E. and Aitken, R.J., 2001, Male mediated developmental defects and childhood disease. *Reprod Med Rev.*8:107-126.

Saxena, R., Brown, L.G., Hawkins, T., Alagappan, R.K., Skaletsky, H., Reeve, M.P., Reijo, R., Rozen, S., Dinulos, M.B., Disteche, C.M. and Page, D.C., 1996, The DAZ gene cluster on the human Y chromosome arose from an autosomal gene that was transposed, repeatedly amplified and pruned. *Nature Genetics.* 14:292-299.

Sayers, T.J., Brooks, A.D., Seki, N., Smyth, M.J., Yagita, H., Blazar, B.R. and Malyguine, A.M., 2000, T cell lysis of murine renal cancer: multiple signaling pathways for cell death via Fas. *J Leuk Biol.* 68:81-86.

Schull, W.J., Neel, J,V. and Hashizume, A. 1966, Some further observations on the sex ratio among infants born to survivors of the atomic bombings of Hiroshima and Nagasaki. *Am J Hum Genet.* 18:328-338.

Schwenn, M.R., and Brill A.B., 1997, Childhood cancer 10 years after the Chernobyl accident. *Curr Opin Pediatr.* 9:51-54.

Selevan, S.G., Borkovec, L., Slott, V.L., Zudova, Z., Rubes, J. Evenson, D.P. and Perreault, S.D., 2000, Semen quality and reproductive health of young Czech men exposed to seasonal air pollution. *Environ Health Perspect.* 108:887-894.

Sharpe, R.M. and Skakkebaeck, N.E., 1993, Are oestrogens involved in falling sperm counts and disorders of the male reproductive tract. Lancet. 341:1392-1395.

Swan, S.H., Elkin, E.P. and Fenster, L., 1998, Have sperm densities declined? A reanalysis of global trend data. *Environ Health Perspect.* 106:A420-421.

Vogt, P.H., Edelmann, A., Kirsch, S., Henegariu, O., Hirschmann, P., Kiesewetter, F., Kohn, F.M., Schill, W.B., Farah, S., Ramos, C., Hartmann, M., Hartschuh, W., Meschede, D., Behre, H.M., Castel, A., Nieschlag, E., Weidner, W., Grone, H.J., Jung, A., Engel, W. and Haidl., G., 1996, Human Y chromosome azoospermia factors (AZF) mapped to different subregions in Yq11. *Hum Molec Genet.* 5:933-943.

Wang, S.L., Wang, X.R., Chia, S.E., Shen, H.M., Song, L., Xing, H.X., Chen, H.Y. and Ong, C.N., 2001, A study on occupational exposure to petrochemicals and smoking on seminal quality. *J Androl.* 22:73-78.

Waters, P.D., Duffy, B., Frost, C.J., Delbridge, M.L. and Graves, J.A., 2001, The human Y chromosome derives largely from a single autosomal region added to the sex chromosomes 80-130 million years ago. *Cytogenet Cell Genet.* 92:74-79.

World Health Organization, 1999, *World Health Organization laboratory manual for the examination of human semen and sperm-cervical mucus interaction*, 4th edn. Cambridge University Press, Cambridge.

Xiao, G., Pan, C., Cai, Y., Lin, H. and Fu, Z., 2001, Effect of benzene, toluene, xylene on the semen quality and the function of accessory gonad of exposed workers. *Ind Health.* 39:206-210.

Zini, A., Buckspan, M., Jamal, M. and Jarvi, K., 1999, Effect of varicocelectomy on the abnormal retention of residual cytoplasm by human spermatozoa. *Human Reprod.* 14:1791-1793.

Zini, A., Defreitas, G., Freeman, M., Hechter, S. and Jarvi, K., 2000a, Varicocele is associated with abnormal retention of cytoplasmic droplets by human spermatozoa. *Fertil Steril.* 74:461-464.

Zini, A., O'Bryan, M.K., Israel, L. and Schlegel, P.N., 1998, Human sperm NADH and NADPH diaphorase cytochemistry: correlation with sperm motility. *Urology.* 51:464-468.

MODEL SYSTEMS FOR STUDYING GERM CELL MUTAGENS: FROM FLIES TO MAMMALS

Ekkehart W. Vogel and Madeleine J. M. Nivard

Leiden University Medical Centre
Department of Radiation Genetics and Chemical Mutagenesis
MGC Sylvius Laboratories
Wassenaarseweg 72, 2300 RA Leiden
The Netherlands

INTRODUCTION

The animal models used to analyze heritable effects induced in male germ cells by genotoxic agents are the fruit fly *Drosophila melanogaster*, the mouse and, to a lesser extent, the zebrafish *Brachydanio rerio*. More than five decades of research on germline mutagenesis have led to the identification of the major variables in the complex relationship between mutagenic insult and germ-cell stage response. One key observation was that the magnitude of genetic damage induced in distinct cell stages by a mutagen can vary by several orders of magnitude (Vogel et al., 1998). A systematical comparison of hereditary damage in Drosophila males and in male mice revealed, nevertheless, common features of germ-cell responses across both species (Vogel and Natarajan, 1995). There is now abundant evidence that the presence and efficiency of DNA repair has a chief function for what is phenomenologically described as germ-cell specificity in the response to mutagen exposure. Other cellular conditions with a major function are the capacity of the cell for enzymatic conversion of stable chemicals into DNA-interactive metabolites, and germinal selection against large structural chromosome aberrations during and prior to meiosis. The interaction of these three complex biological processes largely determines the mutational spectrum finally observed in each individual germ-cell stage.

After a brief consideration of spermatogenesis and genetic endpoints, we will discuss the current state of knowledge regarding the manifestation of genetic damage induced by alkylating agents in defined germ-cell stages of mouse and Drosophila. We will further briefly deal with the question of how malformation and malfunctioning of mature spermatozoa due to non-genotoxic damage could possibly be measured in germ cells.

Advances in Male Mediated Developmental Toxicity, edited by Bernard Robaire and Barbara F. Hales.
Kluwer Academic/Plenum Publishers, 2003.

FEATURES OF SPERMATOGENESIS

Drosophila

The adult testis represent a steady state system in which all stages of spermatogenesis are present (Lindsley and Tokuyasy, 1980). Gonial cells are generated from a small group of stem cells. Following four mitotic and two meiotic divisions each pre-spermatogonial cell gives rise to groups of 64 spermatids which develop synchronously into mature spermatozoa. The transit time from a stem cell to mature sperm available for insemination at 25° C is about 250 hours or 10.4 days (Figure 1). It takes cells nearly 5 days to pass through the various premeiotic stages before they are ready to undergo meiosis. The subsequent development of post-meiotic cells into mature spermatozoa lasts approximately 5.5 days (Lindsley and Tokuyasy, 1980). By the time the early spermatids embark on a program of rapid elongation, male germ cells lose repair competence. Thus for a period of about 3 days, there is no functioning DNA repair in these cells (Smith et al., 1983; Lee and Kelley, 1986).

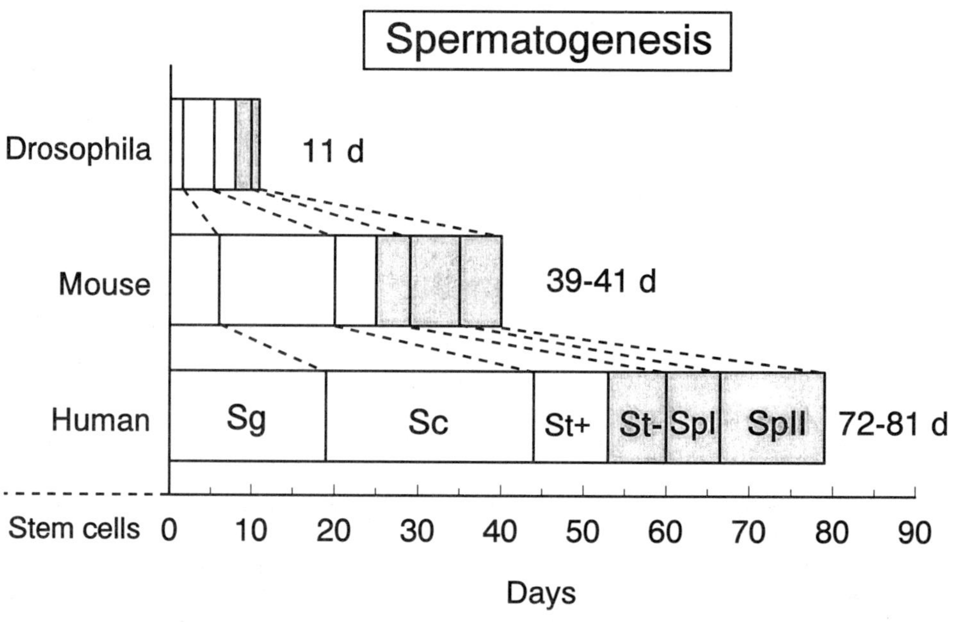

Figure 1. The time-course of spermatogenesis in Drosophila, mouse and man. Sg, spermatogonia; Sc, spermatocytes; St, spermatids; SpI testicular, SpII epididymal spermatozoa. There is no repair in spermatozoa and late spermatids (St-).

Mouse

The testes of adult mice contain germ cells in many distinct stages of development, ranging from undifferentiated spermatogonia to mature spermatozoa (Ehling and Favor, 1984; Sharpe, 1994; Adler, 1996). Slowly cycling stem-cell spermatogonia, which are always present, generate rapidly cycling differentiating spermatogonia. Differentiating spermatogonia go through cycles of mitotic cell divisions and are designated A-spermatogonia, intermediate and B-spermatogonia. This phase lasts 6 days (men - 19 days) (Oakberg, 1956a, b; Heller and Clermont, 1963; Adler, 1996). The following meiotic process takes 14 days (men - 25 days).

During this phase each primary spermatocyte gives rise to four spermatids. Differentiation of spermatids into mature sperm lasts 14-15 days in the mouse and 23 days in man (Adler, 1996). Spermatids are divided into acrosomal (or early spermatid) and elongated stages. The consensus view is that all male germ cell stages prior to elongated stages have active DNA repair but that this repair competence ends during the elongation stage (Sega, 1990; Adler, 1996). Thus, there is no repair of DNA damage in elongated spermatids and testicular spermatozoa - for a time period of 10 days in the mouse (men - 13.2 days) (Fig. 1). Total spermatogenesis takes 35 days in the mouse (Oakberg, 1956a, b) and 64 d in men (Heller and Clermont, 1963). Release of sperm in the ejaculate is 39-41 days (mouse) and 72-81 days (men) after the beginning of spermatogenesis (Adler, 1996).

If the time course of spermatogenesis is not affected by the treatment, the response of distinct germ-cell stages to a mutagen can be examined by the brood fractionation technique. Spermatozoa are sampled 1-7 days, spermatids 8-21 days, spermatocytes 22-35 days, and differentiating spermatogonia from 36 days post-treatment onwards, followed by gonial cells (Ehling and Favor, 1984).

GENETIC METHODS AND DATABASE

Drosophila

The broad range of suitable reporter genes and balancer chromosomes, and the short generation time of 10-12 days, makes it possible to detect in male germ cells of Drosophila the entire spectrum of genetic damage from recessive lethal (or visible) mutations, deletions, reciprocal translocations, chromosome loss, dominant lethals to aneuploidy (Table 1). However, it was realized early that it would not be practicable to assay for all types of genetic damage. In the period prior to 1980, this led to the recommendation of the multiple-loci X-linked recessive lethal (DMX) test to serve as a screen for potential mammalian genotoxins (Abrahamson and Lewis, 1971; Sobels, 1974). The majority of data on mutagenicity of chemicals in Drosophila germ cells has therefore been collected by tests for X-linked recessive lethals in males. Most of these studies were conducted in the period from 1968 - 1980 when the assay was in optimal use. The capability of the DMX test to serve as a screen for the identification of mammalian mutagens and carcinogens was evaluated by a Work Group of the U.S. Gene-Tox Program (Lee et al., 1983) and, more recently, in the context of an IARC activity (Vogel et al., 1999). However, identification of a mutagen is only the first step in the assessment of genetic risk. Thus, if the purpose of a study is to investigate mechanistic aspects of germline mutagenesis, measurement of single endpoints (mutagenicity testing) is less suitable, and more sophisticated parameters are necessary. These are the comparisons of two endpoints with each other, the use of repair mutants and the analysis of mutational changes at the DNA sequence level (Vogel and Nivard, 1994; Vogel et al., 1998).

Mouse

Since the purpose of this paper is to discuss heritable damage induced by covalent DNA binding agents in male germ cells, we consider only the three germline assays for which there is a sufficient database for mechanistic comparisons across species. These three assays are the specific-locus test (SLT), the heritable translocation test (MHT), and the dominant-lethal test (DLT). The technical aspects of these assays have been described by several authors (Russell et al., 1981; Bateman, 1984; Léonard and Adler, 1984).

Table 1. Genetic assays in male germ cells

Abbreviation[1]	Genetic endpoint or test	Types of DNA damage
Drosophila		
DMX	● X-linked recessive lethal mutations	● Base-substitutions, frame-shift mutations, deletions, small insertions
DMA	● Autosomal recessive lethal mutations	● As DMX
DMS	● Specific locus test	● As DMX, molecular analysis feasible
DMH	● Heritable translocation test	● Reciprocal chromosome aberrations
DMC	● Chromosomal aberrations; loss of ring-X, rod-X or Y-chromosome	● Chromosomal aberrations, SCE's
DML	● Dominant lethal test	● Non-viable chromosomal aberrations; (detects also non-genetic damage)
DMN	● Numerical aneuploidy	● Chromosome aberration, non-disjunction
Mouse		
SLT	● Specific locus test	● Base-substitutions, frame-shift mutations, deletions, small aberrations molecular analysis feasible
MHT	● Heritable translocation test	● Reciprocal chromosome aberrations
DLM	● Dominant lethal test	● Non-viable chromosomal aberrations (non-genetic damage: pre-implantation deaths?)
AVA	● Numerical aneuploidy	● Chromosome aberration, non-disjunction

[1]Abbreviations from Jackson *et al.*, 1996.

Database

A review paper by Shelby (1996) lists 29 chemicals that have been tested for induction of specific locus mutations in the SLT until 1996. All but one of these chemicals were also examined for dominant lethals, 16 in the male heritable translocation test. Five of these 29 chemicals are not interesting for our considerations because they act through other mechanisms than DNA-binding (adriamycin, azidothymidine, vincristine sulfate) or because there is insufficient information regarding their mode of interaction with DNA (acrylamide; glycidol). Together with published data on the two nitrosourea compounds BCNU (no. 9) and CCNU (no. 17) (Ehling et al., 1997), on NDEA (no. 24) and DMBA (no. 25), there are 28 chemicals, all covalent DNA binding agents, with test results for one or more of the three mouse germline assays (Table 2). Except for the untested PMS (no. 5) and PNU (no. 15), all these DNA-interactive chemicals are established germ cell mutagens in Drosophila (reviewed in Vogel *et al.*, 1996; 1998). This material provides a solid database for interspecies comparisons of germ cell mutagenesis in mouse, Drosophila and, for two chemicals only, the zebrafish.

GERM-CELL SPECIFICITY: OBSERVATIONS AND CAUSES

How to classify germ-cell mutagens ?

One possibility is to use a classification criterion for each individual chemical based on the quantity of its positive response. Following this strategy, Russell et al. (1990) assigned 11 chemicals that had given positive specific-locus test results to three distinct patterns of maximum response. Most of the chemicals (CPA, DES, EMS, MMS, PCH and TEM?; see Table 2) had their maximum response in early spermatozoa and late spermatids (pattern 1).

With CAB and MEL (pattern 2) the highest rate of mutations was found in early spermatids, and MNU and ENU had their greatest effect in differentiating spermatogonia (pattern 3). The drawback of this classification scheme is that SLTs often rely on one or two doses approaching the LD_{50}. However, the shape of the dose-response relationships and the resulting maximum response can be quite different for distinct germ-cell stages. Well documented examples are, for Drosophila, the dose-related increase of recessive lethal mutations induced by MMS in spermatozoa and spermatid stages compared to the absence of a dose response in younger stages (Vogel et al., 1982; Vogel and Natarajan, 1995) and, for the mouse, the MNU case. With 75 mg/kg body weight MNU, an extremely high frequency of 90.8×10^{-5} mutations per locus was found in progeny conceived in matings made 36-42 days after treatment (Russell and Rinchik, 1993). The corresponding mutation frequency in a study with 70 mg/kg b.wt. MNU (Ehling and Neuhäuser-Klaus, 1991) was lower, 30.7×10^{-5}, whereas a second peak (MF $= 21.8 \times 10^{-5}$), seen in spermatids, had not been reported in the first study.

For two main reasons, we prefer an arrangement of alkylating agents into three categories (see column in Table 2), based upon their DNA alkylation pattern, rather than a classification on individual test responses (Vogel and Natarajan, 1995). The first is the documented lesion-dependence of the rate of adduct removal on adduct size, e.g. ethyl < methyl (Brent et al., 1988; Beranek, 1990), meaning that if DNA repair is a major determinant of germ-cell stage specificity, chemicals modifying DNA in a similar fashion are expected to show a similar pattern. The second is that negative data on individual chemicals, which may be due to trivial rather than mechanistic reasons (metabolic clearance, compound not reaching the target), will be less disturbing in such comparisons. The most illustrative example is the negative response of NDEA (nr. 24; Table 2) in the specific locus test of the mouse. NDEA ethylates DNA in a similar fashion as ENU (about 80% of all modifications are on oxygens; Singer and Grunberger, 1983), and would therefore be expected to be a 'super-mutagen' similar to ENU. This was not found to be the case (Russell, 1984). One possible explanation for this discrepancy might be that the tissue dose in the testis did not reach the critical level needed to see an effect in the SLT.

Thus, 25 of the 28 chemicals can be assigned to three broader categories of alkylating agents. Prototypes of category 1 are MMS and EO, monofunctional agents predominantly reacting with ring nitrogens in DNA, whereas interaction with oxygen positions in DNA is <1% (Singer and Grunberger, 1983; Beranek, 1990). DES and EMS, two ethylating agents producing substantial amounts of O^6-ethylguanine, are nevertheless put into category 1 because they fail to generate measurable levels of O^4-ethylthymine (Beranek, 1990). As will be later shown, this has far-reaching consequences for their *in vivo* genotoxicity. All nitroso compounds, and further PCH and iPMS, belong to the second category composed of agents which, in addition to other positions, can alkylate oxygen sites in pyrimidine bases. Agents capable of cross-linking DNA form category 3. Thirteen of the 28 chemicals displayed in Table 2 are of this type. In a previous analysis, this 3-category classification scheme identified common features for three vastly different experimental indicators of genotoxicity: hereditary damage in Drosophila males *versus* genetic damage in male mice *versus* tumour incidences in rodents (Vogel et al, 1998).

Genotoxic effects in pre- versus postmeiotic cells

The usefulness of our classification scheme becomes also apparent when the 28 mutagens of Table 2 are grouped according to their general activity pattern in post-spermatogonial cells (post-SgS) of the mouse as opposed to spermatogonial stem cells (SgS).

All five category-1 mutagens and six cross-linking agents (category 3) failed to induce specific-locus mutations in stem-cell spermatogonia, but they are all highly active in postmeiotic male germ cells of the mouse. These agents are also powerful mutagens in postmeiotic germ cell stages of Drosophila, as was evident from mutation induction by, for instance, MMS and EMS over a dose range of up to three logarithmic units (Vogel and Natarajan, 1995). Most remarkable is that all attempts using any mutagen of category-1 failed to induce significant numbers of specific-locus mutations in mouse spermatogonial stem-cells

Table 2. Test results with DNA-binding chemicals in germline assays with male mice.

Chemical[1]	DLM[2]	MHT[2]	SLT[2] Post SgS	SLT[2] SgS	Category	Ref.[3]
Group I: Direct alkylating agents active in Post SgS but not active in SgS						
1. Methyl methanesulfonate; MMS	positive	positive	positive	negative	1	a
2. Ethyl methanesulfonate; EMS	positive	positive	positive	negative	1	a
3. Diethyl sulfate; DES	positive	ND	positive	negative	1	a
4. Ethylene oxide; EO positive	positive	positive	negative[4]		1	a, c
5. n-Propyl methanesulfonate; PMS	positive	ND	positive	negative[4]	1	a
6. Chlorambucil; CABpositive	positive	positive	negative		3	a
7. Mechlorethamine; MEC	positive	positive	positive	negative	3	a
8. Myleran; MYL	positive	....ND	positive	negative	3	a
9. Bis(2-chloroethyl)-1-nitrosourea; BCNU	positive	ND	positive	negative	3	b
10. Trophosphamide*; TPA *	positive	positive	positive	negative	3	a
11. Cyclophosphamide *; CPA *	positive	positive	positive	negative	3	a
Group II: Direct alkylating agents active in both Post SgS and in SgS						
12. *N*-Nitroso-*N*-ethylurea; ENU	positive	positive	positive	positive	2	a
13. *N*-Nitroso-*N*-methylurea; MNU	positive	positive	positive	positive	2	a
14. Procarbazine*; PCH *	positive	positive	positive	positive	2	a
15. *N*-Propyl-*N*-nitrosourea; PNU	ND	ND	ND	positive	2	a
16. i-Propyl methanesulfonate; iPMS	positive	positive	positive	positive	2	a
17. 1-(2-Chloroethyl)-3-cyclohexyl-1-nitrosourea; CCNU	positive	....ND	positive	positive	3	b
18. Melphalan; MEL	positive	positive	positive	positive	3	a
19. Mitomycin C; MMC	positive	positive	positive	positive	3	a
20. Triethylenemelamine; TEM	positive	positive	positive	positive	3	a
Group III: Carcinogens inactive in the specific-locus test						
21. Cisplatin; DDP	negative	....ND	negative	negative	3	a
22. Benzo[a]pyrene*; B[a]P*	positive[5]	ND	....ND	negative	uncl.	a
23. 1,2-Dibromo-3-chloropropane *;	negative	....ND	....ND	negative	3	a, c
24. *N*-Nitrosodiethylamine*; NDEA	negative	....ND	?	negative	2	c
25. 7,12-Dimethyl benz[a]anthracene*; DMBA*	ND	ND	....ND	negative[4]	uncl.	c
26. Ethylene dibromide *; EDB *	negative	....ND	....ND	negative[4]	1	a, c
27. Hexamethylphosphoramide *	negative	....ND	ND	negative	3	c
28. Urethane*; UT *	negative	....ND	negative	negative	uncl.	a, c

[1] Chemical marked by * requires enzymatic activation.

[2] DLM, Dominant lethal test; MHT, Heritable translocation test; SLT, specific-locus test in post-spermatogonial cells (Post-SgS) or in spermatogonial stem cells (SgS).

[3] References: [a]Shelby 1996 and references therein; [b]Ehling *et al.*, 1997; [c] Russell, 1984.

[4] Test results are not definite by USEPA criteria of Russell *et al.*, 1981.

[5] Positive in the sperm morphology assay (Wyrobek *et al.*, 1981).

(Ehling and Neuhäuser-Klaus, 1988; 1990; Russell et al., 1990). For Drosophila, it was shown that the resistance of premeiotic cells to MMS and other category-1 mutagens is due to the efficient error-free repair of N-alkylation DNA damage (Vogel and Zijlstra, 1987; Vogel and Natarajan, 1995).

The nine direct-alkylating agents listed as second group in Table 2 display mutagenic activity in a wide range of both post- and premeiotic germ-cell stages. These genotoxic agents either belong to category 2 or 3. Among these germ-cell mutagens is ENU, most likely the most potent monofunctional germline mutagen across species, as shown for the mouse (Russell et al., 1982; Favor, 1998), Drosophila (Vogel and Natarajan, 1995) and zebrafish (Solnica-Krezel et al., 1994). This poses the question why EMS, which like ENU also transfers ethyl groups to DNA, is inactive (mouse, zebrafish) or only weakly active (Drosophila) in immature premeiotic male germ cell stages (Vogel et al., 1982; Ehling and Neuhäuser-Klaus, 1989b), and an unclassified rodent carcinogen (Barbin and Bartsch, 1989). The answer came from a DNA sequence analysis of specific-locus mutations induced by EMS and ENU. In Drosophila (no data for the mouse), 76% of *vermilion* mutations induced by EMS in spermatozoa and spermatid stages were GC→AT transitions derived from mispairing of O^6-ethylguanine (Pastink et al., 1991). The approximately 20-fold lower induction of X-chromosomal mutations in spermatogonia exposed to EMS (Vogel et al., 1982) made it impossible to isolate *vermilion* mutants from this stage. By contrast, ENU is also a powerful mutagen in premeiotic cells of Drosophila. There is, however, a remarkable cell-stage related shift in the types of base-pair substitutions obtained for ENU. In postmeiotic Drosophila germ cells (no molecular data for the mouse), 61% of *vermilion* mutations induced by ENU were GC->AT transitions generated by mispairing of O^6-ethylguanine, and only 18% were AT->GC transitions (Pastink et al., 1988). In premeiotic cells of both Drosophila and the mouse, the mutation spectra of ENU were dominated by changes at A:T sites most likely resulting from O^4-EtT and O^2-EtT, i.e., 85% in Drosophila (Tosal et al., 1998) and 85 to 89% in the mouse (Favor, 1998; 1999 and refs. therein). O^4-EtT and O^2-EtT are minor, apparently slowly repaired adducts which represent high genetic risk lesions. Thus, the failure of EMS and DES to generate measurable levels of O^4-EtT and O^2-EtT (Beranek, 1990) explains their inactivity in mouse spermatogonia, as well as their weak activity in premeiotic Drosophila cells. MNU, PCH, PNU and iPMS, all category-2 mutagens, also display activity in premeiotic cells of both mouse and, so far tested, also in Drosophila. It is not yet known whether their activity in premeiotic cells, like for ENU, is due to O-alkylation at pyrimidines. As all of these agents have a low nucleophilic selectivity (Singer and Grunberger, 1983; Beranek, 1990; Vogel and Nivard, 1994), they should be able to do so.

Another DNA modification representing a risk to premeiotic and spermatogonial stem cells are alkylating agents which can cross-link the DNA. The antineoplastic drugs CCNU, MMC, MEL and TEM (no longer used) are of this type. A characteristic feature of the mutation spectra induced by cross-linking agents is the nearly exclusive occurrence of deletion mutations in postmeiotic cells exposed to these agents. Of the drugs (cisplatin, MEC, hexamethylmelamine, HMPA and BCNU) analysed in Drosophila SLTs, the sum of intra-locus and multi-locus deletions was 92 to 95% for four of the five agents (reviewed in Vogel et al., 1998). In the mouse, the sum of intra-locus (unknown) and multi-locus deletions (60 to 87%) could amount to a similarly high percentage of more than 90%. The only cross-linking agent with data from DNA sequence analysis of specific-locus mutations induced in both postmeiotic and premeiotic cells is chlorambucil (CAB). Of 14 *vermilion* mutants induced in Drosophila spermatozoa or spermatids, 9 were multi-locus and 4 intra-locus deletion mutations (J. Wijen et al., personal communication). In agreement with this dominance of deletions is that, in experiments with mice, all 11 specific locus mutants recovered from post-spermatogonial stages exposed to CAB were deletions (9) or other rearrangements (2) (Favor, 1999). Molecular characterization of 14 Drosophila mutants isolated from premeiotic cells exposed to CAB revealed two multi-locus and 10 intra-locus deletions. Thus, we may conclude that the major effect of DNA-cross-linkage in male germ cells is loss of genetic information. This is in general agreement with the well-documented high clastogenicity of cross-linking agents both in germ cells and in somatic tissue (Vogel *et al.*, 1998).

The third group in Table 2 is composed of 8 rodent carcinogens which are inactive in the mouse SLT. Except for cisplatin (DDP), all are procarcinogens requiring enzymatic conversion to become genotoxic. Since five of the seven procarcinogens were also negative in the mouse dominant lethal test (Shelby, 1996), repair can hardly be the cause of the failure to induce heritable damage in male mice. This conclusion is supported by the fact that cisplatin, 1,2-dibromo-3-chloropropane and hexamethylphosphoramide have cross-linking properties similar to seven agents (CAB, MEC, TPA, CPA, MEL, MMC, TEM) positive in the DLM assay. The reason for the inactivity of most procarcinogens in tests on heritable genetic damage is yet unknown, but the essence is that it largely protects the mouse germline against adverse effects of procarcinogens. This is decidedly different from the situation in Drosophila where all procarcinogens listed in Table 2 are established germ-cell mutagens (reviewed in Vogel et al., 1996; 1999).

DNA repair

Category 1 mutagens, typified by MMS and EO, display major genotoxic effects in male germ-cell stages that have passed the meiotic division; this has been shown for both Drosophila and the mouse. The high resistance of premeiotic stages against this type of mutagen must be due to the efficient error-free repair of N-alkylation - the nearly exclusive DNA modification produced by these agents (Beranek, 1990). The ability of category 2 genotoxins (ENU, MNU) to induce genetic damage in premeiotic stages correlates with their enhanced ability for alkylations at oxygen sites of pyrimidine bases, which are the O^4 T-, O^2 T-, O^2 C- positions (Beranek, 1990). It is this property which distinguishes category-2 from category-1 mutagens. The second type of DNA modification that poses a high hazard to immature germ cells, including spermatogonial stem cells, is DNA cross-linkage. Four members of category 3 display activity in stem-cell spermatogonia, indicating that DNA crosslinks can generate heritable genetic alterations in practically any germ-cell stage. The negative test results obtained for MEC, *cis*DDP, HMPA, CPP and TPP in stem-cell spermatogonia is not in disagreement with such a conclusion because the cause for their inactivity could be rather trivial (e.g. metabolic clearance).

Thus, differences between germ-cell stages in their over-all capacity for DNA repair, together with the low ability of all stages to repair DNA-crosslinks and alkylated pyrimidine bases, are the causes why even structurally closely related chemicals can differ vastly in their genotoxic potency. Experimental evidence for this concept came from studies with Drosophila mutants having a deficiency in the nucleotide excision repair (NER) pathway. The two Drosophila mutants extensively used were the X-linked *DmXPF* (*mei9*) and the chromosome 2 mutant *DmXPG* (*mus201*). Both Drosophila genes encode endnucleases and are homologs of the mammalian XPF and XPG genes, respectively (Araj and Smith, 1996; Sekelsky, *et al.*, 1995; 2000; Calléja et al., 2001).

It should be recalled that in Drosophila continuous repair in germ cells of DNA damage takes place until the formation of mid-spermatids and further in the egg after fertilization, while late spermatids and spermatozoa have lost their capability of repair (Smith et al., 1983; Lee and Kelley, 1986). This lack of repair in the latest stages of spermatogenesis, together with the fact that efficient DNA repair occurs in the developing embryo, provided the basis for a genetic DNA repair assay first applied by Smith et al. (1982). In this assay, postmeiotic (spermatozoa and spermatids) germ-cell stages of P_1 males are mutagenized and forward mutations, after their fixation in a NER$^+$ or a NER$^-$ background of the maternal oocyte, are measured in the F_2 as X-linked recessive lethal (RL) frequencies. The RL frequency (M_{NER-}) estimated from crosses of *mei9* or *mus201* NER$^-$ females with NER$^+$ males is divided by the mutant frequency (M_{NER+}) derived from repair-proficient conditions (cross NER$^+$ ♀ X NER$^+$ ♂), providing the M_{NER-}/M_{NER+} mutability ratio (Smith et al., 1982; 1983; Vogel, 1989; Vogel et al., 1998).

The blockage of the NER pathway may or may not potentiate mutation induction, depending on the type of mutagen. Drastic increases in M_{NER-}/M_{NER+} mutability ratios have been observed only with category-1 mutagens. Examples are ethylene oxide (EO) and propylene imine (Fig. 2). In NER proficient conditions (NER$^+$♀ X NER$^+$♂), both chemicals, like many other category-1 mutagens, are only weakly mutagenic in spermatozoa and spermatid stages, unless these cells are exposed to high doses at nearly toxic levels. However, when the female NER system is disrupted, they become powerful mutagens over a wide dose range (Fig. 2). This is a most impressive illustration of the efficiency of repair of N-alkylation damage in germ cells. On the contrary, the NER system of germ cells cannot repair mono-adducts at oxygen sites and DNA crosslinks. Therefore, alkylating agents of category 2 or 3 mostly give no response in the repair assay. The sometimes weak responses seen with crosslinking agents of the nitrogen mustard, aziridine or epoxide type, are caused by unrepaired mono-adducts, which are also produced by crosslinking agents. For a more detailed discussion of the three mutagen categories see Vogel and Nivard (2001).

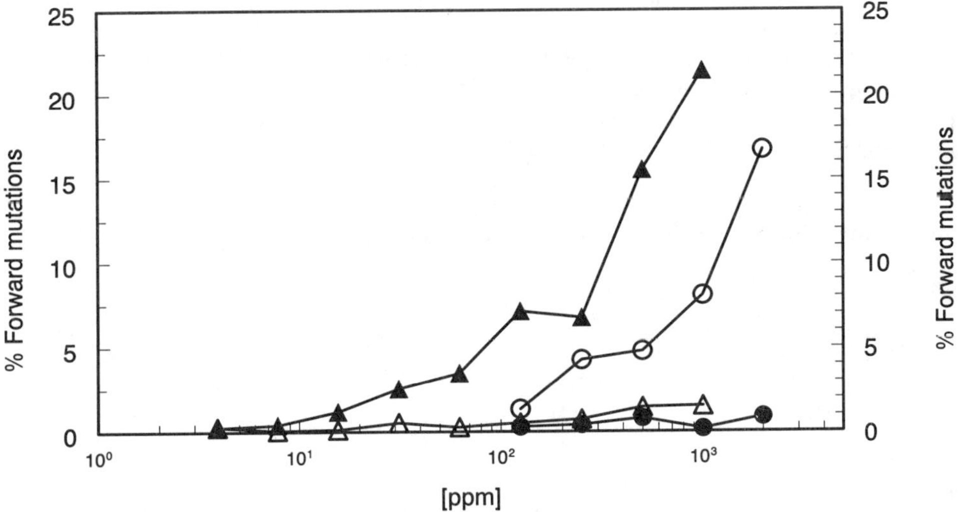

Figure 2. The potentiating effect of a deficiency for nucleotide excision repair in the female repair system on DNA damage induced by EO and PI in spermatozoa. Triangles, EO; circles, PI; open symbols, NER-deficient (DmXPG); closed symbols, wildtype. Data from Vogel and Nivard, 1997; 1998.

Although repair deficient mice have not yet been used in experiments on germ cells, there is nevertheless good evidence for stage-related differences in DNA repair in mice. Sega and his colleagues investigated in detail DNA repair (as measured by unscheduled DNA synthesis, UDS) in meiotic and postmeiotic germ-cell stages after exposure to EMS (Sega, 1974), cyclophosphamide and mechlorethamine (Sotomayor *et al.*, 1978), MMS (Sotomayor *et al.*, 1979), acrylamide (Sega *et al.*, 1989), and ethylene oxide (Sega, 1990). The principle finding was that mutagens induce UDS in meiotic and postmeiotic stages up to midspermatids. However, no UDS was detectable in late spermatids and spermatozoa (Sega, 1990).

The far reaching similarities in the mutational response of mouse and Drosophila to the 3 categories of germ-cell mutagens shows that the cellular conditions defining germ cell specificity and the types of mutations finally observed, must be largely the same. Thus, if the type of DNA damage induced by a mutagen is known, its genotoxicity profile in germ cells can generally be predicted, given that the chemical reaches the target cells.

It is undisputed these days that a ground-level of DNA damage is unavoidable and must be regarded as constitutive to a cell (Gupta and Lutz, 1999). Observations that certain mutagen-induced DNA modifications are more persistent, and therefore more hazardous than others, are also widely accepted scientific facts. But why are primary and secondary lesions at ring nitrogens, induced by mutagens of the EO-type, so efficiently repaired whereas the defence against O^4 alkylT-, O^2 alkylT or DNA-DNA crosslinks is obviously rather poor? Is this because mutagens of category 1 produce DNA modifications also derived from endogenous and unavoidable exogenous sources? All small alkylating agents belonging to category 1 generate N-7 alkylguanine, by far the most abundant DNA adduct produced by these agents, and further N-3 alkyladenine, the next most frequent DNA adduct. Alkylations at ring nitrogens labilize the glycosyl bond and may result in depurination, generating apurinic (AP-) sites in DNA, and other secondary modifications (Singer and Grunberger, 1983; Segerbäck, 1994; Zhao et al., 1999). These abasic sites can trigger mutations and deletions (Segerbäck et al., 1998; Nivard et al. 1992; 1993). Due to the instability of DNA, AP-sites are constantly produced in unexposed cells (Table 3) and must be repaired. It has been estimated that in double-stranded DNA of rat liver cells, the rate of AP-site formation is 10^4 per day and per 10^{10} bases for depurination and 5×10^2 for depyrimidination (Singer and Grunberger, 1983). Apurinic sites form about 40% of the total level of known background DNA adducts (Gupta and Lutz, 1999). Repair competent germ cells are apparently well prepared to deal with these lesions, explaining their low genotoxic potency in premeiotic stages. In analogy, high doses are generally needed to produce tumors in rodents by MMS and related carcinogens (Barbin and Bartsch, 1989; Vogel et al., 1990).

Table 3. Formation and levels of background DNA damage.

Type	Damage level (Lesions / 10^9 nucleotides)	Reference
Formation of AP-sites		
Depurination per day[1]	≈ 1000	Singer and Grunberger, 1983
Depyrimidination per day[1]	≈ 50	Singer and Grunberger, 1983
Levels of DNA damage		
Apurinic sites	4000	Nakamura *et al.*, 1998
Alkylated dG	< 300	Zhao *et al.*, 1999
Alkylphosphotriesters	> 2500	Gupta and Lutz, 1999; and references therein
Total level of background DNA adducts	$\approx 10,000$	Gupta and Lutz, 1999

[1] Events / day / 10^{10} bases in double-stranded DNA (rat liver cell) are $\approx 10^4$ for depurination and 5×10^2 for depyrimidination (Singer and Grunberger, 1983; and ref. therein)

For DNA-DNA crosslinks no quantitative estimates are available. Apart from human exposure to antineoplastic drugs such as BCNU and MEC, which in principle is avoidable (Gupta and Lutz, 1999), there is no substantial and general exposure of living organisms to crosslinking agents, although chemicals (e.g. pyrrolizidine alkaloids) with this property do occur in nature (Culvenor and Jago, 1978). Thus, although a potential source of secondary tumours in cancer patients exposed to alkylating drugs, crosslinking agents form no apparent threat to the survival of whole species. It would therefore seem plausible that evolution has not resulted in efficient repair systems against category-3 chemicals.

NON-GENETIC DAMAGE

Drosophila

For more than twenty years, Drosophila germ cells were exclusively used to investigate aspects related to the formation and repair of DNA damage by physical and chemical agents. Due to a general lack of interest in the scientific community, adverse effects on male fertility caused by other than genotoxic mechanisms were neglected. Therefore, it is largely unknown that two assays which can monitor toxic effects due to non-genetic damage and resulting in reduction of male fertility and/or viable offspring, were developed and described in detail (Würgler et al., 1984).

A first indication on the potentially sterilizing action of a chemical can be obtained by running a combined toxicity and sterility test. Groups of males are exposed to a range of concentrations including toxic doses. Acute toxic effects causing male mortality will be seen within 1-3 days. By mating surviving males to unexposed females in brood fractionation experiments for a period of 10-14 days, reduced fertility or complete sterility can be monitored by counting the number of F_1 offspring in the different broods (Würgler et al., 1984). It should be noticed that this simple test does not allow distinguishing non-genetic from genetic damage.

A slightly more sophisticated technique is the dominant lethal assay in which adult males are exposed to the chemical under test. The term 'dominant lethality' is generally used as an operational term. True dominant lethality is due to DNA damage, mostly structural chromosomal aberrations, which in heterozygous condition do not allow a zygote to develop into an adult fly. Dominant lethals can be mimicked by other phenomena such as unfertilized eggs. Thus to separate true chromosomal damage from other adverse but non-genetic effects, the eggs deposited by the females have to be examined for developmental potential. As a checkpoint for progeny survival one can choose either the hatching of the larvae from the egg or the hatching of the adult fly from the puparium. In the first type of experiment, 'white' lethal eggs and 'brown' lethal embryos are counted separately. The class of white lethal eggs represents embryos killed during the first few hours of embryonic development (cleavage division, gastrulation) but also includes unfertilized eggs and eggs inseminated by non-functional sperm. Brown lethal embryos represent the induction of genetic damage leading to late death and are comparable to the post-implantation lethality in the mouse dominant lethal test. An increase of only white lethal eggs indicates that the substance under test does not induce DNA damage but interferes either with fertilization, gametogenesis or physiological functions of the sperm (Würgler *et al.,* 1984). Final confirmation as to whether or not a test chemical can damage DNA should come from a 'DNA-interaction' test. This does not necessarily have to be one of the laborious germline assay, such as the recessive lethal test. One of the rapid and sensitive assays measuring recombination events in somatic cells of Drosophila (Vogel and Nivard, 2000) would fully meet this purpose.

There is yet another reason why a dominant lethal test should always be combined with a 'true' system for genotoxicity. Parallel runs on both dominant lethals and forward mutations (recessive lethals) revealed a higher sensitivity of the latter method. This point is exemplified by two examples (Figure 3). After exposure of males to Cl_3PDMT (1-[2,4,6,trichlorophenyl]-3,3-dimethyltriazene), a methylating agent after enzymatic activation, dominant lethal mutations are not enhanced at a dose <1.0 mM, which produces approximately one forward mutation per cell. Even more drastic is the second case, NDEA. This nitrosamine, one of the most potent point mutagens in Drosophila, was entirely inactive in the DML and other assays measuring gross chromosomal damage (Vogel and Leigh, 1975). Since even some cross-linking agents gave problems in the DLM (Vogel, 1975), the system was rejected and no longer used in studies on genotoxicity. But the test might provide a simple method to monitor non-genotoxic damage in male germ cells.

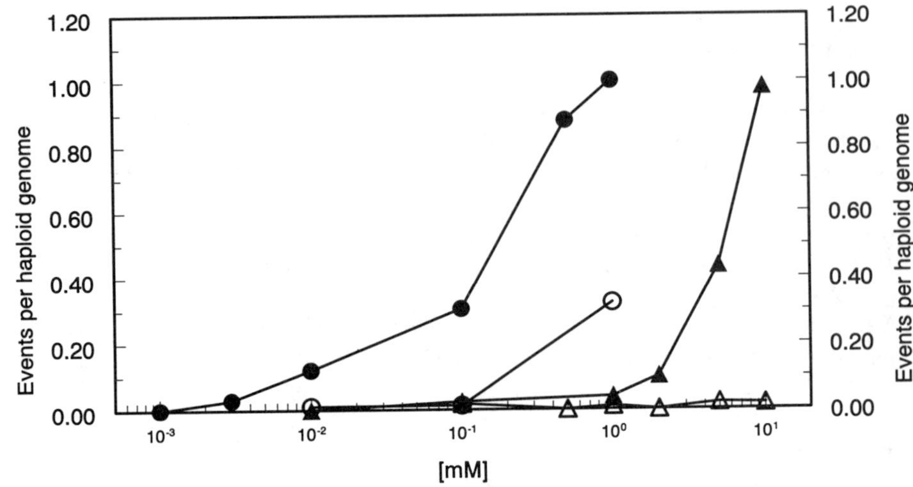

Figure 3. The low sensitivity of the Drosophila dominant lethal test to monitor DNA damage induced by (1-[2,4,6,trichlorophenyl]-3,3-dimethyltriazene (Cl₃PDMT) and *N*-nitrosodiethylamine (NDEA) in spermatozoa. Open symbols, dominant lethal test; closed symbols, recessive lethal test for forward mutations. Circles, Cl₃PDMT; triangles, NDEA. Data from Vogel and Leigh, 1975.

Mouse

Similar to the situation in Drosophila, studies on the effects of chemicals in germ cells of mice focussed on endpoints and methods which could detect DNA damage, such as the dominant lethal test. Up to date, the consensus view seems to be that the assay detects chromosome breakage rather than point mutations (Bateman, 1984). Dominant lethal mutations induced in gametes of male mice will reduce litter size (Bateman, 1984). This reduction in litter size can be due to failure of the fertilized egg either to implant or to develop after implantation. A decrease in implantations per female from the control value indicates induced pre-implantation death or an excess of unfertilized eggs (Ehling and Neuhäuser-Klaus, 1990). Bateman pointed out that pre-implantation losses through dominant lethal mutation cannot be distinguished from non-genetic factors, including inadequate quality or quantity of sperm, unless there had been microscopic examination of the pre-implantation eggs. He therefore recommended to "safely ignore pre-implantation deaths and concentrate on the fate of implantations", classified as live embryos, late deaths and early deaths. This seems to be the reason why the outcome of dominant lethal assays is judged on the occurrence of post-implantation losses. Thus, although the corpora lutea are counted in tests on chemically induced dominant lethal mutations (cf. Ehling and Neuhäuser-Klaus, 1988; 1989a; 1990), the contribution of non-genetic damage to pre-implantation loss was not systematically evaluated and remains therefore unknown. For more-detailed considerations of non-genetic damage see the papers by S. Perreault, F. Marchetti and other contributions to this issue.

CONCLUSIONS

The two animal models extensively used in studies on heritable effects of genotoxic chemicals in male germ cells are Drosophila and the mouse. Evaluation of the database on germ cell mutagenesis revealed that the presence and efficiency of DNA repair has a chief function for what is phenomenologically described as germ-cell specificity in the response to mutagen exposure. Based on the types of DNA adducts produced, 25 of 28 carcinogens with data on germline assays in mice and Drosophila can be assigned to three sub-categories of alkylating agents. For these three classes, the principle response of male germ cells of both Drosophila and mouse can be predicted, provided the chemical reaches the cellular target.

Category-1 germ-cell mutagens show peak mutagenic activity in postmeiotic male germ cells. This mutagen type predominantly alkylates ring nitrogens in DNA, whereas interaction with oxygen positions of pyrimidine bases is below detection limits. Alkylations at ring nitrogens may result in depurination. These abasic sites can trigger mutations and deletions. Due to the instability of DNA, abasic sites are also constantly produced in unexposed cells and must be repaired. Apurinic sites form about 40% of the total level of known background DNA adducts (Gupta and Lutz, 1999) and represent therefore an unavoidable source of adverse genotoxic effects in both germ cells and somatic cells of living organisms. Therefore, it seems logical that during evolution cells developed efficient repair systems against this type of damage, otherwise life would not be possible. It further seems no surprise that all category-1 germ-cell mutagens are also weak carcinogens in rodents (Vogel et al., 1998).

Alkylation of oxygen positions of pyrimidines and the formation of DNA-DNA crosslinks are the action principles of category-2 and category-3 mutagens, respectively. The defence against these types of DNA damage is clearly less efficient than against that generated by category-1. We find the most potent germ-cell mutagens (and carcinogens) among these chemicals. Interestingly, nitroso compounds (NDEA) and crosslinking agents are not considered to form a major source of background DNA damage (Gupta and Lutz, 1999). However, the extensive use of crosslinking agents as antineoplastic drugs in cancer chemotherapy should not be ignored is such a consideration. Several of these alkylating drugs have been implicated in causing secondary tumours in cancer patients (Penn, 1986; Haas et al., 1987). It is entirely unknown whether these drugs induce mutations in human stem cells, like they do in stem cells of male mouse and Drosophila.

REFERENCES

Abrahamson, S. and Lewis, E.B., 1971, The detection of mutations in *Drosophila melanogaster*, in: *Chemical Mutagens, Principles and Methods for their Detection*, A. Hollaender, A. ed., Plenum Press, New York, pp. 461-487.

Adler, I.D., 1996, Comparison of the duration of spermatogenesis between male rodents and humans. *Mutat Res.* 352:169-172.

Araj, H., and Smith, P.D., 1996, Positional cloning of the *Drosophila melanogaster mei-9* gene, the putative homolog of the *Saccharomyces cerevisiae RAD1* gene. *Mutat Res.* 364:209-215.

Barbin, A., and Bartsch, H., 1989, Nucleophilic selectivity as a determinant of carcinogenic potency (TD_{50}) in rodents: a comparison of mono- and bi-functional alkylating agents and vinyl chloride metabolites. *Mutation Res.* 215:95-106.

Bateman, A.J., 1984, The dominant lethal assay in the mouse, in: *Handbook of Mutagenicity Test Procedures*, B.J. Kilbey, M. Legator, W. Nichols and C. Ramel (eds.), Elsevier Amsterdam, 471.

Beranek, D.T., 1990, Distribution of methyl and ethyl adducts following alkylation with monofunctional alkylating agents. *Mutat Res.* 231:11-30.

Brent, T.P., M.E. Dolan, M.E. Fraenkel-Conrat, H., Hall, J., Karran, P., Laval, F., Margison, G.P., Montesano, R., Pegg, A.E., Potter, P.M., Singer, B., Swenberg, J.A., and Yarosh, D.B., 1988, Repair of O-alkylpyrimidines in mammalian cells: a present consensus. *Proc Natl Acad Sci USA.* 85:1759-1762.

Calléja, F.M.G.R. Nivard, M.J.M. and Eeken, J.C.J., 2001, Induced mutagenic effects in the nucleotide excision repair deficient *Drosophila* mutant *mus201^{D1}*, expressing a truncated XPG protein. *Mutat Res.* 461:279-288.

Culvenor, C.C.J. and Jago, 1978, Carcinogenic plant products and DNA, in: *Chemical Carcinogens and DNA*, Vol. I, P.L. Grover, ed., CRC Press, Boca Raton, Florida, pp. 161-186.

Ehling, U.H., and Favor, J., 1984, Recessive and dominant mutations in mice, in: *Mutation, Cancer and Malformation*, Chu, E.H.Y. and Generoso, W.M., ed., Plenum Publishing Corporation, New York, pp. 389-428.

Ehling, U.H., and Neuhäuser-Klaus, A., 1988, Induction of specific-locus and dominant-lethal mutations in male mice by diethyl sulfate (DES). *Mutat Res.* 199:191-198.

Ehling, U.H., and Neuhäuser-Klaus A., 1989a, Induction of specific-locus and dominant lethal mutations in male mice by chlormethine. *Mutat Res.* 227:81-89.

Ehling, U.H., and Neuhäuser-Klaus< A., 1989b, Induction of specific-locus mutations in male mice by ethyl methanesulfonate (EMS). *Mutat Res.* 227:91-95.

Ehling, U.H., and Neuhäuser-Klaus, A., 1990, Induction of specific-locus and dominant lethal mutations in male mice in the low dose range with methyl methanesulfonate (MMS). *Mutat Res.* 230:61-70.

Ehling, U.H. and Neuhäuser-Klaus, A., 1991, Induction of specific-locus mutations and dominant lethal mutations in male mice by 1-methyl-1-nitrosourea (MNU). *Mutat Res.* 250:447-456.

Ehling, U.H., Adler, I.D., Favor, J., and Neuhäuser-Klaus, A., 1997, Induction of specific-locus and dominant lethal mutations in male mice by 1,3-bis(2-chloroethyl)-1-nitrosourea (BCNU) and 1-(2-chloroethyl)-3-cyclohexyl-1-nitrosourea (CCNU). *Mutat Res.* 379:219-231.

Favor, J., 1998, The mutagenic activity of ethylnitrosourea at low doses in spermatogonia of the mouse as assessed by the specific-locus test. *Mutat Res.* 405:221-226.

Favor, J., 1999, Mechanisms of mutation in germ cells of the mouse as assessed by the specific locus test. *Mutat Res.* 428:227-236.

Gupta, R.C. and Lutz, W.K., 1999, Background DNA damage from endogenous and exogenous carcinogens: a basis for spontaneous cancer incidence? *Mutat Res.* 424:1-8.

Haas, J.F., Kittelmann, B, Mehnert, W.H., Staneczek, W., Mohner, M, Kaldor, J.M., and Day, N.E., 1987, Risk of leukaemia in ovarian tumour and breast cancer patients following treatment by cyclophosphamide. *Br J Cancer.* 55:213-218.

Heller, C.G. and Clermont, Y., 1963, Spermatogenesis in man: an estimate of its duration. *Science.* 140:184-186.

Jackson, M.A., Stack, H.F. and Waters, M.D., 1996, Genetic activity profiles of anticancer drugs. *Mutat Res.* 355:171-208.

Lee,W.R., Abrahamson, S.,Valencia, R., von Halle,. E.S., Würgler, F.E. and Zimmering, S., 1983, The sex-linked recessive lethal test for mutagenesis in *Drosophila melanogaster. Mutat Res.* 123:183-279.

Lee, W.R., and Kelley, M.R., 1986, Correction for differences in germ cell stage sensitivity in risk assessment, in: *Risk Assessment in Relation to Environmental Mutagens and Carcinogens*, Alan R. Liss, Inc., New York, pp. 99-102.

Lindsley, D.L., and Tokuyasu, K.T., 1980, Spermatogenesis, in: *The Genetics and Biology of Drosophila,* M. Ashburner and T.R.F. Wright, ed., Academic Press, London and New York, Vol. 2d, pp 225-294..

Léonhard, A. and Adler, I.D., 1984, Test for heritable translocations in male mammals, in: *Handbook of Mutagenicity Test Procedures*, B.J. Kilbey, M. Legator, W. Nichols and C. Ramel (eds.), Elsevier Amsterdam, 471.

Nakamura, J., Walker, V.E., Upton, P.B., Chiang, S.-Y., Kow, Y.W., and Swenberg, J.A., 1998, Highly sensitive apurinic / apyrimidinic site assay can detect spontaneous and chemically induced depurination under physiological conditions. *Cancer Res.* 58:222-225.

Nivard, M.J.M., Pastink, A., Vogel, E.W., 1992, Molecular analysis of mutations in the *vermilion* gene of Drosophila melanogaster by methyl methanesulfonate. *Genetics.* 131:673-682.

Nivard, M.J.M., Pastink, A., Vogel, E.W., 1993, Impact of DNA nucleotide excision repair on methyl methanesulfonate-induced mutations in *Drosophila melanogaster. Carcinogenesis.* 14:1585-1590.

Oakberg, E.F., 1956a, A description of spermatogenesis in the mouse and its use in analysis of the cycle of the seminiferous epithelium and germ cell renewal. *Am J Anat.* 99:391-413.

Oakberg, E.F., 1956b, Duration of spermatogenesis in the mouse and timing of stages of the cycle of the seminiferous epithelium. *Am J Anat.* 99:507-516.

Pastink, A, Vreeken, C, Nivard, M.J.M., Searles, L., and Vogel, E.W., 1989, Sequence analysis of N-ethyl-N-nitrosourea induced *vermilion* mutations on *Drosophila melanogaster. Genetics.* 123:123-129.

Pastink, A., Heemskerk, E., Nivard, M.J.M., van Vliet, C.J., and Vogel, E.W., 1991, Mutational specificity of ethyl-methanesulfonate in excision-repair proficient and -deficient strains of *Drosophila melanogaster. Mol Gen Genetics.* 229:213-218.

Penn, I., 1986, Malignancies induced by drug therapy: a review. IARC Sci. Publ. No. 78, International Agency for Research on Cancer, Lyon, pp. 13-27.

Russell, L.B., Selby, P.B., von Halle, E., Sheridan, W., and Valcovic, L., 1981, The mouse specific-locus test with agents other than radiations, Interpretations of data and recommendations for future. *Mutat Res.* 86:329-354.

Russell, L.B., Russell, W.L., Rinchik, E.M., and Hunsicker, P.R., 1990, Factors affecting the nature of induced mutations, in: *Biology of Mammalian Germ Cell Mutagenesis*, Banbury Report 34, J.W. Allen, B.A., Bridges, F.M., Lyon, M.J., Moses, and Russell, L.B.,ed., Cold Spring Harbor Laboratory Press, pp. 271-289.

Russell, L.B., and Rinchik, E.M., 1993, Structural differences between specific-locus mutations induced by different regimes in mouse spermatogonial stem cells. *Mutat Res.* 288:187-195.

Russell, W.L., Hunsicker, P.R., Raymer, G.D., Steele, M.H., Stelzner K.F., and Thomson, H.M., 1982, Dose-response curve for ethylnitrosourea-induced specific-locus mutations in mouse spermatogonia. *Proc Natl Acad Sci USA.* 79:3589-3591.

Russell, W.L., 1984, Dose response, repair, and no-effect dose levels in mouse germ-cell mutagenesis, in: *Problems of Threshold in Chemical Mutagenesis*, Y. Tazima, S., Kondo, and S. Kuroda, eds., The Environmental Mutagen Society of Japan, 153.

Sega, G.A., 1974, Unscheduled DNA synthesis in the germ cells of male mice exposed in vivo to the chemical mutagen ethyl methanesulfonate. *Proc Natl Acad Sci USA.* 71:4955-4959.

Sega, G.A., Alcota, V.R.P., Tancongco, C.P., and Brimer, P.A., 1989, Acrylamide binding to the DNA and protamine of spermiogenic stages in the mouse and its relationship to genetic damage. *Mutat Res.* 216:221-230.

Sega, G.A., 1990, Molecular targets, DNA breakage, and DNA repair: their roles in mutation induction in mammalian germ cells, in: *Biology of Mammalian Germ Cell Mutagenesis,* Banbury Report 34, J.W. Allen, B.A., Bridges, F.M., Lyon, M.J. Moses, and J. and Russell, L.B., eds., Cold Spring Harbor Laboratory Press, pp. 79-91.

Segerbäck, D., 1994, DNA alkylation by ethylene oxide and some monosubstituted epoxides, in: *DNA Adducts Identification and Biological Significance*, K. Hemminki, A. Dipple, D.E.G. Shuker, F.F. Kadlubar, D. Segerbäck, and H. Bartsch, ed., IARC Scientific Publ. No. 125, IARC Lyon, pp. 37-47.

Segerbäck, D., Plná, K., Faller, T., Kreuzer, P.E., Håkansson, K., Filser, J.G. and Nilsson, R., 1998, Tissue distribution of DNA adducts in male Fischer rats exposed to 500 ppm of propylene oxide: quantitative analysis of 7-(2-hydroxypropyl)guanine by ^{32}P-postlabelling. *Chem Biol Interact.* 115:229-246.

Sekelsky, J.J., McKim, K.S., Chin, G.M. and Hawley, R.S., 1995, The Drosophila meiotic recombination gene *mei-9* encodes a homologue of the yeast excision repair protein rad1. *Genetics.* 141:619-627.

Sekelsky, J.J., Hollis, K.J., Eimerl, A.I., Burtis, M.R. and Jawley, R.S., 2000, Nucleotide excision repair endonuclease genes in *Drosophila melanogaster. Mutat Res.* 459:219-228.

Sharpe, R.M., 1994, Regulation of spermatogenesis, in: *The Physiology of Reproduction*, E. Knobil and J.D. Neil, eds, Vol. 2, Raven Press, pp. 1363-1434.

Shelby, M.D., 1996, Selecting chemicals and assays for assessing mammalian germ cell mutagenicity. *Mutat Res.* 352:159-167.

Singer,B., and Grunberger, D., 1983, Molecular Biology of Mutagens and Carcinogens. Plenum Press, New York.

Smith, P.D., Dusenbery, R.L., Cooper, S.F., and Baumen C.F., 1982, Examining the mechanism of mutagenesis in repair-deficient strains of *Drosophila melanogaster*, in: *Environmental Mutagens and Carcinogens*, T. Sugimura, S. Kondo, and H. Takebe ,eds., University of Tokyo Press, Tokyo, pp. 147-155.

Smith, P.D., Baumen, C.F., and Dusenbery, R.L., 1983, Mutagen sensitivity of *Drosophila melanogaster*, VI. Evidence from the excision-defective *mei-9^{AT1}* mutant for the timing of DNA repair activity during spermatogenesis. *Mutat Res.* 108:175-184.

Sobels, F.H., 1974, The advantage of Drosophila for mutation studies. *Mutat Res.* 26:277-284.

Solnica-Krekel, L., Schier, A.F. and W. Driever, W., 1994, Efficient recovery of ENU-induced mutations from the zebrafish germline. *Genetics.* 136:1401-1420.

Sotomayor, R.E., Sega, G.A., and Cumming, R.B., 1978, Unscheduled DNA synthesis in spermatogenic cells of mice treated in vivo with the indirect alkylating agent cyclophosphamide and mitomen. *Mutat Res.* 50:229-240.

Sotomayor, R.E., Sega, G.A., and Cumming, R.B., 1979, An autoradiographic study of unscheduled DNA synthesis in the germ cells of mice treated with X-rays and methyl methanesulfonate. *Mutat Res.* 62:293-309.

Tosal, L., Comendador, M.A. and Sierra, L.M., 1998 *N*-Ethyl-*N*-nitrosourea predominantly induces mutations at AT base pairs in pre-meiotic germ cells of Drosophila males. *Mutagenesis.* 13:375-380.

Vogel, E.W., 1975, Some aspects of the detection of potential mutagenic agents in Drosophila. *Mutat Res.* 29:241-250.

Vogel, E.W. and Leigh, B., 1975, Concentration-effect studies with MMS, TEB, 2,4,6-triCl-PDMT, and DEN on the induction of dominant lethals, recessive lethals, chromosome loss and translocations in Drosophila sperm. *Mutat Res.* 29:383-396.

Vogel, E.W., Blijleven, W.G.H., Kortselius, M.J.H., and Zijlstra J.A., 1982, A search for some common characteristics of chemical mutagens in Drosophila. *Mutat Res.* 92:69-87.

Vogel, E.W. and Zijlstra, J.A., 1987, Somatic cell mutagenesis in Drosophila melanogaster in comparison with genetic damage in early germ-cell stages. *Mutat Res.* 180:189-200.

Vogel, E.W., 1989, Nucleophilic selectivity of carcinogens as a determinant of enhanced mutational response in excision repair-defective strains in *Drosophila*: effects of 30 carcinogens. *Carcinogenesis.* 10:2093-2106.

Vogel, E.W., Barbin, A., Nivard, M.J.M. and Bartsch, H., 1990, Nucleophilic selectivity of alkylating agents and their hypermutability in *Drosophila* as predictors of carcinogenic potency in rodents. *Carcinogenesis*. 11:2211-2217.

Vogel, E.W. and Nivard, M.J.M., 1994, International Commission for Protection Against Environmental Mutagens and Carcinogens. The subtlety of alkylating agents in reactions with biological macromolecules. *Mutat Res*. 305:13-32.

Vogel, E.W. and Natarajan, A.T., 1995, DNA damage and repair in somatic and germ cells *in vivo*. *Mutat Res*. 330:183-208.

Vogel, E.W., Nivard, M.J.M., Ballering, L.A.B., Bartsch, H., Barbin, A., Nair, J., Comendador, M.A., Sierra, L.M., Aguirrezabalaga, I., Tosal, L., Ehrenberg, L., Fuchs, R.P.P., Janel-Bintz, R., Maenhaut-Michel, G., Montesano, R., Hall, J., Kang, H., Miele, M., Thomale, J., Bender, K., Engelbergs, J., and Rajewsky, M.F., 1996, DNA damage and repair in mutagenesis and carcinogenesis: implications of structure-activity relationships for cross-species extrapolation. *Mutat Res*. 353:177-218.

Vogel, E.W. and Nivard, M.J.M., 1997, The response of germ cells to ethylene oxide, propylene oxide, propylene imine and methyl methanesulfonate is a matter of cell-stage related DNA repair. *Environ Mol Mutagen*. 29:124-135.

Vogel, E.W. and Nivard, M.J.M., 1998, Genotoxic effects of inhaled ethylene oxide, propylenen oxide and butylene oxide on germ cells: Sensitivity of genetic endpoints in relation to dose and repair status. *Mutat Res*. 405:259-271.

Vogel, E.W., Barbin, A., Nivard, M.J.M., Stack, H.F., Waters, M.D. and Lohman, P.H.M., 1998, Heritable and cancer risks of exposure to anticancer drugs: inter-species comparisons of covalent deoxyribonucleic acid-binding agents. *Mutat Res*. 400:509-540.

Vogel, E.W., Graf, U., Frei, H.-J. and Nivard, M.J.M., 1999, The results of assays in Drosophila as indicators of exposure to carcinogens, in: *The Use of Short- and Medium-term Tests for Carcinogens and Data on Genetic Effects in Carcinogenic Hazard Evaluation*, D.B. McGregor, J.M. Rice and S. Venitt, eds., IARC Scientific Publ. No. 146, International Agency for Research on Cancer, Lyon, pp. 427-469.

Vogel, E.W. and Nivard, M.J.M., 2000, Parallel monitoring of mitotic recombination, clastogenicity and teratogenic effects in eye tissue of *Drosophila*. *Mutat Res*. 455:141-153.

Vogel, E.W. and Nivard, M.J.M., 2001, Phenotypes of *Drosophila* homologs of human *XPF* and *XPG* to chemically-induced DNA modifications. *Mutat Res*. 476:149-165.

Wijen, J.P.H., van Lange, R.E.E., Nivard, M.J.M. and Vogel, E.W., Differences in the proportion of multi-locus deletions induced by chlorambucil in post- and premeiotic germ cells of male Drosophila, paper submitted.

Würgler, F.E., Sobels, F.H. and Vogel E.W., 1984, Drosophila as assay system for detecting genetic changes, in: *Handbook of Mutagenicity Testing Procedures*, B. Kilbey, M. Legator, W. Nichols and C. Ramel ,eds., Elsevier/North Holland Biomedical Press, Amsterdam, pp. 555-602

Wyrobek, A., Gordon, L., and Watchmaker, G., 1981, Evaluation of 17 chemical agents including 6 carcinogen/noncarcinogen pairs on sperm shape abnormalities in mice, in: *Evaluation of Short-Term Tests for Carcinogens*, F.J. de Serres and J. Ashby, ed., Elsevier/North Holland, 712.

Zhao, C., Tyndyk, M., Eide, I. and Hemminki, K., 1999, Endogenous and background DNA adducts by methylating and 2-hydroxylating agents. *Mutat Res*. 424:117-125.

GERMLINE MUTATION INDUCTION AT MOUSE AND HUMAN TANDEM REPEAT DNA LOCI

Yuri E. Dubrova

Department of Genetics
University of Leicester
Leicester, LE1 7RH
United Kingdom

INTRODUCTION

The ability to predict the genetic consequences for humans of exposure to ionising radiation has certainly been one of the most important goals of human genetics in the past fifty years. However, despite numerous experimental studies, little is known about the effects of radiation exposure on germline mutation in humans. For example, data collected in Hiroshima and Nagasaki during the past 40 years on children of atomic bomb survivors using standard monitoring systems have not provided evidence of any statistically signify-cant differences in mutation rate between exposed and control families (Neel at al., 1990). Similarly, a survey of survivors treated with radiotherapy showed that the occurrence of genetic diseases in their offspring was similar to that in control families (Byrne et al., 1998). For this reason, germline mutation induction in mice still remains the main source of experimental data used to evaluate the genetic risk of human exposure to ionising radiation (UNCEAR, 1993; Sankaranarayanan and Chakraborty, 2000).

Estimating the genetic hazards of radiation and other mutagens depends on extrapolation from experimental systems. The obvious shortcoming of current approaches for monitoring radiation-induced mutation is the necessity to use very large numbers of individuals (more than 100,000) to detect increases in mutation rate. Furthermore, the question of adequate scoring of radiation-induced mutations by these approaches remains uncertain (Sankaranarayanan, 1999). It is therefore clear that new experimental approaches for monitoring radiation-induced germline mutation in human populations need to be developed.

We have recently proposed that hypervariable tandem repeat loci may provide a new experimental approach to the evaluation of germline mutation induction in mice and humans by low-dose exposure to ionising radiation (Dubrova et al., 1993, 1996). These loci have a very high spontaneous mutation rate both in humans (Jeffreys et al., 1988; Vergnaud and Denoeud, 2000) and mice (Kelly et al., 1989; Gibbs et al., 1993), and therefore they are capable of detecting changes in mutation rates in relatively small

Advances in Male Mediated Developmental Toxicity, edited by Bernard Robaire and Barbara F. Hales.
Kluwer Academic/Plenum Publishers, 2003.

population samples. Here I present a summary of recent publications analysing mutation induction at mouse and human tandem repeat loci.

MUTATION INDUCTION IN MICE

Tandem repeat loci in the mouse genome are represented by relatively short microsatellites (<500 bp) with repeat size of 1-4 bp, long expanded simple tandem repeats (ESTR, 0.5-16 kb, repeat size 4-6 bp) and true minisatellites (0.5-10 kb) with repeat size of 14-47 bp (Blake et al., 2000; Kelly et al., 1989; Gibbs et al., 1993; Bois et al., 1998a,b). To date, germline mutation induction has only been studied at the mouse ESTR loci (Dubrova et al., 1993, 1998, 2000a; Sadamoto et al., 1994; Fan et al., 1995; Barber et al., 2000). These loci were originally termed minisatellites but have recently been renamed to distinguish them from the much more stable true minisatellites in the mouse genome (Bois et al., 1998a,b). Unstable ESTRs consist of homogenous arrays of relatively short repeats and show a very high spontaneous mutation rate both in germline and somatic cells (Kelly et al., 1989; Gibbs et al., 1993; Bois et al., 1998b), whereas human GC-rich minisatellites have substantially long (10-100 bp) repeat units and primarily mutate in the germline (Jeffreys et al., 1994; May et al., 1996; Jeffreys and Neumann, 1997; Tamaki et al., 1999; Buard et al., 2000; Stead and Jeffreys, 2000). In contrast, little is known about the true mouse minisatellites, and the only attempt to isolate them has revealed a set of relatively stable loci with a mutation rate well below 10^{-3} per gamete (Bois et al., 1998a).

Several initial studies have shown that ionising radiation can substantially increase the germline mutation rate at mouse ESTR loci (Dubrova et al., 1993; Sadamoto et al., 1994; Fan et al., 1995). However, these studies produced conflicting results and failed to establish any reliable relationship between radiation dose and mutation frequency. To analyse mutation induction at ESTR loci in mice in more detail, we have measured mutation rate in male mice exposed to various sources of ionising radiation (Dubrova et al., 1998, 2000a; Barber et al., 2000).

Mutation Induction At Mouse ESTR Loci By Low- And High-LET Irradiation

Mutation induction at ESTR loci was analysed in CBA/H male mice exposed to acute low-LET X-rays delivered at 0.5 Gy min^{-1}, chronic low-LET γ-irradiation (^{60}Co source, 1.66×10^{-4} Gy min^{-1}) and chronic high-LET neutron irradiation (^{252}Cf source, 0.003 Gy min^{-1}). Males were mated to unexposed CBA/H females 6 or 10 weeks post-irradiation, ensuring that litters were derived from irradiated pre-meiotic A_s spermatogonia and stem cells (Searle, 1974). ESTR mutations in offspring were scored using the mouse-specific single-locus probes Ms6-hm and Hm-2 (Kelly et al., 1989; Gibbs et al., 1993).

Acute exposure to X-rays caused a statistically significant increase in the paternal ESTR mutation rate and yielded a linear dose response in the 0-1 Gy dose range (Figure 1a) with the doubling dose of 0.33 Gy (Dubrova et al., 1998). The doubling dose estimate obtained for paternal ESTR mutation is close to those obtained using traditional mutation scoring systems in mice, including the specific locus method (Russell 7-locus test), previously the most efficient system of studying point mutations in mice (Searle, 1974). This remarkable similarity between the estimates of the doubling dose suggests that the effects of ionising radiation on ESTR germline mutation rates may accurately reflect mutation induction at protein coding genes; it provides the basis for the possible application of this system for monitoring germline mutation. Most importantly, statistically significant evidence for mutation induction at ESTR loci was obtained by profiling only 252 offspring, whereas other systems required the analysis of thousands or even hundreds of thousands of mice to detect significant increases in mutation rate (Searle, 1974), highlighting the unique advantage of ESTR loci for the detection of radiation-induced germline mutations in mice. Furthermore, in all previous studies male mice were exposed to very high doses (3-6 Gy) of acute irradiation (Figure 1b), far greater than the doubling dose itself. Therefore, doubling doses were estimated by extrapolation assuming a linear dose response. In contrast, high frequencies of spontaneous and induced mutations at ESTR loci permitted us to evaluate mutation induction at much lower doses of exposure, allowing a more robust estimation of the doubling dose without extrapolation from high doses of radiation.

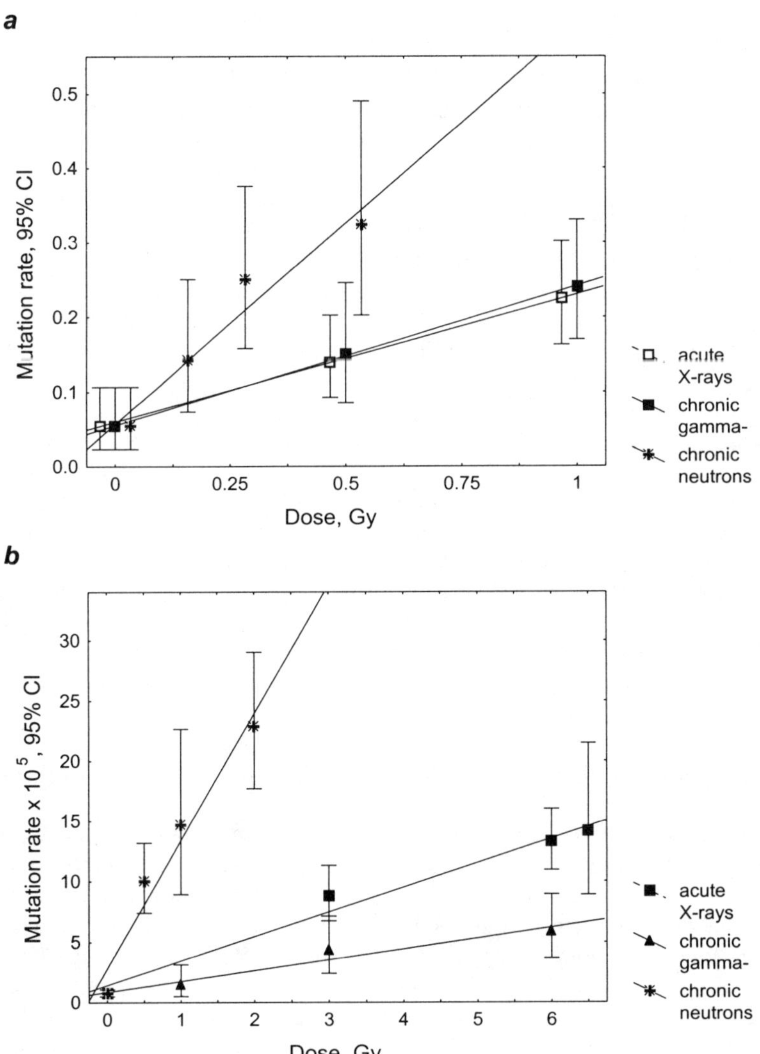

Figure 1. Dose-response curves for paternal mutation rates in mice. (a) Dose-response curves for ESTR mutations. Data from Dubrova et al. (1998, 2000a). (b) Dose-response curves for mutations scored by the Russell 7-locus test. Data from Searle (1974) and Russell and Kelly (1982). 95% confidence intervals for mutation rates are given.

Chronic exposure to low-LET γ-rays (Dubrova et al., 2000a) also caused a statistically significant increase in the paternal mutation rate and a linear dose response (Figure 1a). Comparison of the results for chronic γ-ray (100 hrs) and acute X-ray (2 min) exposure shows an identical pattern of mutation induction within the dose range 0.5-1 Gy.

A direct comparison of mutation induction in the mouse germline exposed to such low doses of acute and chronic low-LET radiation has not previously been reported. Therefore, this is the first experimental data to show that the mutagenic potential of chronic and acute low-LET radiation at low doses is very similar. In contrast, results using the Russell 7-locus test, and predominantly high total doses, show that acute exposure results a higher germline mutation rate than chronic exposure (Figure 1b). Russell (1965) proposed that acute exposure to high-dose irradiation saturates the capacity of DNA repair, so that only a fraction of the premutational damage can be repaired before fixation. With chronic exposure, however, a higher proportion of the premutational damage is repaired. On the other hand, at low doses of acute exposure, DNA repair may not be fully saturated. Therefore the acute response could approach the chronic at the low end of the dose-response curve. Our results on the unexpectedly similar mutagenicity of chronic and acute low-LET exposure at relatively low doses raise important issues with regard to the extrapolation of genetic risk from high-dose estimates. Current estimates of genetic hazard of low-dose radiation have been extrapolated from data obtained using high doses of radiation under the assumption that chronic radiation exposure induces ~25-33% as many germline mutations as an acute dose rate (UNSCEAR, 1993). In contrast, our data show that the evaluation of genetic risk of low-dose chronic exposure, at least for ESTR loci, does not require any correction as the risk factors for chronic and acute irradiation for this dose range should be regarded as similar.

Chronic exposure to high-LET fission neutrons is more effective than acute or chronic low-LET exposure in the induction of mouse germline ESTR mutations (Figure 1a) with the relative biological effectiveness of neutrons, RBE as 3.36 (Dubrova et al., 2000a). Previous studies in mice (Figure 1b) have also shown a marked increase in mutation rates after chronic or acute paternal exposure to neutrons, but a higher RBE value (RBE =5.40) was obtained (Russell, 1965; Batchelor et al., 1966, 1967). It should be stressed that in those studies the efficiency of mutation induction was compared for different dose-ranges for the two radiations - 0.5-2 Gy neutrons and 3-6 Gy low-LET radiation. In contrast, our data were obtained for neutron doses from 0.125 to 0.5 Gy and, importantly, were compared to a similar dose range of 0.5 and 1 Gy of X-rays.

Mechanisms Of Mutation Induction At Mouse ESTR Loci

It is generally accepted that DNA double strand breaks are the most potentially mutagenic type of DNA damage induced by ionising radiation (Frankenberg-Schwager, 1990). The observed radiation-induced increases in ESTR mutation rates cannot be attributed to the direct effects of radiation-induced DNA double strand breaks at these small genomic loci (Sadamoto et al., 1994; Fan et al., 1995; Dubrova et al., 1998, 2000a; Barber et al., 2000), and similar conclusions have been obtained from studies of somatic and germline mutation rates at other DNA repeat sequences (Schiestl et al., 1994; Dubrova et al., 1996, 1997; Kovalchuk et al., 2000). The main argument for the non-targeted mechanisms of mutation induction at DNA repeat loci is that if mutations found in the offspring of irradiated males were initiated by direct targeted events, this would require an unrealistically high number of extra double-strand breaks or other damage per genome (details of estimates are given in Schiestl et al., 1994 and Dubrova et al., 1998). It would appear that the observed radiation-induced increase in ESTR mutation rates results from radiation-induced damage elsewhere in the genome/cell, raising the possibility that a cell's response to untargeted structural damage induces mutation at ESTR loci.

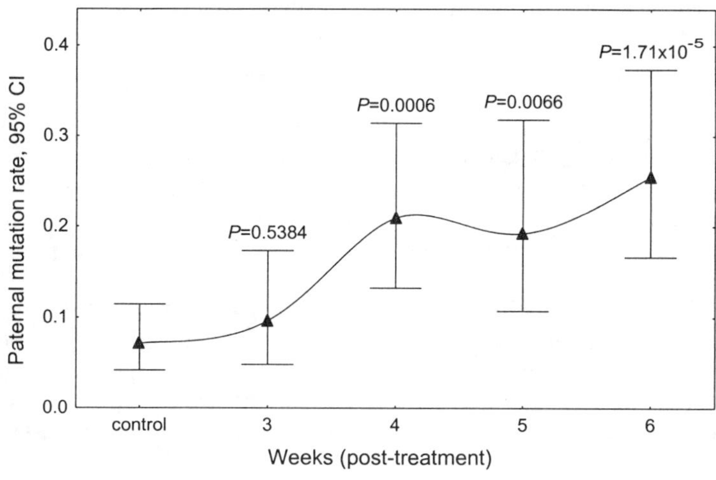

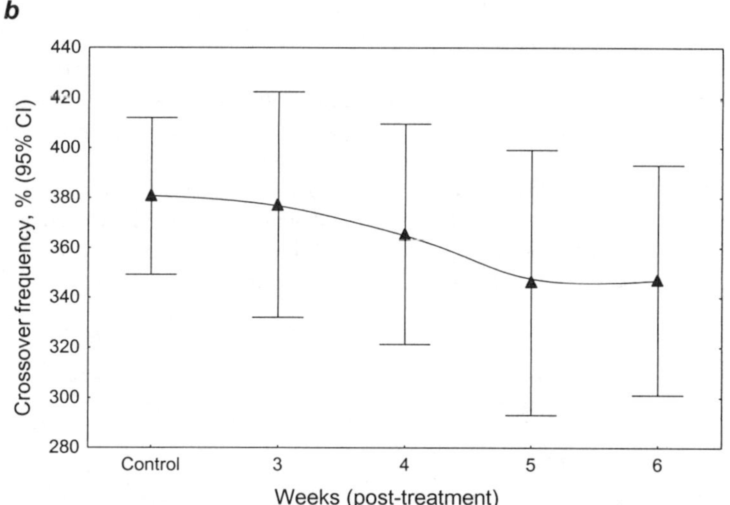

Figure 2. ESTR mutation rates and crossover frequencies in male mice following irradiation by 1Gy of X-rays. (a) Paternal ESTR mutation rate. The probabilities of difference from the control group (Fisher's exact test, two-tailed) are given. (b) Crossover frequencies. 95% confidence intervals for mutation rate and crossover frequency are given. Data from Barber et al. (2000).

We have recently tested the hypothesis that ESTR mutation induction may be associated with radiation-induced increases in meiotic crossing-over (Barber et al., 2000). Hybrid (C57BL/6 x CBA/H) F_1 male mice were exposed to acute 1 Gy X-rays, and backcrossed to unexposed CBA/H females. A panel of 25 polymorphic chromosome-specific microsatellite (six mouse chromosomes, 422 cM) were used to genotype the offspring to estimate crossover frequencies in the irradiated F_0 male germline. The same offspring were analysed for paternal ESTR mutations (single-locus probes Ms6-hm and Hm-2) for comparison.

Figure 2a presents the frequencies of ESTR mutation in offspring derived from germ cells exposed at the different stages of mouse spermatogenesis. Offspring conceived 3 weeks after exposure showed no increase in ESTR mutation rates, whereas a highly significant increase in paternal mutation rate was found in offspring conceived 4, 5 or 6 weeks after treatment. These data, together with the results of our previous study (Dubrova et al., 1998), show that radiation-induced ESTR mutations can occur at all stages of spermatogenesis prior to metaphase I (3 weeks). In contrast, crossover frequencies in the same irradiated males remained unchanged over the whole period of exposure and were similar to those seen in the same mice before irradiation (Figure 2b), therefore there is no obvious relationship between ESTR mutation rates and meiotic crossover frequencies. In theory, DSBs induced by radiation at the crossover-proficient stages of spermatogenesis could lead to an increase in meiotic crossover rates. However, our experimental design included these stages of mouse spermatogenesis (mice bred 4 weeks after exposure), and no evidence for crossover induction was found.

While radiation induced-DNA double strand breaks (DSBs) may stimulate recombination, such DSBs in pre-meiotic cells are likely to be repaired by non-homologous end-joining and recombination, without crossover between homologous chromosomes (Weaver, 1996). Alternatively, cells containing DSBs may be lost by apoptosis. Mutation induction at ESTR loci therefore cannot be attributed to a genome-wide increase in meiotic recombination rates.

The persistence of instability into the germline of offspring of irradiated mice, resulting in transgenerational mutagenesis (Dubrova et al., 2000b), provides further evidence for a non-targeted process at mouse ESTR loci. Figure 3 presents the ESTR mutation rates in CBA/H male mice directly exposed to 0.5 Gy of fission neutrons and their non-exposed offspring. Following paternal irradiation, breeding from unexposed F_1 male and female mice showed that the F_1 germline mutation rates were very similar to those of the irradiated F_0 males and unexpectedly did not return to normal unexposed mutation rates. The increase was seen in most F_1 offspring and was found at alleles derived from both the F_0 irradiated father and the unexposed F_0 mothers, indicating that this instability can affect both irradiated paternal and non-irradiated maternal F_0 genomes.

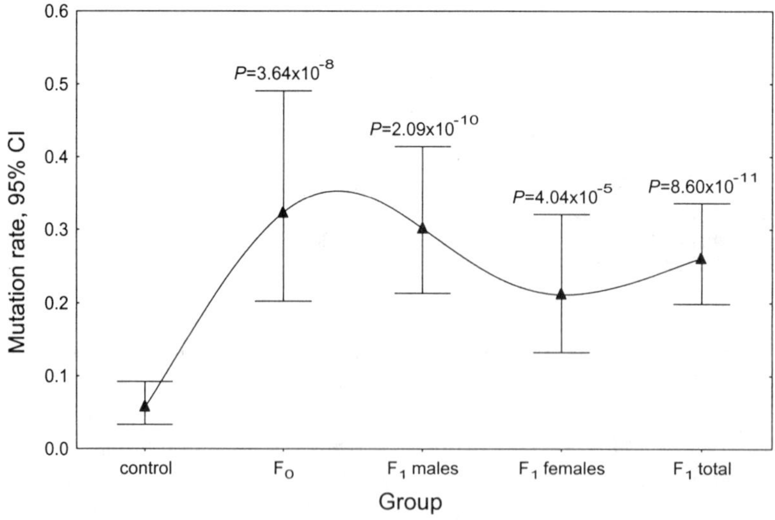

Figure 3. ESTR germline mutation rates in male mice exposed to 0.5 Gy of fission neutrons (F_0 males) and in the descendants of irradiated males (F_1 males and females). The probabilities of difference from the control group (Fisher's exact test, two-tailed) are given. Data from Dubrova et al. (2000a, 2000b).

This remarkable stimulation of mutation in the F_1 germline is reminiscent of the phenomenon of delayed radiation-induced genome instability in somatic cells, where radiation can induce an increase in non-clonal genetic changes in the clonal offspring of a single irradiated cell, an instability that can persist for many cell generations (Morgan et al., 1996). In our studies, each F_1 offspring is the clonal offspring of a single irradiated F_0 spermatogenic stem cell. By analogy, radiation-induced instability in the F_0 germline is transmitted via a single sperm to the clonal F_1 offspring in whom instability arises many cell generations later in the developing germline.

Our results do not support the hypothesis of direct targeting of any specific set of genes in the F_0 male, causing for example mutational inactivation of a DNA repair gene and consequent destabilisation of ESTRs. An elevated ESTR mutation rate appears to be uniform over all F_1 offspring derived from ten independent sperm cells from the irradiated F_0 male and, as neutron irradiation results in random structural damage to DNA, it is unlikely that the same genes were affected in the different sperm cells. Thus, we are led to consider the possibility that the radiation exposure signal is inherited through sperm in an epigenetic fashion.

One hypothesis is that genomic instability may be transmitted via changes in DNA methylation. Methylation is known to be transmissible through many cell divisions (Holliday, 1986) and can influence DNA repair functions, as in the case of microsatellite instability in colorectal carcinomas resulting from loss of *hMLH1* mismatch repair activity due to promoter hypermethylation (Herman et al., 1998; Wheeler et al., 1999). Radiation-induced DNA damage triggers a cascade of events, ranging from the repair of damage to programmed cell death. Surviving cells will bear the marks of these activities, and may therefore show changes in methylation status of genes responsible for maintaining genomic integrity. Theoretically, such alterations may not be cell autonomous, since A_s spermatogonia in mice are connected by intercellular bridges (De Rooij, 1998) which could allow them to share different biochemical pathways including DNA repair and methylation activities. If so, then the initial response of one cell to radiation-induced damage could spread to neighbouring unaffected cells. If this response results in an epigenetic signal influencing DNA integrity systems that can survive the reprogramming of DNA methylation during spermatogenesis and early development (Roemer et al., 1997; Constancia et al., 1998), then subsequent activation of this signal in the germline could lead to transgenerational mutagenesis. Future work should address the issue of whether induction can persist over more than two generations, and whether radiation causes detectable shifts in DNA methylation patterns in the male germline.

Conclusions

The results of our studies highlight the unique advantage of ESTR loci for detection of radiation-induced germline mutation in mice. The dose response of ESTR germline mutation induction in male mice exposed to high- and low-LET ionising radiation is remarkably similar to that for protein-coding genes. ESTR loci have also revealed new and unexpected features of germline mutation induction in mice. An elevated ESTR mutation rate can be robustly detected at doses that were previously inaccessible using standard approaches for monitoring germline mutation in mice, and will provide new insights into the estimation of genetic hazard of low-dose radiation exposure for humans. Moreover, this model reveals a novel form of radiation-induced genetic instability which is readily transmitted to the next generation. Further studies are clearly needed to analyse mechanisms of radiation-induced mutation at ESTR loci.

MUTATION INDUCTION IN HUMANS

Minisatellites are the most unstable loci in the human genome. In contrast to the mouse ESTRs, they consist of longer repeats (10-60 bp), which show considerable sequence variation along the array (Jeffreys et al., 1991, 1994; May et al., 1996; Buard et al., 1998; Tamaki et al., 1999; Vergnaud and Denoeud, 2000; Stead and Jeffreys, 2000). Mutation at these loci is almost completely restricted to the germline, with very rare and simple muta-tional events occurring in the somatic cells (Jeffreys at al., 1994; May et al., 1996; Jeffreys and Neumann, 1997; Tamaki et al., 1999; Buard et al., 2000; Stead and Jeffreys, 2000). Germline mutation at human minisatellites is attributed to complex gene-conversion like events altering repeat unit copy number. Previous studies have shown very high spontaneous germline mutation rates for some minisatellite loci, ranging from 0.5 up to 13% per gamete (Jeffreys et al., 1998; Vergnaud and Denoeud, 2000), which potentially make these loci useful markers for monitoring germline mutation in humans. To date, the frequency of minisatellite mutation has been analysed in families exposed to several different types of ionising radiation (Kodaira et al., 1995; Dubrova et al., 1996, 1997; Livshits et al., 2000). Mutation rate has also been examined by quantitative small-pool PCR (SP-PCR) methods in human sperm collected from the limited number of cancer patients exposed to therapeutic mutagens and radiation (Armour et al., 1999; May et al., 2000; Zheng et al., 2000). These studies have generated conflicting results on the effects of radiation on human germline minisatellite instability.

Mutation Analysis In Human Pedigrees

We analysed the frequency of minisatellite mutation in human families inhabiting the heavily polluted rural areas of the Mogilev district of Belarus following the Chernobyl accident and compared it to the mutation frequency in a control data set from the UK (Dubrova et al., 1996, 1997). The exposed cohort consisted of 127 children born between February and September 1994 for whom both parents were continuously resident in the Mogilev district from the time of the Chernobyl accident. For all parents the local level of surface contamination by ^{137}Cs exceeded 1 Ci/km^2.

Minisatellite germline mutations were scored using two multi-locus minisatellite probes 33.6 and 33.15 and eight hypervariable single-locus minisatellite probes B6.7, MS1, CEB1, CEB15, CEB25, CEB36, MS31 and MS32, chosen for their high spontaneous mutation rate (Jeffreys et al., 1988, 1990; Olaisen et al., 1993; Vergnaud and Denoeud, 2000). We produced three different estimates of mutation rate based on these independent sets of minisatellites (Figure 4a). For all these sets of loci, mutation rates in the exposed families were higher than in the control group. Moreover, six of the eight single-locus minisatellites showed an increase in mutation rate in the exposed group (Figure 4b). Finally, we estimated the total frequency of mutant bands in offspring using all independent probes (Figure 4a). A highly statistically significant 1.9 fold increase in the mutation rate was found in the exposed group.

We also estimated the individual radiation doses for chronic exposure for families in the exposed group, using published data on the annual external and internal exposure to ^{137}Cs in soil, milk and vegetables in different localities of the Mogilev district (Drosdovitch et al., 1989) and family histories after the Chernobyl accident. A marginally-significant positive correlation between the parental dose and the number of mutant bands in offspring was found within the exposed group (Kendall's τ=0.1111, P=0.0672), consistent with a steady increase of mutation rate with radiation dose.

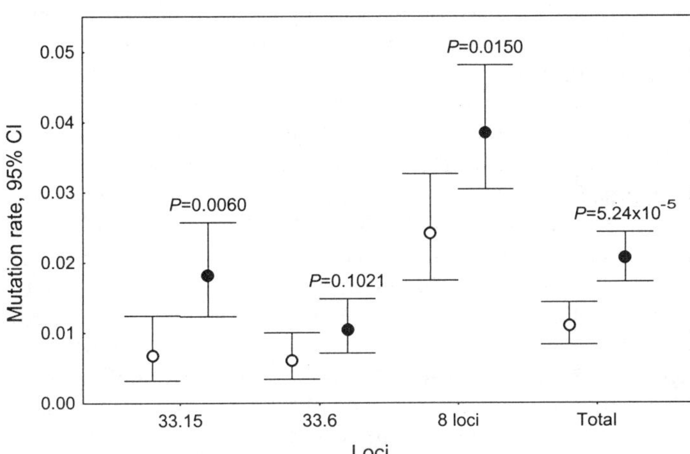

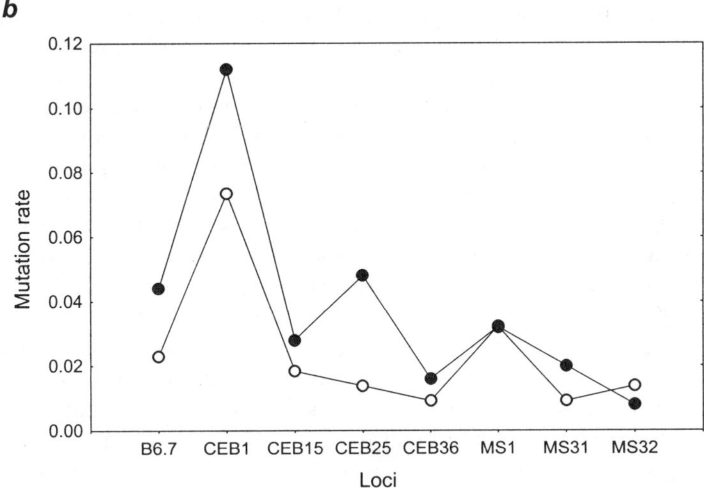

Figure 4. Mutation rates in control and exposed human populations. (a) Total mutation rates scored by three independent sets of minisatellite probes. 95% confidence intervals for mutation rate and probabilities of difference between control and exposed group are given. (b) Mutation rates for eight independent single-locus minisatellite loci (Wilcoxon Matched Pairs Test, P=0.0357). Open circles, control families; closed circles, exposed families. Data from Dubrova et al. (1996, 1997).

The observed correlation of mutation rate within the exposed group with parental radiation dose suggests that the observed increase in mutation rate may be directly caused by ionising radiation. Alternatively, the increased mutation rate found in the exposed group might be explained either by genetic or environmental factors. Scoring of mutations by three independent sets of minisatellite loci detected separately by 33.15, 33.6 and eight different single-locus probes revealed an elevated mutation rate in the exposed group for most of the minisatellite systems, indicating that increased mutability cannot be attributed to a single locus which has accumulated unusually unstable alleles in the Mogilev population. Thus, the most probable cause of increased mutation rate in the exposed group is the influence of environmental mutagens. Furthermore, the correlation of mutation rate within the exposed group with parental radiation dose suggests that this observed increase in mutation rate may be directly caused by ionising radiation. If correct, then these data provide the first experimental evidence for radiation-induction of human germline mutation. It should be noted that we have recently extended this analysis to another cohort of irradiated families chronically exposed to radioactive fallout near the Semipalatinsk nuclear test site in Kazakhstan. The results of this study also show an elevated germline mutation rate in the exposed population (Dubrova et al., in preparation).

Using a similar approach, two other studies do not provide any evidence for radiation-induced mutation in the offspring of irradiated parents (Kodaira et al., 1995; Livshits et al., 2000). This discrepancy could result from totally different types of exposure to ionising radiation in these studies. The atomic bomb survivors from Hiroshima and Nagasaki studied by Kodaira et al. (1995) were externally exposed to a considerable dose of acute irradiation, whereas chronic internal and external exposure was the main source of radiation hazard after the Chernobyl disaster. The results of recent publications show that the effects of low-dose chronic exposure to ionising radiation may be substantially more mutagenic than previously thought (Vilenchik and Knudson, 2000; Kovalchuk et al., 2000). In addition, most of the children in the Japanese study were born more than ten years after the single acute parental irradiation, which means that at least some radiation-induced DNA alterations could have been repaired over this period of time. In contrast, the population of the post-Chernobyl affected areas has been constantly irradiated following the accident. Finally, the Japanese data are derived from families in which mainly one parent received A-bomb radiation, whereas our data were obtained by profiling families with both parents exposed to chronic irradiation after the Chernobyl accident.

Another study of germline minisatellite mutation rate was curried out on the offspring of Chernobyl clean-up workers (Livshits et al., 2000). This group of workers is thought to be extremely heterogeneous, in terms of exposure and doses. Most of the participants involved in the decontamination work around the Chernobyl nuclear power plant, sarcophagus construction and other clean-up operations received doses less than 0.25 Gy (Pitkevitch et al., 1997). The main exposure to this cohort was due to external and relatively uniform γ-irradiation, with a relatively minor contribution from the intake of radionuclides. Most importantly, this group was exposed to repeated small daily doses of ionising radiation. Previous studies in male mice have clearly shown that the yield of germline mutations after such an acute external fractionated exposure was less than when the same dose was given in a single exposure (Lyon et al., 1972). Given the fact that the maximum reported dose to the Chernobyl clean-up workers of 0.25 Gy is below any known estimates of the doubling dose for mice (Searle, 1974; Dubrova et al., 1998) and assuming the dose-fractionation effects, the expected increase in mutation rate in this group may be quite small (less than 25%) and therefore cannot be statistically detected within 183 children from the exposed group.

Small-pool PCR Analysis of Sperm DNA

A novel small-pool PCR approach for detection of minisatellite mutations has recently been used to quantify germline mutation rate in the cancer patients treated with therapeutic mutagens and radiation (Armour et al., 1999; May et al., 2000; Zheng et al., 2000). This approach initially developed for the analysis of spontaneous mutation at human minisatellite loci (Jeffreys et al., 1994), is based on the amplification of multiple diluted aliquots of sperm DNA and allows the detection of a large number of de novo mutants in a single male. Compared to the pedigree approach, this technique dramatically reduces the number of individuals needed for the measurement of germline mutation frequencies. However, the major shortcoming of the SP-PCR approach is a very high variation between spontaneous mutation rates of individual alleles at a single locus (Jeffreys et al., 1994; May et al., 1996; Buard et al., 1998; Tamaki at al., 1999). This variation between alleles may be as high as 10-fold, effectively preventing comparison of mutation rates between non-exposed and exposed men. Therefore, this technique can only be used to evaluate mutation rate in the same man before and after mutagenic treatment. Moreover, SP-PCR does not allow amplification of very large minisatellite alleles (longer than 5 kb), thus restricting mutation scoring to a subset of relatively small alleles.

Using SP-PCR of the MS205 minisatellite, the germline mutation rate was determined in cancer patients treated with therapeutic mutagens (Armour et al., 1999; Zheng et al., 2000). It should be noted that to date, little is known about germline mutation induction at tandem repeat loci by chemical mutagens, although in vitro mutagenicity studies in somatic cells have demonstrated induction of minisatellite mutations by chemicals (Ledwith et al., 1995; Kaplanski et al., 1997). The only germline evidence comes from studies of urban herring gulls exposed to chemical pollutants (Yauk and Quinn, 1996; Yauk et al., 2000) and male mice treated with polychlorinated biphenyls (Hedenskog et al., 1997). The results of our recent study show that the anticancer drug cisplatin does not affect germline mutation rate at the mouse ESTR loci (Barber et al., 2000). Consequently, the minisatellite data on cancer patients can only be compared with the results of numerous studies of mutagenicity of anticancer drugs in the mouse germline, where the different end-points such as Russell 7-locus test, dominant lethality and chromosome translocations were used (Witt and Bishop, 1996).

In the first study, sperm samples of two men exposed to the anticancer drugs cyclophosphamide, etoposide and vincristine were analysed (Armour et al., 1999). The alkylating agent cyclophosphamide induces a variety of germ cell effects in post-meiotic stages only, with no effects on pre-meiotic stages (Witt and Bishop, 1996). Our previous analysis has shown that radiation-induced ESTR mutations in mice cannot occur at the post-meiotic stages, and the same may also hold true for exposure to chemicals. The topoisomerase-II inhibitor etoposide affects only meiotic germ cells and is not mutagenic in pre- and post-meiotic cells (Russell et al., 1998; Marchetti et al., 2001). If correct, then the window of time for mutation induction by this drug is very short and may be difficult to analyse. Vincristine prevents the assembly of tubulin into spindle fibres, and there is no indication of germ cell mutagenicity for this drug in mice (Witt and Bishop, 1996). Therefore, the two men analysed in the above-mentioned study (Armour et al., 1999) were exposed to the anticancer drugs which, because of their stage-specificity of mutation induction in the male germline, may not induce mutations at minisatellite loci at all or, alternatively, could affect a very small subset of germ cells over a relatively short period of time.

In the second study, sperm samples from ten men treated for Hodgkin's disease were analysed (Zheng et al., 2000). Nine patients, treated either with vinblastine or adriamycin and bleomycin, did not show any increases in mutation rate after cancer chemotherapy. Vinblastine binds to tubulin, and exposure of male mice to this drug results in aneuploidy rather than chromosome breakage or gene mutation (Witt and Bishop, 1996). Adriamicyn is an intercalating agent and an inhibitor of topoisomerase-II. Exposure to adriamicyn predominantly results in cell toxicity in mouse germ cells, without mutation induction (Witt and Bishop, 1996). Bleomycin, a radiomimetic drug, selectively targets mouse oocytes; mutation induction has not been observed in male germ cells (Witt and Bishop, 1996). Therefore, judging from the mouse data, the negative results for the cancer patients are not unexpected. Interestingly, the only patient treated with procarbazine showed a significant increase in mutation rate after chemotherapy (Zheng et al., 2000). In contrast to all above-mentioned anticancer drugs, exposure of male mice to procarbazine does induce specific locus mutations in stem cell spermatogonia (Witt and Bishop, 1996), the stage where radiation-induced ESTR mutations were also detected (Dubrova et al., 1998; Barber et al., 2000).

The analysis of sperm DNA from three seminoma patients before and after radiotherapy has also failed to detect any increases in mutation rate at the hypervariable minisatellites B6.7 and CEB1 (May et al., 2000). These men were repeatedly exposed to 15 fractions of acute X-rays with a total testicular dose ranging between 0.4 and 0.8 Gy, a value close to the estimates of doubling dose in male mice (Searle, 1974; Dubrova et al., 1998). The negative results of this study could be attributed to various factors, including the assumption that the yield of germline mutations after fractionated exposure is less than when the same dose is given in a single exposure (Lyon et al., 1972). A direct comparison of the results of this study with our data on the Chernobyl families (Dubrova et al., 1996, 1997) is limited by the differences in doses and, most importantly, by types of exposure, being acute external for the seminoma patients and chronic internal for the Belarus families.

Conclusions

In summary, it should be stressed that all above-mentioned data on mutation induction in humans were derived from relatively small numbers of people exposed to a wide range of potential mutagens. This variety of environmental factors, together with the different types of exposure ranging from high-dose acute to low-dose chronic, currently prevents any reliable comparisons of the results of these pilot studies. More work is clearly needed to validate the potential applications of minisatellite loci for monitoring mutation rate in human populations.

ACKNOWLEDGMENTS

This work was supported by grants from the Wellcome Trust.

REFERENCES

Armour, J.A.L., Brinkworth, M.H., and Kamischke, A., 1999, Direct analysis by small-pool PCR of MS205 minisatellite mutation rates in sperm after mutagenic therapies. *Mutat Res.* 445:73-80.

Barber, R., Plumb, M.A., Smith, A.G., Cesar, C.E., Boulton, E., Jeffreys, A.J., and Dubrova, Y.E., 2000, No correlation between germline mutation at repeat DNA and meiotic crossover in male mice exposed to X-rays or cisplatin. *Mutat Res.* 457:79-91.

Batchelor, A.L., Phillips, R.J.L., and Searle, A.G., 1966, A comparison of the mutagenic effectiveness of chronic neutron- and γ-irradiation of mouse spermatogonia. *Mutat Res.* 3:218-229.

Batchelor, A.L., Phillips, R.J.L., and Searle, A.G., 1967, The reversed dose-rate effect with fast neutron irradiation of mouse spermatogonia. *Mutat Res.* 4:229-231.

Blake, J.A., Eppig, J.T., Richardson, J.E., Davisson, M.T., and The Mouse Genome Database Group, 2000, The Mouse Genome Database (MGD): expanding genetic and genomic resources for the laboratory mouse. *Nucl Acids Res.* 28:108-111.

Bois, P., Stead, J.H.D., Bakshi, S., Williamson, J., Neumann, R., Moghadaszadeh, B., and Jeffreys, A.J., 1998a, Isolation and characterization of mouse minisatellites. *Genomics.* 50:317-330.

Bois, P., Williamson, J., Brown, J., Dubrova, Y.E., and Jeffreys, A.J., 1998b, A novel unstable mouse VNTR family expanded from SINE B1 element. *Genomics.* 49:122-128.

Buard, J., Bourdet, A., Yardley, J., Dubrova, Y.E., and Jeffreys, A.J., 1998, Influence of array size and homogeneity on minisatellite mutation. *EMBO J.* 17:3495-3502.

Buard, J., Collick A.J., Brown, J., and Jeffreys, A.J., 2000, Somatic versus germline mutation process at minisatellite CEB1 (*D2S90*) in humans and transgenic mice. *Genomics.* 65:95-103.

Byrne, J., Rasmussen, S.A., Steinhorn, S.C., Connelly, R.R., Myers, M.H., Lynch, C.F., Flannery, J., Austin, D.F., Holmes, F.F., Holmes, G.E., Strong, L.C., and Mulvihill, J.J., 1998, Genetic diseases in offspring of long-term survivors of childhood and adolescent cancer. *Am J Hum Genet.* 62:45-52.

Constancia, M., Pickard, B., Kelsey, G., and Reik, W., 1998, Imprinting mechanisms. *Genome Res.* 8:881-900.

De Rooij, D.G., 1998, Stem cells in the testis. *Int J Exp Path.* 79:67-80.

Drosdovitch, V.V., Minenko, M.F., Ulanovski, A.V., and Shemyakina, E.V., 1989, *Prognosis of Doses of Exposure for Population of BSSR from the Caesium Radionuclides*, Ministry of Health of Belarus, Minsk.

Dubrova, Y.E., Jeffreys, A.J., and Malashenko, A.M., 1993, Mouse minisatellite mutations induced by ionizing radiation. *Nature Genet.* 5:92-94.

Dubrova, Y.E., Nesterov, V.N., Krouchinsky, N.G., Ostapenko, V.A., Neumann, R., Neil, D.L., and Jeffreys, A.J., 1996, Human minisatellite mutation rate after the Chernobyl accident. *Nature.* 380:683-686.

Dubrova, Y.E., Nesterov, V.N., Krouchinsky, N.G., Ostapenko, V.A., Vergnaud, G., Giraudeau, F., Buard, J., and Jeffreys, A.J., 1997, Further evidence for elevated human minisatellite mutation rate in Belarus eight years after the Chernobyl accident. *Mutat Res.* 381:267-278.

Dubrova, Y.E., Plumb, M., Brown, J., Fennelly, J., Bois, P., Goodhead, D., and Jeffreys, A.J., 1998, Stage specificity, dose response, and doubling dose for mouse minisatellite germ-line mutation induced by acute radiation. *Proc Natl Acad Sci USA.* 95:6251-6255.

Dubrova, Y.E., Plumb, M., Brown, J., and Jeffreys, A.J., 1998, Radiation-induced germline instability at minisatellite loci. *Int J Radiat Biol.* 74:689-696.

Dubrova, Y.E., Plumb, M., Brown, J., Boulton, E., Goodhead, D., and Jeffreys, A.J., 2000a, Induction of minisatellite mutations in the mouse germline by low-dose chronic exposure to γ-radiation and fission neutrons. *Mutat Res.* 453:17-24.

Dubrova, Y.E., Plumb, M., Gutierrez, B., Boulton, E., and Jeffreys, A.J., 2000b, Transgenerational mutation by radiation. *Nature.* 405:37.

Fan, Y.J., Wang, Z., Sadamoto, S., Ninomiya, Y., Kotomura, N., Kamiya, K., Dohi, K., Kominami, R., and Niwa, O., 1995, Dose-response of radiation induction of a germline mutation at a hypervariable mouse minisatellite locus. *Int J Radiat Biol.* 68:177-183.

Frankenberg-Schwager, M., 1990, Induction, repair and biological relevance of radiation-induced DNA lesions in eukaryotic cell. *Radiat Environ Biophys.* 29:273-292.

Gibbs, M., Collick, A., Kelly, R., and Jeffreys, A.J., 1993, A tetranucleotide repeat mouse minisatellite displaying substantial somatic instability during early preimplatation development. *Genomics.* 17:121-128.

Hedenskog, M., Sjogren, M., Cederberg, H., and Rannug, U., 1997, Induction of germline-length mutations at the minisatellites PC-1 and PC-2 in male mice exposed to polychlorinated biphenyls and diesel exhaust emissions. *Environ Mol Mutagen.* 30:254-259.

Herman, J.G., Umar, A., Polyak, K., Graff, J.R., Ahuja, N., Issa, J.-P.J., Markowitz, S., Willson, J.K.V., Hamilton, S.R., Kinzler, K.W., Kane, M.F., Kolodner, R.D., Vogelstein, B., Kunkel, T.A., and Baylin, S.B., 1998, Incidence and functional consequences of *hMLH1* promoter hypermethylation in colorectal carcinoma. *Proc Natl Acad Sci USA.* 95:6870-6875.

Holliday, R., 1987, The inheritance of epigenetic defects. *Science.* 238:163-170.

Jeffreys, A.J., Royle, N.J., Wilson, V., and Wong, Z., 1988, Spontaneous mutation rate to new length alleles at tandem-repeat hypervariable loci in human DNA. *Nature.* 332:278-281.

Jeffreys, A.J., Turner, M., and Debenham, P., 1990, The efficiency of multi-locus DNA fingerprint probes for individualization and establishment of family relationships, determined from extensive casework. *Am J Hum Genet.* 48:824-840.

Jeffreys, A.J., MacLeod, A., Tamaki, K., Neil, D.L., and Monckton, D.G., 1991, Minisatellite repeat coding as a digital approach to DNA typing. *Nature.* 354:204-209.

Jeffreys, A.J., Tamaki, K., MacLeod, A., Monckton, D.G., Neil, D.L., and Armour, J.A.L., 1994, Complex gene conversion events in germline mutation at human minisatellites. *Nature Genet.* 6:136-145.

Jeffreys, A.J., and Neumann, R., 1997, Somatic mutation process at a human minisatellite. *Hum Mol Genet.* 6:129-136.

Kaplanski, C., Chisari, F.V., and Wild, C.P., 1997, Minisatellite rearrangements are increased in liver tumours induced by transplacental alfatoxin B_1 treatment of hepatitis B virus transgenic mice, but not in spontaneously arising tumours. *Carcinogenesis.* 18:633-639.

Kelly, R., Bulfield, G., Collick, A., Gibbs, M., and Jeffreys, A.J., 1989, Characterization of a highly unstable mouse minisatellite locus: Evidence for somatic mutation during early development. *Genomics.* 5:844-856.

Kodaira, M., Satoh., C, Hiyama, K., and Toyama, K., 1995, Lack of effects of atomic-bomb radiation on genetic instability of tandem-repetitive elements in human germ-cells. *Amer J Hum Genet.* 57:1275-1283.

Kovalchuk, O., Dubrova, Y.E., Arkhipov, A., Hohn, B., and Kovalchuk, I., 2000, Wheat mutation rate after Chernobyl. *Nature.* 407:583-584.

Ledwith, B.J., Joslin, D.J., Troilo, P., Leander, K.R., Clair, J.H., Soper, K.A., Manam, S., Prahalada, S., van Zwieten, J., and Nichols, W.W., 1995, Induction of minisatellite DNA rearrangements by genotoxic carcinogens in mouse liver tumors. *Carcinogenesis.* 16:1167-1172.

Livshits, L.A., Malyarchuk, S.G., Lukyanova, E.M., Antipkin, Y.G., Arabskaya, L.P., Kravchenko, S.A., Matsuka, G.H., Petit, E., Giraudeau, F., Gourmelon, P., Vergnaud, G., and Le Guen, B., 2001, Children of Chernobyl cleanup workers do not show elevated rates of mutations in minisatellite alleles. *Radiat Res.* 155:74-80.

Lyon, M.F., Phillips, R.J.S., and Bailey, H.J., 1972, Mutagenic effects of repeated small doses to mouse spermatogonia. I. Specific-locus mutation rates. *Mutat Res.* 15:185-190.

Marchetti, F., Bishop, J.B., Lowe, X., Generoso, W.M., Hozier, J., and Wyrobek, A.J., 2001, Etoposide induces heritable chromosome aberrations and aneuploidy during male meiosis in the mouse. *Proc Natl Acad Sci USA.* 98:3952-3957.

May, C.A., Jeffreys, A.J., and Armour, J.A.L., 1996, Mutation rate heterogeneity and the generation of allele diversity at the human minisatellite MS205 (*D16S309*). *Hum Mol Genet.* 5:1823-1833.

May, C.A., Tamaki, K., Neumann, R., Wilson, G., Zagars, G., Pollack, A., Dubrova, Y.E., Jeffreys, A.J., and Meistrich, M.L., 2000, Minisatellite mutation frequency in human sperm following radiotherapy. *Mutat Res.* 453:67-75.

Morgan, W.F., Day, J.P., Kaplan, M.I., McGhee, E.M., and Limoli, C.L., 1996, Genomic instability induced by ionizing radiation. *Radiat Res.* 146:247-258.

Neel, J.V., Schull, W.J., Awa, A.A., Satoh, C., Kato, H., Otake, M., and Yoshimoto, Y., 1990, The children of parents exposed to atomic bombs: estimates of the genetic doubling dose of radiation for humans. *Am J Hum Genet.* 46:1053-1072.

Olaisen, B., Bekkemoen, M., Hoff-Olsen, P., and Gill, P., 1993, Human VNTR and sex, in: *DNA Fingerprinting: State of the Science,* S.D.J. Pena, R. Chakraborty, J.T. Epplen and A.J. Jeffreys, eds., Birkhauser, Basel, 63-69.

Pitkevitch, V.A., Ivanov, V. K., Tsyb, A. F., Maksyoutov, M. A., Matiash V. A., and Shchukina, N. V., 1997, Exposure levels for persons involved in recovery operations after the Chernobyl accident. Statistical analysis based on the data of the Russian National Medical and Dosimetric Registry (RNMDR). *Radiat Environ Biophys.* 36:149-160.

Roemer, I., Reik, W., Dean, W., and Klose, J., 1997, Epigenetic inheritance in the mouse. *Current Biol.* 7:277-280.

Russell, W.L., 1965, Studies in mammalian radiation genetics. *Nucleonics.* 23:53-62.

Russell, W.L., and Kelly, E.M., 1982, Mutation frequencies in male mice and the estimation of genetic hazards of radiation in men. *Proc Natl Acad Sci USA.* 79:542-544.

Russell, L.B., Hunsicker, P.R., Johnson, D.K., and Shelby, M.D., 1998, Unlike other chemicals, etoposide (a topoisomerase-II inhibitor) produces peak mutagenicity in primary spermatocytes of the mouse. *Mutat Res.* 400:279-286.

Sadamoto, S., Suzuki, S., Kamiya, K., Kominami, R., Dohi, K., and Niwa, O., 1994, Radiation induction of germline mutation at a hypervariable mouse minisatellite locus. *Int J Radiat Biol.* 65:549-557.

Sankaranarayanan, K., 1999, Ionizing radiation and genetic risks X. The potential "disease phenotypes" of radiation-induced genetic damage in humans: perspectives from human molecular biology and radiation genetics. *Mutat Res.* 429:45-83.

Sankaranarayanan, K., and Chakraborty, R., 2000, Ionizing radiation and genetic risks XI. The doubling dose estimates from the mid- 1950s to present and the conceptual change to the use of human data on spontaneous mutation rates and mouse data on induced mutation rates for doubling dose calculations. *Mutat Res.* 453:107-127.

Searle, A.G., 1974, Mutation induction in mice. *Adv Radiat Biol.* 4:131-207.

Schiestl, R.H., Khogali, F., and Carls, N., 1994, Reversion of the mouse *pink-eyed unstable* mutation induced by low doses of x-rays. *Science.* 266:1573-1576.

Stead, J.D.H., and Jeffreys, A.J., 2000, Allele diversity and germline mutation at the insulin minisatellite. *Hum Mol Genet.* 9:713-723.

Tamaki, K., May, C.A., Dubrova, Y.E., and Jeffreys, A.J., 1999, Extremely complex repeat shuffling during germline mutation at human minisatellite B6.7. *Hum Mol Genet.* 8:879-888.

UNSCEAR, 1993, *Sources and Effects of Ionizing Radiation*, United Nations, New York.

Vergnaud, G., and Denoeud, F., 2000, Minisatellites: mutability and genome architecture. *Genome Res.* 10:899-907.

Vilenchik, M. M., and Knudson, A.G., 2000, Inverse radiation dose-rate effects on somatic and germ-line mutations and DNA damage rates, *Proc. Natl. Acad. Sci. U.S.A.* 97: 5381-5386.

Weaver, D.T., 1996, Regulation and repair of double-strand DNA breaks. *Crit Rev Euk Gene Exp.* 64:345-375.

Wheeler, J.M.D., Beck, N.E., Kim, H.C., Tomlinson, I.P.M., Mortensen, N.J.M., and Bodmer, W.F., 1999, Mechanisms of inactivation of mismatch repair genes in human colorectal cancer cell lines: The predominant role of hMLH1. *Proc Natl Acad Sci USA.* 96:10296-10301.

Witt, K.L., and Bishop, J.B., 1996, Mutagenicity of anticancer drugs in mammalian germ cells. *Mutat Res.* 355:209-234.

Yauk, C.L., and Quinn, J.S., 1996, Multilocus DNA fingerprinting reveals high rate of heritable genetic mutation in herring gulls nesting in an industrialized urban site. *Proc Natl Acad Sci USA.* 93:12137-12141.

Yauk, C.L., Fox, G.A., McCarry, B.E, and Quinn, J.S., 2000, Induced minisatellite mutations in herring gulls (*Larus argentatus*) living near steel mills. *Mutat Res.* 452:211-218.

Zheng, N., Monckton, D.G., Wilson, G., Hagemeister, F., Chakraborty, R., Connor, T.H., Siciliano, M.J., and Meistrich, M.L., 2000, Frequency of minisatellite repeat number changes at the MS205 locus in human sperm before and after cancer chemotherapy. *Environ Mol Mutagen.* 36:134-145.

PAINT/DAPI ANALYSIS OF MOUSE ZYGOTES TO DETECT PATERNALLY TRANSMITTED CHROMOSOMAL ABERRATIONS

Francesco Marchetti and Andrew J. Wyrobek

Biology and Biotechnology Research Program
Lawrence Livermore National Laboratory
Livermore, CA 94550, USA

INTRODUCTION

Chromosomal abnormalities transmitted through gametes are associated with pregnancy loss, infant mortality, developmental and morphological defects, infertility and genetic diseases including cancer (Chandley, 1991; Wyrobek, 1993; Hassold et al., 1996; McFadden and Friedman, 1997; Wyrobek et al., 2000; Hassold and Hunt, 2001). These abnormalities are typically *de novo* events that originate in the germ cells of either parent. The relative parental contribution to transmitted chromosomal aberrations has been shown to depend on the specific chromosomes and defects that are involved. Autosomal aneuploidies are predominantly maternal in origin (Hassold et al., 1996; Hassold and Hunt, 2001), while sex chromosomal aneuploidies have a substantial paternal contribution (Hassold et al., 1993; Hassold, 1998). In addition, more than 80% of *de novo* structural chromosomal abnormalities among livebirths appear to be paternally derived (Olson and Magenis, 1988; Overhauser et al., 1990; Chandley, 1991; Dallapiccola et al., 1993; Cody et al., 1997). There is epidemiological evidence linking paternal occupational or environmental exposure to abnormal reproductive outcomes (Narod et al., 1988; Savitz and Chen, 1990; Olshan, 1995); however, despite the health risk to the developing embryo and offspring, little is known about the etiology of paternally-derived chromosomal abnormalities. Research with laboratory animals is essential for identifying and characterizing reproductive toxicants and for assessing the human risk of exposure to such agents.

MOUSE MODELS FOR INVESTIGATING PATERNALLY TRANSMITTED CHROMOSOMAL ABNORMALITIES

The mouse is an important animal model in reproductive toxicology, and it has been widely used to investigate the mechanisms of paternally transmitted genetic damage

Advances in Male Mediated Developmental Toxicity, edited by Bernard Robaire and Barbara F. Hales.
Kluwer Academic/Plenum Publishers, 2003.

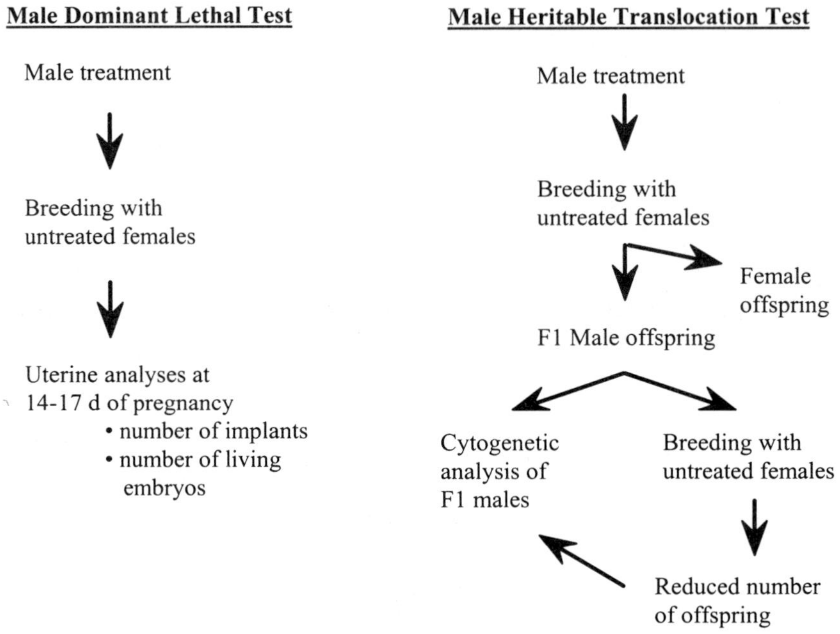

Male Dominant Lethal Test

Male treatment

↓

Breeding with
untreated females

↓

Uterine analyses at
14-17 d of pregnancy
• number of implants
• number of living
embryos

Male Heritable Translocation Test

Male treatment

↓

Breeding with
untreated females

F1 Male offspring

Female
offspring

Cytogenetic
analysis of
F1 males

Breeding with
untreated females

↓

Reduced number
of offspring

Figure 1. General experimental schemes of Male Dominant Lethal and Male Heritable Translocation tests.

(Preston, 1994). It is known that when male mice are treated with a germinal mutagen and mated with unexposed females, the deleterious effects on reproduction can be dramatic; these include embryonic lethality, heritable translocations, malformations and cancer in the offspring (Nomura, 1982; Kirk and Lyon, 1984; Nomura, 1988; Shelby, 1996). Two of the most common methods used for identifying and characterizing male mammalian germ cell mutagens have been the dominant lethal (DL) and the heritable translocation (HT) tests (Shelby et al., 1993). Figure 1 shows the basic protocols used for conducting these tests.

The DL test measures embryonic death of the progeny of treated males, while the HT test measures the induction of chromosomal rearrangements in the offspring of treated males. These tests can be used to determine the sensitivity of the various spermatogenic cell types to the induction of mutations by controlling the time between male exposure and mating. Matings within the first 3 weeks after exposure measure mutagenic effects on postmeiotic germ cells (spermatids), while matings occurring 3-5 weeks and more than 5 weeks after exposure measure effects on meiotic (spermatocytes) and mitotic (spermatogonia) cells, respectively.

Approximately 30 chemicals have been tested for the induction of DL and/or HT mutations in the mouse (Shelby, 1996). Figure 2 shows 11 chemicals that produced positive results in both tests. The majority of chemicals produced the highest DL response during, if not exclusively, the postmeiotic window of spermatogenesis. Two of the chemicals, melphalan and trophosphamide, also produced effects in meiotic cells. Interestingly, etoposide, a common chemotherapeutic agent, produced a distinctly different pattern of germ cell-stage sensitivity by affecting meiotic cells only. The high sensitivity of postmeiotic cells to chemical exposure is probably related to the diminished DNA repair capacity of spermatozoa and late spermatids, as compared to early spermatids and the other spermatogenic cell types (Sega, 1979). It is thus possible that damage induced in these postmeiotic cell types accumulates and is transmitted to offspring, while damage induced in earlier spermatogenic cell types is repaired.

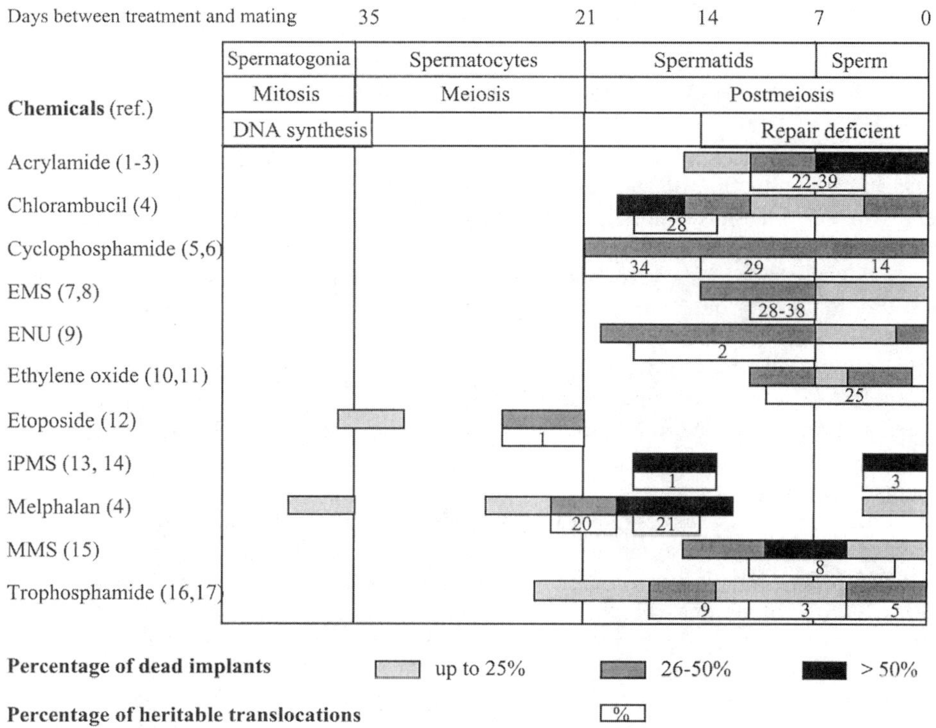

Figure 2. Dominant Lethal and Heritable Translocation data for 11 chemicals that have produced positive results in both tests. A scheme of mouse spermatogenesis with the time from fertilization (day 0) is shown at the top. For each chemical the mating intervals that produced significant increases with respect to control values are indicated. Color bars represent the amount of dominant lethality observed. Numbers in white bars report the percentage of translocation carriers in the offspring. Original data was presented in the following papers: 1) Shelby et al., 1986; 2) Shelby et al., 1987; 3) Adler et al., 1994; 4) Generoso et al., 1995; 5) Ehling and Neuhäuser-Klaus, 1988; 6) Sotomayor and Cumming, 1975; 7) Ehling et al., 1968; 8) Cattanach et al., 1968; 9) Generoso et al., 1984; 10) Generoso et al., 1986; 11) Generoso et al., 1990; 12) Shelby et al., 2001; 13) Ehling and Neuhäuser-Klaus, 1995; 14) Generoso et al., 1979; 15) Lang and Adler, 1977; 16) Ehling and Neuhäuser-Klaus, 1994; 17) Adler et al., 1994. Adapted from Marchetti et al. (2001a).

In addition to the spermatogenic pattern of sensitivity, the 11 chemicals of Figure 2 can be grouped based upon the rates at which DL and HT effects were induced. For the majority of the chemicals there was a close relationship between the rates of DL and of HT mutations, however, for 3 of them (ENU, etoposide and isopropyl methanesulphonate) a strong induction of DL was associated with a weak HT induction. The reasons for these differing responses are not fully understood. Our knowledge of the mechanisms involved in the conversion of chemically induced DNA lesions in male germ cells into chromosomal aberrations and their transmission to the offspring is very limited. There is evidence that the nature of mutations is dependent upon the germ cell stage rather than upon the chemical administered (Russell, 1994) and that there are differences in the ability of the eggs of different strains to repair the same type of sperm DNA lesion (Generoso et al., 1979).

However, because the DL test cannot provide information on the mechanism(s) responsible for the observed effects, more direct methods are required to understand the relationship between nature of the mutagen, germ cell stage exposed and its repair status and reproductive genetic outcome. The cytogenetic analysis of mouse zygotes is a method that can be used to achieve this goal.

CYTOGENETIC ANALYSIS OF MOUSE ZYGOTES

The cytogenetic analysis of mouse first-cleavage (1-Cl) zygotes has been used for measuring the induction of aneuploidy after treatment of female germ cells (Mailhes and Marchetti, 1994) and the induction of chromosomal structural aberrations after treatment of male germ cells (Tanaka et al., 1981; Albanese, 1982; Matsuda and Tobari, 1988; Matsuda et al., 1989; Matsuda et al., 1989; Pacchierotti et al., 1994). An advantage of this method is the ability to identify the parental origin of the abnormalities because paternal and maternal chromosomes do not join until metaphase of the first mitotic division (McGaughey and Chang, 1969) and because maternal chromosomes show a higher degree of condensation with respect to paternal chromosomes (Donahue, 1972). The development of chromosome composite probes for the mouse (Breneman et al., 1993; Boei et al., 1994; Breneman et al., 1995) has opened the possibility of using chromosome painting for the study of heritable chromosome aberrations. We have developed a methodology, which we called PAINT/DAPI analysis, that combines DAPI staining, to detect unstable aberrations such as dicentrics and acentric fragments, with chromosome-specific painting probes to detect stable aberrations such as translocations and insertions in mouse zygotes (Marchetti et al., 1996). Stable aberrations, which in general do not affect cell viability, may persist long after exposure in stem cells and are directly implicated in heritable mutagenesis and carcinogenesis (Mitelman et al., 1991).

In the remainder of the chapter, we describe the PAINT/DAPI analysis, the various endpoints that can be measured with it, and the progress in our studies of whether chromosomal aberrations in mouse 1-Cl zygotes are quantitatively and temporally predictive of the risk for spontaneous abortions and birth of offspring with chromosomal abnormalities.

PAINT/DAPI ANALYSIS OF MOUSE ZYGOTES

The basic protocol for performing the PAINT/DAPI analysis of mouse zygotes can be divided in 4 parts: 1) animal treatment and zygote collection; 2) slide scanning; 3) Fluorescence In Situ Hybridization (FISH); and 4) PAINT/DAPI analysis.

Animal Treatment and Zygote Collection

Various mouse strains have been used in the past for obtaining 1-Cl zygote metaphases, including inbred and outbred strains, hybrids and strains with chromosomal markers (Mailhes and Marchetti, 1994). In our laboratory, we generally use B6C3F1 mice because the males are good breeders and the females respond well to superovulation. The animals are usually between 8 and 12 weeks of age and maintained under a 14 hr light/10 hr dark photoperiod (light from 6.00 am to 8:00 pm) at room temperature (21-23° C) with a relative humidity of $50 \pm 5\%$. Food and water are provided *ad libitum*.

To collect 1-Cl zygotes, a group of twelve males is treated with a test agent and used for breeding with untreated females at various times from treatment to sample sperm that were at different phases of spermatogenesis at the moment of treatment (Figure 2). Males are kept in single cages throughout the duration of the experiment. Additional groups of twelve males are used until a sufficient number of 1-Cl zygote metaphases are collected (see below).

Females are allowed to acclimate to the light/dark photoperiod for at least one week before they are used for superovulation. Two days before mating, females are induced to superovulate by an intraperitoneal (i.p.) injection of 7.5 I.U. of pregnant mare's serum. On the day of mating they receive an i.p. injection of 5.0 I.U. of human chorionic gonadotrophin (HCG) to induce ovulation. Immediately after the HCG injection (10:00 am), females are caged with males (1:1) and checked for vaginal plugs 8 hr later. This is done to ensure that sperm are present in the female reproductive tract when the eggs are ovulated (~12 h after HCG) and to synchronize zygotic development among the various females. Mated females receive an i.p. injection of 0.08 mg of colchicine 24 hr after HCG to prevent the union of the two parental pronuclei and arrest zygote development at the metaphase of the first cleavage division.

Mated females are euthanized by CO_2 inhalation 6 hr after colchicine injection, i.e., 30 hr after HCG (the timing of collection may vary depending on the strain of mice used). Zygotes are harvested from all mated females, pooled and processed according to the mass harvest procedure (Mailhes and Yuan, 1987). Briefly, zygotes are treated with 150 I.U./ml hyaluronidase in Hank's balanced salt solution (HBSS) for 15 min at room temperature to remove residual cumulus cells still surrounding the zygotes and to weaken the zona pellucida. Zygotes are then washed twice in HBSS, and transferred to a hypotonic solution of 0.3% sodium citrate for 30 min at room temperature. They are then gradually fixed in microcentrifuge tubes with 3:1 (methanol:acetic acid) fixative and finally with 1:1 fixative before being dropped onto a wet, cold slide. One slide per experiment containing 50-100 1-Cl metaphases is typically made. Particular care is taken to ensure that all the cells are dropped within an area that can be covered by a 22x22 mm coverslip. Each slide is analyzed twice: first without staining to measure pre- and post-fertilization toxicity, and second after FISH to detect chromosomal abnormalities.

Slide Scanning

To assess pre- and post-fertilization toxicity, each slide is scanned with a 20x objective under phase contrast and each cell is classified into the following groups according to its appearance:

(1) unfertilized oocytes - oocytes with meiotic chromosomes or degenerating chromatin and without a sperm head or tail;

(2) developmentally arrested zygotes - zygotes showing female meiotic chromosomes and a sperm head or tail, or occasionally male meiotic chromosomes;

(3) degenerated zygotes – zygotes with degenerating chromatin and a sperm head or tail, or fragmented pronuclei;

(4) pronuclei – zygotes with two well defined pronuclei showing the difference in size between paternal (larger) and maternal (smaller) pronuclei;

(5) zygotes – zygotes with mitotic chromosomes, which are further categorized in diploid zygotes (containing ~40 chromosomes), polyploid zygotes (containing ~60 chromosomes) and haploid zygotes (containing ~20 chromosomes).

Examples of these cells have been published elsewhere (Marchetti et al., 1992; Marchetti and Mailhes, 1994). The data collected during this analysis are used for determining the fertilization rate (sum of groups 2, 3, 4, and 5 divided by the total number of cells), the rate of zygotic development (group 5 divided by the sum of groups 2, 3, 4 and 5) and the metaphase rate (group 5 divided by the total number of cells). These endpoints are used to assess treatment-induced toxic effects on male germ cells. A reduction in the metaphase rate in the treated group with respect to controls is an indication that the chemical has produced pre- and/or post-fertilization toxicity. Pre-fertilization toxicity manifests itself as a reduction in the number of fertilized eggs and is an indication of impaired sperm physiology. Post-fertilization toxicity is manifested as a reduction in the number fertilized eggs that reach the metaphase stage of the first cleavage division and is an indication that paternal exposure affected pronuclear formation and the proper progression of the first cell cycle of development.

During the slide scanning, the coordinates of each 1-Cl metaphase amenable to cytogenetic analysis are recorded. These coordinates are used to relocate the metaphases after FISH. After scanning has been completed, the slides are kept in a nitrogen atmosphere at -20°C until hybridization.

Fluorescence In Situ Hybridization (FISH)

The protocol for performing FISH in mouse zygotes has been described in detail elsewhere (Marchetti et al., 1996; Marchetti et al., 1997; Marchetti et al., 2001). Here, our discussion is limited to factors that may influence the quality of hybridization.

It is our experience that performing FISH in zygotes is more difficult than in somatic cells. Probes that hybridize well in somatic cells may not work in zygotes. This is probably due to quantitative and qualitative differences in the cytoplasm of the two cell types. The zygotic cytoplasm is much larger in volume than that of a typical somatic cell, and it is extremely rich in mRNAs and nutrients that must sustain the early stages of embryonic development. Both these factors can affect FISH. It is extremely important that as much cytoplasm as possible is removed during the fixation process since there is an inverse correlation between the amount of cytoplasm in the cytogenetic preparations and the quality of FISH. In addition, longer hybridization times (at least 36 hr) rather than short times (i.e., overnight) are preferred, as are probes labeled with fluorochromes whose signals can be amplified.

PAINT/DAPI Analysis

PAINT analysis. Each metaphase is relocated using the coordinates obtained during the slide scan. A double-band filter is used to detect chromosome aberrations involving only painted chromosomes. These are identified by the presence of color junctions, i.e., locations along a chromosome where one color ends and another color begins. A triple-bandpass filter, which allows the observation of bright DAPI staining of the heterochromatic region near the centromere, is used to discriminate between translocations, dicentrics and acentric fragments. Scoring of these aberrations is done using the PAINT nomenclature system (Tucker et al., 1995). Briefly, each aberrant chromosome is classified according to the type of aberration (translocation, dicentric, insertion, etc.). This is followed by parentheses enclosing letters that describe the aberration in detail. Uppercase letters denote the presence of the centromeric region, while lowercase letters denote the absence of the centromeric region. The letter A refers to an unpainted chromosome region; additional letters (B, C, etc.) are assigned to the different fluorochromes beginning with the one that paints the greater fraction of the genome. For example, if biotin-labeled probes cover 20% of the genome, and digoxigenin-labeled probes cover 15% of the genome, the letter B is assigned to the former and the letter C to the latter. In this context, t(Ab) indicates a chromosome with a piece of unpainted chromosome containing the centromeric region associated with a piece of a chromosome labeled with biotin without the centromeric region.

The objective of the PAINT analysis is to score as many metaphases as required to analyze at least 100 cell-equivalents. This is dependent on the percentage of chromosomal exchanges that can be detected with the probe combination used. For example, a painting probe combination that allows the detection of ~50% of all chromosomal exchanges will require the analysis of at least 200 metaphases to reach 100 cell-equivalents. The initial probe combination used for the validation studies contained four composite probes specific for chromosomes 1, 2, 3 and X labeled with biotin and a probe specific for chromosome Y labeled with digoxigenin (Marchetti et al., 1996). This probe combination detects ~37% of all possible exchanges. We have recently developed a new probe combination that allows the detection of ~60% of all chromosomal exchanges. The new PAINT/DAPI assay uses a combination of five composite probes specific for chromosomes 1, 3, 5, X and Y, each labeled with FITC, plus five composite probes specific for chromosomes 2, 4, 6, X and Y labeled with biotin. With this new probe combination only 170 metaphases are needed to reach the goal of 100 cell-equivalents thereby reducing the number of animals needed.

DAPI analysis. Once all chromosomal aberrations involving painted chromosomes are characterized, a DAPI filter is used to detect chromosome aberrations regardless of whether they involve painted or unpainted chromosomes. The brighter DAPI stain of the centromeric heterochromatin region is used to identify the centromere of each chromosome. Chromosomes with a bright DAPI band at each end are classified as dicentrics. Chromosomes lacking a bright DAPI band are classified as acentric fragments. This analysis, referred to as DAPI analysis, gives information similar to that obtained by C-banding (Arrighi and Hsu, 1971).

VALIDATION OF PAINT/DAPI ASSAY

We validated the PAINT/DAPI assay for the detection of stable and unstable chromosomal aberrations in 1-Cl zygotes after paternal treatment with either acrylamide (Marchetti et al., 1997) or etoposide (Marchetti et al., 2001b). These two chemicals were selected because of their different patterns of spermatogenic sensitivity and mechanisms of action. The objectives were to determine whether PAINT/DAPI analysis provided reliable estimates of DL and HT frequencies compared with results obtained using standard breeding methods.

Acrylamide (AA) is extensively used in the paper industry, waste water treatment plants and in research laboratories (Dearfield et al., 1995). It is an alkylating agent that binds mostly to protamines (Sega et al., 1989), basic proteins that replace histones during mid-spermiogenesis and are the predominant nuclear proteins during the late postmeiotic stages (Meistrich et al., 1978). Consistent with this proposed mechanism of action, AA induced the highest DL response in late spermatids and early sperm (Shelby et al., 1986). AA also induced high frequencies of translocation carriers (Shelby et al., 1987; Adler, 1990; Adler et al., 1994) with a DL/HT ratio of ~2, that is, there were ~2 embryos with dominant lethality for every reciprocal translocation at birth.

Etoposide (ET) is a topoisomerase II inhibitor widely used in cancer chemotherapy (Liu, 1989; Smith, 1990). Topoisomerase II (topo II) enzymes function by transiently introducing DNA double strand breaks, allowing the passage of one double helix through another, and resealing the double strand break (Wang et al., 1990). Topo II activity is needed for removing regions of DNA catenation during DNA replication and prior to chromosome segregation (DiNardo et al., 1984; Rose et al., 1990; Downes et al., 1991), and for chromosome condensation (Adachi et al., 1991). ET inhibits topo II activity by forming a ternary complex, DNA-topo II-ET, that prevents the ligation of the double strand breaks (Smith, 1990; Anderson and Berger, 1994; Ferguson and Baguley, 1994; Ferguson and Baguley, 1996). ET induced DL only in early and late meiotic stages and was a weak inducer of heritable translocations with a DL/HT ratio of ~19 (Shelby et al., 2001).

In our studies, male mice were treated either with 5 consecutive daily doses of 50 mg/kg AA dissolved in distilled water or with a single dose of 80 mg/kg ET dissolved in dimethyl sulphoxide. Males were mated with untreated females at intervals chosen to allow comparisons with published DL and HT data (Figure 2). Concurrent controls were treated with the matching solvent. 1-Cl metaphases were analyzed using a painting probe combination that detected ~37% of all possible chromosomal exchanges (Marchetti et al., 1996).

Pre- and post-fertilization toxicities

Figure 3 shows the effects of paternal exposure to AA and ET on fertilization rate and on proper development to the 1-Cl metaphase stage. The number of 1-Cl metaphases was significantly reduced (P<0.001 vs. concurrent controls, Chi-square) at all time points after paternal exposure to AA. Both fertilization rate and zygotic development (i.e., number of fertilized eggs that reached the metaphase stage) were affected. Conversely, ET treatment of male germ cells reduced the number of 1-Cl metaphases (P<0.01 vs. concurrent controls, Chi-square) only after exposure of meiotic cells (24.5 days). This was due mostly to a reduction in the number of eggs that were fertilized. As shown in Table 1, the majority of unfertilized eggs were represented by oocytes with degenerating chromatin and no recognizable chromosomes. The remaining unfertilized eggs were represented by oocytes arrested at the metaphase stage of the second metiotic division (only one oocyte arrested at the metaphase stage of the first meiotic division was found).

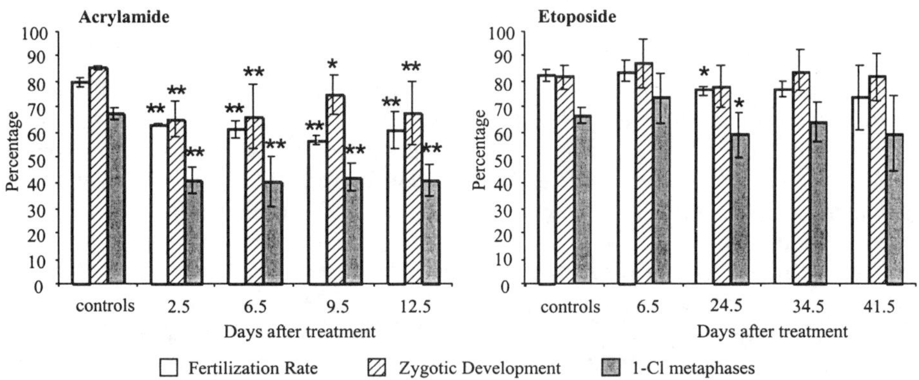

Figure 3. Pre- and post-fertilization toxicities induced by paternal treatment with acrylamide or etoposide. Bars represent the standard error of the mean. *P<0.01 vs. concurrent controls. **P<0.001 vs. concurrent controls.

Among the fertilized eggs, the majority of those that did not reach the metaphase stage consisted of eggs in which pronuclear formation did not take place (Table 1). Some of these had no identifiable pronuclei or chromosomes, however, they contained at least a sperm tail, which is known to be incorporated in the mouse egg during fertilization (Simerly et al., 1993). Others contained a female complement of meitoic chromosomes associated with a sperm head in various stages of decondensation, a sperm tail or, occasionally, a group of male-derived chromosomes that did not undergo DNA synthesis (Clarke and Masui, 1986). As shown in Table 1 and Figure 3, AA increased the number of eggs that failed to form pronuclei at all time points, while ET did so only at the time point with the highest level of chromosomal aberrations (24.5 days). It is possible that these cells represent eggs that were fertilized by highly damaged sperm. Following fertilization a series of events takes place in the egg that leads to the formation of male and female pronuclei. These events include completion of the second meiotic division and formation of the female pronucleus for the egg, and decondensation of the male chromatin followed by pronuclear formation for the sperm (Garagna and Redi, 1988; Nonchev and Tsanev, 1990; Newport and Dunphy, 1992; Perreault, 1992). Both gametes must undergo these changes in a coordinated fashion to insure normal zygotic development. If the timing of these critical events is disrupted and one pronucleus does not form, the development of the other pronucleus will also be affected (Perreault, 1990).

Table 1. Frequencies and types of eggs recovered after paternal exposure to 5x50 mg/kg acrylamide or 80 mg/kg etoposide.

| Chemical[1] | Total | Unfertilized eggs[2] | | Fertilized eggs[3] | | | |
	eggs	Meiotic oocytes	Degen[4] oocytes	Meiotic oocytes	Degen[4] oocytes	Pronuclei	Zygotes
Acrylamide							
controls	680	6 (4.2)	136 (95.8)	3 (0.6)	38 (7.1)	18 (3.3)	479 (89.0)
2.5	479	73 (42.2)	104 (58.3)	5 (1.7)	53 (17.6)[5]	34 (11.2)[5]	210 (69.5)[6]
6.5	822	65 (20.4)	254 (79.6)	7 (1.4)	102 (20.4)[5]	45 (9.0)[5]	349 (69.4)[6]
9.5	712	62 (20.2)	245 (79.8)	14 (3.4)	68 (16.8)[5]	8 (2.0)	315 (77.8)[6]
12.5	690	61 (22.5)	210 (77.5)	4 (0.9)	99 (23.7)[5]	22 (5.2)	294 (70.2)[6]
Etoposide							
controls	698	18 (13.8)	112 (86.2)	3 (0.5)	75 (13.2)	15 (2.7)	475 (83.6)
6.5	627	6 (5.7)	99 (94.3)	4 (0.8)	48 (9.2)	14 (2.7)	456 (87.3)
24.5	835	18 (8.8)	187 (91.2)	3 (0.5)	115 (18.2)[5]	17 (2.7)	495 (78.6)
34.5	318	3 (4.0)	73 (96.0)	2 (0.8)	30 (12.4)	6 (2.5)	204 (84.3)
41.5	321	27 (30.7)	61 (69.3)	3 (1.3)	29 (12.4)	10 (4.3)	191 (82.0)

[1]Days from exposure.
[2]Numbers of cells and percentages among unfertilized eggs are shown.
[3]Number of cells and percentages among fertilized eggs are shown.
[4]Degenerated oocytes. Eggs with degenerating chromatin were classified as unfertilized if no sperm head or tail was present, and fertilized if sperm head or tail was present. Degenerated fertilized eggs included also cells with fragmented or morphologically abnormal pronuclei.
[5]P<0.05 vs. concurrent controls (Chi-square).
[6]P<0.001 vs. concurrent controls (Chi-square).

Zygotes with two well-defined pronuclei represented ~3% of control zygotes. These cells do not necessarily represent damaged or delayed cells. In fact, because fertilization occurs over a period of ~8 hr (Donahue, 1972; Marchetti and Mailhes, 1994), pronuclear zygotes may have originated from oocytes that were fertilized toward the end of the fertilization period. ET did not increase the frequencies of zygotes at the pronuclear stage; however, there was a significant increase in the percentages of zygotes at the pronuclear stage after treatment of epididymal (2.5 d) and early spermatozoa (6.5 d) with AA. This suggests that AA may have induced cell-cycle delay in some of the zygotes, similar to the effect found after AA treatment of bone marrow cells (Gassner and Adler, 1996).

As shown in Table 2, the majority of zygotes that reached the metaphase stage were diploid. Neither polyploid nor haploid zygotes were increased by paternal exposure to AA and ET. This is because polyploid zygotes generally originate from fertilization of a diploid egg while haploid zygotes are typically an artifact of the mass harvest procedure, i.e., zygotes in which one group of chromosomes was lost during the fixation process.

Because it is not always possible to identify the parental origin of the single pronucleus, only diploid and polyploid zygotes were analyzed for the presence of chromosomal aberrations. The results of the specific types of chromosomal aberrations detected by PAINT/DAPI analysis of these metaphases have been presented in detail elsewhere (Marchetti et al., 1997; Marchetti et al., 2001b).

Table 2. Frequencies and ploidy status of first-cleavage (1-Cl) zygotes recovered after paternal exposure to 5x50 mg/kg acrylamide or 80 mg/kg etoposide.

Chemical[1]	1-Cl zygotes[2]		
	Total	Haploid (~20 chrs.)	Diploid (~40 chrs.)	Diploid (~60 chrs.)
Acrylamide				
controls	479	20 (4.2)	449 (93.7)	10 (2.1)
2.5	210	13 (6.2)	193 (91.9)	4 (1.9)
6.5	349	16 (4.6)	331 (94.8)	2 (0.6)
9.5	315	13 (4.1)	299 (94.9)	3 (1.0)
12.5	294	11 (3.7)	279 (94.9)	4 (1.4)
Etoposide				
controls	475	16 (3.4)	452 (95.1)	7 (1.5)
6.5	456	3 (0.6)	452 (99.1)	1 (0.2)
24.5	495	9 (1.8)	484 (97.8)	2 (0.4)
34.5	204	3 (1.5)	200 (98.0)	1 (0.5)
41.5	191	3 (1.6)	187 (97.9)	1 (0.5)

[1]Days from exposure.

[2]Numbers of cells and percentages are shown.

Predictive value of PAINT/DAPI analysis for dominant lethality and heritable translocations

The main objective of these studies was to determine whether PAINT/DAPI analysis of chromosomal aberrations at 1-Cl could be used to predict subsequent embryonic development. As shown in Figure 4, for both AA and ET, the number of zygotes with unstable aberrations provided estimates of dead implants that agreed both in magnitude and in kinetics with the results obtained in the DL tests. Additionally, PAINT/DAPI analysis provided cytogenetic data for the mechanism leading to DL. While AA-induced dominant lethality was due mainly to chromosome structural aberrations, ET induced dominant lethality by a dual mechanism: chromosome structural aberrations and aneuploidy. To date, ET is the only chemical for which paternal exposure leads to an increased incidence of aneuploidy in the offspring (Marchetti et al., 2001b). The finding of the aneugenic activity of ET in male germ cells has been confirmed in cancer patients treated with ET (De Mas et al., 2001).

As shown in Figure 5, there was also agreement between the frequencies of zygotes with stable aberrations and the frequencies of offspring with heritable translocations reported using the standard HT method (28% versus 22% for AA; 0% vs 1.4% for ET). These results suggest that essentially all of the reciprocal translocations induced by paternal exposure in zygotes are compatible with embryonic survival.

Finally, PAINT/DAPI analysis has provided an explanation for the differing DL/HT ratios produced by the two chemicals. A significant proportion of AA-induced chromosomal aberrations consisted of chromosomal exchanges (Marchetti et al., 1997). Some of these exchanges were unstable (i.e., dicentrics) resulting in dominant lethality, and some

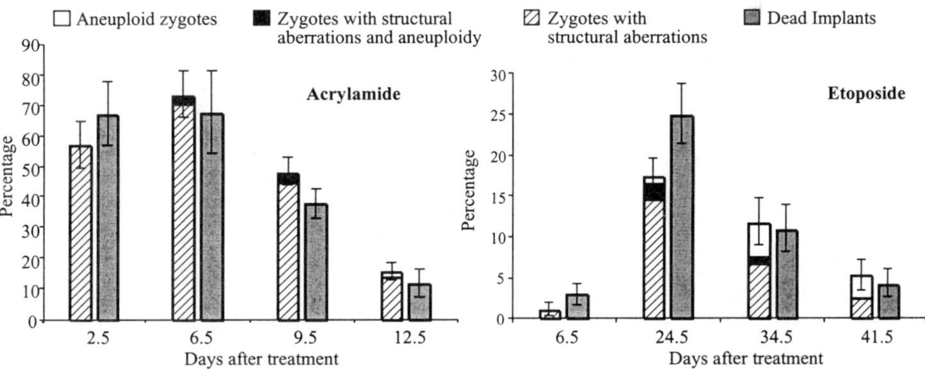

Figure 4. Comparison of the percentage of zygotes with chromosomal abnormalities that are expected to result in embryonic lethality vs. the percentages of dead implants. Bars represent the standard error of the sample. Adapted from Marchetti et al. (1997; 2001b).

were stable (translocations) resulting in reciprocal translocations. Conversely, ET induced mostly chromosome fragmentation and deletions (Marchetti et al., 2001). These types of aberrations are expected to result in embryonic lethality, because they represent the loss of genetic material. Chromosomal exchanges after paternal exposure to ET, which would produce reciprocal translocations, were very rare.

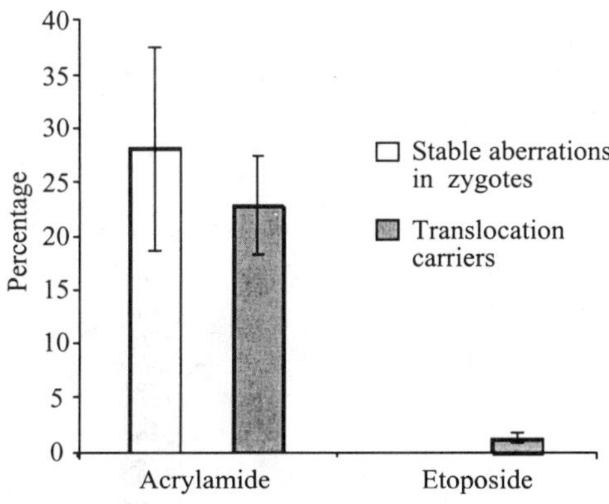

Figure 5. Comparison between the frequencies of zygotes with reciprocal translocations vs. the frequencies of translocation carriers in the offspring. Bars represent the standard errors of the samples.

CONCLUSIONS

Our validation studies indicate that cytogenetic abnormalities at 1-Cl are critical intermediates between paternal exposure and abnormal reproductive outcome and that PAINT/DAPI analysis of mouse zygotes may predict future embryonic development.

Potential advantages of PAINT/DAPI analysis of zygotes for assessing abnormal embryonic outcomes would be the simultaneous collection of data predictive of DL and HT outcomes, the elucidation of the cytogenetic mechanisms responsible for the observed outcomes, and also a dramatic reduction in the number of animals required for conducting the tests. Additional experiments in zygotes are being conducted to determine whether the correlation between unstable aberrations and dominant lethality and stable aberrations and heritable translocations holds true for other chemicals with differing patterns of germ cell specificity and differing DL and HT outcomes.

In conclusion, the results presented here show that PAINT/DAPI analysis is a powerful tool for investigating the correlation between chemical exposure, maternal and paternal germ cell genotoxicity and embryonic outcomes.

ACKNOWLEDGMENTS

We are indebted to Dr. Jack B. Bishop for his support of the work described in this book chapter. We also thank Drs. Lidia Cosentino, Matt Coleman, Eddi Sloter and Lisa Tomascik-Cheeseman for helpful comments and discussions. This work was done under the auspices of the U.S. Department of Energy by the University of California, Lawrence Livermore National Laboratory, contract W-7405-ENG-48, with funding support from NIEHS Interagency Agreement Y01-ES-10203-00.

REFERENCES

Adachi, Y., Luke, M. and Laemmli, U.K., 1991, Chromosome assembly in vitro: topoisomerase II is required for condensation. *Cell.* 643:137-148.

Adler, I.D., 1990, Clastogenic effects of acrylamide in different germ-cell stages of male mice, in: *Banbury Report 34: Biology of Mammalian Germ Cell Mutagenesis*, Cold Spring Harbor, Cold Spring Harbor Laboratory Press, pp. 115-131.

Adler, I.D., Reitmeir, P., Schmoller, R. and Schriever-Schwemmer, G., 1994, Dose response for heritable translocations induced by acrylamide in spermatids of mice. *Mutat Res.* 309:285-291.

Adler, I.D., Schriever-Schwemmer, G. and Kliesch, U., 1994, Clastogenicity of trophosphamide in somatic and germinal cells of mice. *Mutat Res.* 307:237-243.

Albanese, R., 1982, The use of fertilized mouse eggs in detecting potential clastogens. *Mutat Res.* 97:315-326.

Anderson, R.D. and Berger, N.A., 1994, Mutagenicity and carcinogenicity of topoisomerase-interactive agents. *Mutat Res.* 309:109-142.

Arrighi, F.E. and Hsu, T.C., 1971, Localization of heterochromatin in human chromosomes. *Cytogenetics.* 10:81-86.

Boei, J.J.W.A., Balajee, A.S., de Boer, P., Rens, W., Aten, J.A., Mullenders, L.H. and Natarajan, A.T., 1994, Construction of mouse chromosome-specific DNA libraries and their use for the detection of X-ray-induced aberrations. *Int J Rad Biol.* 65:583-590.

Breneman, J.W., Ramsey, M.J., Lee, D.A., Eveleth, G.G., Minkler, J.L. and Tucker, J.D., 1993, The development of chromosome-specific composite probes for the mouse and their application to chromosome painting. *Chromosoma.* 102:591-598.

Breneman, J.W., Swiger, R.R., Ramsey, M.J., Minkler, J.L., Eveleth, G.G., Langlois, R.A. and Tucker, J.D., 1995, The development of painting probes for dual-color and multiple chromosome analysis in the mouse, *Cytogenet. Cell Genet.* 68:197-202.

Cattanach, B.M., Pollard, C.E. and Isaacson, J.H., 1968, Ethyl methanesulphonate-induced chromosome breakage in the mouse. *Mutat Res.* 6:297-307.

Chandley, A.C., 1991, On the paternal origin of de novo mutation in men. *J Med Genet.* 28:217-223.

Clarke, H.J. and Masui, Y., 1986, Transformation of sperm nuclei to metaphase chromosomes in the cytoplasm of maturing oocytes of the mouse. *J Cell Biol.* 102:1039-1046.

Cody, J.D., Pierce, J.F., Brkanac, Z., Plaetke, R., Ghidoni, P.D., Kaye, C. and Leach, R.J., 1997, Preferential loss of the paternal alleles in the 18q- syndrome. *Am J Med Genet.* 69:280-286.

Dallapiccola, B., Mandich, P., Bellone, E., Selicorni, A., Mokin, V., Ajmar, F. and Novelli, G., 1993, Parental origin of chromosome 4p deletion in Wolf-Hirschhorn syndrome. *Am J Med Genet.* 47:921-924.

De Mas, P., Daudin, M., Vincent, M.C., Bourrouillou, G., Calvas, P., Mieusset, R. and Bujan, L., 2001, Increased aneuploidy in spermatozoa from testicular tumor patients after chemotherapy with cisplatin, etoposide and bleomycin. *Hum Reprod.* 16:1204-1208.

Dearfield, K.L., Douglas, G.R., Ehling, U.H., Moore, M.M., Sega, G.A. and Brusick, D.J., 1995, Acrylamide: a review of its genotoxicity and an assessment of heritable genetic risk. *Mutat Res.* 330:71-99.

DiNardo, S., Voelkel, K. and Sternglanz, R., 1984, DNA topoisomerase II mutant of *Saccharomices cerevisiae*: topoisomerase II is required for segregation of daughter molecules at the termination of DNA replication. *Proc Natl Acad Sci USA.* 81:2616-2620.

Donahue, R.P., 1972, Cytogenetic analysis of the first cleavage division in mouse embryos. *Proc Natl Acad Sci USA.* 69:74-77.

Downes, C.S., Mullinger, A.M. and Johnson, R., 1991, Inhibitors of DNA topoisomerase II prevent chromatid separation in mammalian cells but do not prevent from exit from meiosis. *Proc Natl Acad Sci USA.* 88:8895-8899.

Ehling, U.H., Cumming, R.B. and Malling, H.V., 1968, Induction of dominant lethal mutations by alkylating agents in male mice. *Mutat Res.* 5:417-428.

Ehling, U.H. and Neuhäuser-Klaus, A., 1988, Induction of specific-locus and dominant-lethal mutations by cyclophosphamide and combined cyclophosphamide-radiation treatment in male mice. *Mutat Res.* 199:21-30.

Ehling, U.H. and Neuhäuser-Klaus, A., 1994, Induction of specific-locus and dominant lethal mutations in male mice by trophosphamide. *Mutat Res.* 307:229-236.

Ehling, U.H. and Neuhäuser-Klaus, A., 1995, Induction of specific-locus and dominant lethal mutations in male mice by *n*-propyl and isopropyl methanesulphonate. *Mutat Res.* 328:73-82.

Ferguson, L.R. and Baguley, B.C., 1994, Topoisomerase II enzymes and mutagenicity. *Environ Mol Mutagen.* 24:245-261.

Ferguson, L.R. and Baguley, B.C., 1996, Mutagenicity of anticancer drugs that inhibit topoisomerase enzymes. *Mutat Res.* 355:91-101.

Garagna, S. and Redi, C., 1988, Chromatin topology during the transformation of the mouse sperm nucleus into pronucleus in vivo. *J Exp Zool.* 246:187-193.

Gassner, P. and Adler, I.D., 1996, Induction of hypoploidy and cell cycle delay by acrylamide in somatic and germinal cells of male mice. *Mutat Res.* 367:195-202.

Generoso, W.M., Cain, K.T., Cornett, C.C. and Cacheiro, N.L.A., 1984, DNA target sites associated with chemical induction of dominant-lethal mutations and heritable translocations in mice, in: *Genetics: New Frontiers,* V.L. Chopra, B.C. Joshi, R.P. Sharma and H.C. Bansal, eds., New Delhi, Oxford and IBH, 1: 347-356.

Generoso, W.M., Cain, K.T., Cornett, C.V., Cacheiro, N.L.A. and Hughes, L.A., 1990, Concentration-response curves for ethylene-oxide-induced heritable translocations and dominant lethal mutations. *Environ Mol Mutagen.* 16:126-131.

Generoso, W.M., Cain, K.T., Hughes, L.A., Sega, G.A., Braden, P.W., Gosslee, D.G. and Shelby, M.D., 1986, Ethylene oxide dose and dose-rate effects in the mouse dominant-lethal test. *Environ Mutagen.* 8:1-7.

Generoso, W.M., Cain, K.T., Krishna, M. and Huff, S.W., 1979, Genetic lesions induced by chemicals in spermatozoa and spermatids of mice are repaired in the egg. *Proc Natl Acad Sci USA.* 76:435-437.

Generoso, W.M., Witt, K.L., Cain, K.T., Hughes, L., Cacheiro, N.L., Lockhart, A.M. and Shelby, M.D., 1995, Dominant lethal and heritable translocation tests with chlorambucil and melphalan in male mice. *Mutat Res.* 345:167-180.

Hassold, T., Abruzzo, M., Adkins, K., Griffin, D., Merrill, M., Millie, E., Saker, D., Shen, J. and Zaragoza, M., 1996, Human aneuploidy: incidence, origin, and etiology. *Environ Mol Mutagen.* 28:167-175.

Hassold, T. and Hunt, P., 2001, To err (meiotically) is human: the genesis of human aneuploidy. *Nat Rev Genet.* 2:280-291.

Hassold, T., Hunt, P.A. and Sherman, S., 1993, Trisomy in humans: incidence, origin and etiology. *Curr Opin Genet Develop.* 3:398-403.

Hassold, T.J., 1998, Nondisjunction in the human male. *Curr Top Develop Biol.* 37:383-406.

Kirk, K. M. and Lyon, M. F., 1984, Induction of congenital malformations in the offspring of male mice treated with X-rays at pre-meiotic and post-meiotic stages. *Mutat Res.* 125: 75-85.

Lang, R. and Adler, I.D., 1977, Heritable translocation test and dominant lethal-assay in male mice treated with methyl methanesulphonate. *Mutat Res.* 48:75-88.

Liu, L.F., 1989, DNA topoisomerase poisons as antitumor drugs. *Annu Rev Biochem.* 58:351-375.

Mailhes, J.B. and Marchetti, F., 1994, Chemically-induced aneuploidy in mammalian oocytes. *Mutat Res.* 320:87-111.

Mailhes, J.B. and Yuan, Z.P., 1987, Cytogenetic technique for mouse metaphase II oocytes. *Gamete Res.* 18:77-83.

Marchetti, F., Bishop, J.B., Lowe, X., Generoso, W.M., Hozier, J. and Wyrobek, A.J., 2001b, Etoposide induces heritable chromosomal aberrations and aneuploidy during male meiosis in the mouse. *Proc Natl Acad Sci USA.* 98:3952-3957.

Marchetti, F., Lowe, X., Bishop, J. and Wyrobek, A.J., 1997, Induction of chromosomal aberrations in mouse zygotes by acrylamide treatment of male germ cells and their correlation with dominant lethality and heritable translocations. *Environ Mol Mutagen.* 30:410-417.

Marchetti, F., Lowe, X., Moore, D. II, Bishop, J. and Wyrobek, A.J., 1996, Paternally inherited chromosomal structural aberrations detected in mouse first-cleavage zygote metaphases by multicolor fluorescence in situ hybridization painting. *Chromosome Res.* 4:604-613.

Marchetti, F. and Mailhes, J.B., 1994, Variation of mouse oocyte sensitivity to griseofulvin-induced aneuploidy and meiotic delay during the first meiotic division. *Environ Mol Mutagen.* 23:179-185.

Marchetti, F., Sloter, E. and Wyrobek, A. J., 2001a, Advances in understanding paternally transmitted chromosomal abnormalities, in: *Teplice program: Impact of air pollution on human health,* R. J. Sram, ed., Prague, Academia, pp. 193-205.

Marchetti, F., Tiveron, C., Bassani, B. and Pacchierotti, F., 1992, Griseofulvin-induced aneuploidy and meiotic delay in female mouse germ cells. II. Cytogenetic analysis of one-cell zygotes. *Mutat Res.* 266:151-162.

Matsuda, Y., Seki, N., Utsugi-Takeuchi, T. and Tobari, I., 1989, Change in X-ray sensitivity of mouse eggs from fertilization to the early pronuclear stage, and their repair capacity. *Int J Rad Biol.* 55:233-256.

Matsuda, Y., Seki, N., Utsugi-Takeuchi, T. and Tobari, I., 1989, X-Ray- and mitomycin C (MMC)-induced chromosome aberrations in spermiogenic germ cells and the repair capacity of mouse eggs for the X-ray and MMC damage. *Mutat Res.* 211:65-75.

Matsuda, Y. and Tobari, I., 1988, Chromosomal analysis in mouse eggs fertilized in vitro with sperm exposed to ultraviolet light (UV) and methyl and ethyl methanesulphonate (MMS and EMS). *Mutat Res.* 198:131-144.

McFadden, D. and Friedman, J., 1997, Chromosome abnormalities in human beings. *Mutat Res.* 396:129-140.

McGaughey, R.W. and Chang, M.C., 1969, Meiosis of mouse egg before and after sperm penetration. *J Exp Zool.* 170:397-410.

Meistrich, M.L., Brock, W.A., Grimes, S.R., Platz, R.D. and Hnilica, L.S., 1978, Nucleoprotein transitions during spermatogenesis. *FASEB J.* 37:2522-2525.

Mitelman, F., Kaneko, Y. and Trent, J., 1991, Report of the committee on chromsome changes in neoplasia. *Cytogenet Cell Genet.* 58:1053-1079.

Narod, S.A., Douglas, G.R., Nestmann, E.R. and Blackey, D.H., 1988, Human mutagens: evidence from paternal exposure. *Environ Mol Mutagen.* 11:401-415.

Newport, J. and Dunphy, W., 1992, Characterization of the membrane binding and fusion events during nuclear envelope assembly using purified components. *J Cell Biol.* 116:295-306.

Nomura, T., 1982, Parental exposure to x rays and chemicals induces heritable tumours and anomalies in mice. *Nature.* 296:575-577.

Nomura, T., 1988, X-Ray- and chemically induced germ-line mutation causing phenotypical anomalies in mice. *Mutat Res.* 188:309-320.

Nonchev, S. and Tsanev, R., 1990, Protamine-histone replacement and DNA replication in the male mouse pronucleus. *Mol Reprod Dev.* 25:72-76.

Olshan, A. F., 1995, Lessons learned from epidemiologic studies of environmental exposure and genetic disease. *Environ Mol Mutagen.* 25, Suppl 26:74-80.

Olson, S.B. and Magenis, R.E., 1988, Preferential paternal origin of de novo structural chromosome rearrangements, in: *The Cytogenetics of Mammalian Autosomal Rearrangements,* A. Daniel, ed., New York, Liss, pp. 583-599.

Overhauser, J., McMahon, J., Oberlender, S., Carlin, M.E., Niebuhr, E., Wasmuth, J.J. and Lee-Chen, J., 1990, Parental origin of chromosome 5 deletions in the cri-du-chat syndrome. *Am J Med Genet.* 37:83-86.

Pacchierotti, F., Tiveron, C., D'Archivio, M., Bassani, B., Cordelli, E., Leter, G. and Spanò, M., 1994, Acrylamide-induced chromosomal damage in male mouse germ cells detected by cytogenetic analysis of one-cell zygotes. *Mutat Res.* 309:273-284.

Perreault, S.D., 1990, Regulation of sperm nuclear rectivation during fertilization, in: *Fertilization in Mammals,* B.D. Bavister, J. Cumming and E.R.S. Roldan, eds., Norwell, MA, Serono Symphosia, pp. 285-296.

Perreault, S.D., 1992, Chromatin remodeling in mammalian zygotes. *Mutat Res.* 296:43-55.

Preston, R.J., 1994, Future of germ cell cytogenetics. *Environ Mol Mutagen.* 23:54-58.

Rose, D., Thomas, W. and Holm, C., 1990, Segregation of recombined chromosomes in meiosis I requires DNA topoisomerase II. *Cell.* 60:1009-1017.

Russell, L., 1994, Effects of spermatogenic cell type on quantity and quality of mutations, in: *Male-Mediated Developmental Toxicity,* A. Olsham and D. Mattison, eds., New York, Plenum Press, pp. 37-48.

Savitz, D.A. and Chen, J., 1990, Paternal occupation and childhood cancer: review of epidemiologic studies. *Environm Health Perspect.* 88:325-358.

Sega, G.A., 1979, Unscheduled DNA synthesis (DNA repair) in the germ cells of male mice: its role in the study of mammalian mutagenesis. *Genetics.* 92:349-358.

Sega, G.A., Valdivia Alcota, R.P., Tancongco, C.P. and Brimer, P., 1989, Acrylamide binding to the DNA and protamine of spermiogenic stages of the mouse and its relationship to genetic damage. *Mutat Res.* 216:221-230.

Shelby, M.D., 1996, Selecting chemicals and assays for assessing mammalian germ cell mutagenicity. *Mutat Res.* 325:159-167.

Shelby, M.D., Bishop, J.B., Hughes, L., Morris, R.W. and Generoso, W.M., 2001, Dominant lethal and heritable translocation effects of etoposide in male mice, *in press.*

Shelby, M.D., Bishop, J.B., Mason, J. M. and Tindall, K.R., 1993, Fertility, reproduction, and genetic disease: studies on the mutagenic effects of environmental agents on mammalian germ cells. *Environm Health Perspect.* 100:283-291.

Shelby, M.D., Cain, K.T., Cornett, C.V. and Generoso, W.M., 1987, Acrylamide: induction of heritable translocations in male mice. *Environ Mol Mutagen.* 9:363-368.

Shelby, M.D., Cain, K.T., Hughes, L.A., Braden, P.W. and Generoso, W.M., 1986, Dominant lethal effects of acrylamide in male mice. *Mutat Res.* 173:35-40.

Simerly, C.R., Hecht, N.B., Goldberg, E. and Schatten, G., 1993, Tracing the incorporation of the sperm tail in the mouse zygote and early embryo using an anti-testicular alpha-tubulin antibody. *Dev Biol.* 158:536-548.

Smith, P.J., 1990, DNA topoisomerase dysfunction: a new goal for antitumor chemotherapy. *BioEssays.* 12:167-172.

Sotomayor, R.E. and Cumming, R.B., 1975, Induction of translocations by cyclophosphamide in different germ cell stages of male mice: cytological characterization and transmission. *Mutat Res.* 27:375-388.

Tanaka, N., Katoh, M. and Iwahar, S., 1981, Formation of chromosome-type aberrations at the first cleavage after MMS treatment in late spermatids of mice. *Cytogenet Cell Genet.* 31:145-152.

Tucker, J.D., Morgan, W.F., Aw, A.A., Bauchinger, M., Blakey, D., Cornforth, M.N., Littlefield, L.G., Natarajan, A.T. and Shasserre, C. 1995, A proposed system for scoring structural aberrations detected by chromosome painting, *Cytogenet. Cell Genet.* 68:211-221.

Wang, J.C., Caron, P.R. and Kim, R.A., 1990, The role of DNA topoisomerases in recombination and genome stability: a double-edged sword? *Cell.* 62:403-406.

Wyrobek, A.J., 1993, Methods and concepts in detecting abnormal reproductive outcomes of paternal origin. *Reprod Toxicol.* 7:3-16.

Wyrobek, A.J., Marchetti, F., Sloter, E. and Bishop, J., 2000, Chromosomally defective sperm and their developmental consequences, in: *Human Monitoring after Environmental and Occupational Exposure to Chemical and Physical Agents,* D. Anderson, A.E. Karakaya and R.J. Sram, eds., Amsterdam, IOS press. 313:134-150.

PATERNAL OCCUPATION AND CHILDHOOD CANCER

Andrew F. Olshan[1] and Edwin van Wijngaarden[1]

Department of Epidemiology
School of Public Health
University of North Carolina
Chapel Hill, NC 27599-7435

INTRODUCTION

Childhood cancer remains an important pediatric disease, not only in terms of disease burden, but also because the epidemiologic and molecular study of childhood cancers has provided a unique insight into the etiology of adult cancer. The two-hit mutational model developed in the 1970s by Knudson based on the descriptive epidemiology and genetics of childhood tumors such as retinoblastoma, Wilms tumor, and neuroblastoma provided an important model for mechanisms of adult cancers (Knudson et al., 1972; Knudson et al., 1975). The environmental (broadly defined) etiology of childhood cancer is uncertain, but paternal occupation has remained as one of the most suggestive associations for a number of years. In this review we will consider the current evidence for paternal occupational exposures and the risk of childhood cancer. We will focus on the promising leads from the most extensively studied childhood cancers, leukemia and brain tumors (central nervous system tumors), and provide directions for future research.

EPIDEMIOLOGY OF CHILDHOOD CANCER

Although childhood cancer is relatively rare, there are still 12,400 new cases and 2,500 deaths among persons less than 15 years of age each year in the United States (Ries et al., 1999). Childhood cancer is the fourth leading cause of death among children and is a significant contributor to years of life lost (Ries et al., 1999). Childhood cancer is not a single disease, but is comprised of diverse malignancies that vary by histology, site, demographic factors, and possibly other risk factors. The five most common childhood cancers are leukemia (31% of all cases under age 15 years), central nervous system tumors (20%), lymphomas and reticuloendothelial neoplasms (11%), sympathetic nervous system (mostly neuroblastoma, 7.8%), and renal tumors (mostly Wilms tumor, 6.3%) (Ries et al., 1999). The incidence of childhood cancer is generally higher for males than females and higher for whites than blacks, with rates among Asians intermediate (Ries et al., 1999).

Advances in Male Mediated Developmental Toxicity, edited by Bernard Robaire and Barbara F. Hales.
Kluwer Academic/Plenum Publishers, 2003.

Worldwide, there is some variability in incidence among some childhood cancers (Parkin et al., 1998). Since the mid-1970s, the incidence rates for some childhood cancers (e.g., brain tumors) have shown an increase, but the overall rate during the last decade has remained stable (Linet et al., 1999).

A number of environmental, lifestyle, and genetic risk factors for childhood cancer have been investigated (Little, 1999, Ries et al., 1999). There are several well-documented genetic conditions, birth defects, and syndromes associated with childhood cancer such as Down syndrome, neurofibromatosis, Bloom syndrome, ataxia telangiectasia, Li-Fraumeni syndrome, and Beckwith-Wiedemann syndrome. Reasonably strong evidence for an association has emerged for factors such as in utero ionizing radiation exposure (leukemia and brain tumors), immunodeficiency, and Epstein-Barr virus (lymphomas) (Ries et al., 1999). Other exposures or factors that have been more inconsistently associated with childhood cancer include high birth weight, maternal history of fetal loss, older maternal age at pregnancy, birth order, parental tobacco use, dietary factors, maternal alcohol use, residential electromagnetic fields, parental recreational drug use, residential pesticide use, products containing N-nitroso compounds, maternal hair dye use, coffee and tea consumption, and high maternal hormone use during pregnancy (Little, 1999; Ries et al., 1999).

POTENTIAL MECHANISMS FOR A MALE-MEDIATED EFFECT

There are several possible mechanisms to explain an epidemiologic association between paternal occupational exposure and cancer in offspring (Olshan and Faustman, 1993). These include: 1) direct effects on the male germ cells resulting in heritable genetic damage, 2) direct effects on male germ cells via epigenetic changes that alter gene function, 3) indirect effects by the transportation of chemical to the mother in seminal fluid, and 4) indirect effects by means of household contamination with substances brought home on the father's clothing. A number of animal studies have provided evidence of a male-mediated effect on cancer in offspring for agents such as ionizing radiation, urethane, and ethylnitrosourea (ENU) (Tomatis et al., 1992). It should be noted that the tumors found in offspring are often those found to be spontaneous in the specific species or strain (lung, lymphoid, liver of mice). In addition, the tumors are analogous to those seen in adult humans based on age and histology, not those seen in childhood (Anderson et al., 2000). Nonetheless, the totality of the animal evidence indicates that parental male exposures can induce neoplasia in offspring. These transgenerational effects may be due to direct germ cell mutation, although epigenetic alterations of gene expression may be another possibility (Anderson et al., 1994).

With respect to indirect mechanisms, there have been several studies demonstrating that fathers can bring home a variety of substances that lead to measurable household contamination (Olshan and Faustman, 1993). Studies of semen transfer indicate that drugs can be transmitted in seminal fluid, but typically at levels that would be expected to have no effect. One animal study did report an increase in fetal loss via seminal fluid transfer of cyclophosphamide (Hales et al., 1986). There is no evidence that these indirect mechanisms can increase the risk of childhood cancers.

EXPOSURE ASSESSMENT IN COMMUNITY-BASED CASE-CONTROL STUDIES

Because of its critical importance in interpreting epidemiologic evidence, prior to discussion of the study findings on paternal occupation and the risk of cancer among offspring we will describe the sources and methods of collecting and assessing occupational exposures in case-control studies. Because of the rarity of childhood cancers, most epidemiologic studies have been of the case-control design. We will also consider the strengths and limitations of the various approaches.

Background. The value and validity of an occupational epidemiology study is judged to a large extent by the quality of the exposure assessment (Checkoway et al., 1996). The ideal exposure assessment would include detailed work history from company records, industrial hygiene data, and personal monitoring which is sometimes available in occupational cohort studies. Nevertheless, community-based case-control studies usually involve a wide variety of workforces, work environments, and agents for each person longitudinally, as well as a wide spectrum of these work-related factors cross-sectionally. The main disadvantage of exposure data in these types of epidemiological studies is the lack of personal monitoring or industrial hygiene data measured in the study population.

Most previous studies of childhood cancer have inferred paternal occupational exposures based on (a) self report of broad exposure groups using checklists; (b) job titles or industries that are presumed to involve certain exposures; and (c) generic job exposure matrices (Tielemans et al., 1999). The approach of using general and job specific questionnaires combined with an evaluation by trained experts has been less commonly used. Each of these methods is briefly described in the remainder of this section.

Self Report. Probably the most simple and inexpensive method of assessing exposure is based on self reports, usually by means of a checklist (Tielemans et al., 1999). Subjects indicate whether they were exposed to any of the agents mentioned on a list. Although workers may know broad categories of exposure (e.g., paints, degreasers, solvents, metals, cutting oils) (McGuire et al., 1998), they are not likely to know the names of individual chemical substances (Teschke et al., 1994). Another potential problem of particular concern when relying on self report is whether parents of cancer cases will be motivated (by guilt or concern) to differentially report more adverse exposures than parents of controls. However, studies of the theoretical impact of such recall bias have shown that even severe recall bias would cause only weak to moderate spurious associations (Little, 1999).

Job Title or Industry. Job title or industry are often used as surrogates for workplace exposure to potentially hazardous agents based on information obtained from death certificates, medical records, disease registries, or questionnaires. Examples of exposure surrogates used in occupational investigations include the longest job held during the career, the usual industry, or the last job. This method is relatively inexpensive, and may provide suggestive leads for further research. However, it has been long recognized that the use of data from occupation or industry does not take into account the variety of exposure levels and work environments within these categories (Stewart and Herrick, 1991), or the change in work practices and levels of exposure over time (McGuire et al., 1998).

Generic Job Exposure Matrix. Job exposure matrices, which are usually constructed a priori (Benke et al., 2001), have been developed in order to address some of the limitations related to the use of job titles alone. In its simplest form, a job exposure matrix is a cross-tabulation of a list of job titles with a list of agents to which workers carrying out the jobs may be exposed. Each cell of a job exposure matrix contains information on exposure to an agent within a category or job. This information on the job level can then be used to aggregate workers on the basis of common exposures. In more complex job exposure matrices, exposure may be inferred on the basis of job title, industry, and/or calendar times of employment (Louik et al., 2000). The measure of exposure may be either dichotomous (e.g., ever vs. never exposed), ordinal (e.g., low, medium, high), or expressed as probability (i.e., proportion of occupationally exposed subjects in the job category). Examples of such generic (i.e., a priori) job exposure matrices are those developed by Hoar et al. (1980), the National Institute of Occupational Safety and Health (NIOSH) (Sieber et al., 1991), and more recently by the Finnish Institute of Occupational Health (FIOH) (Kauppinen et al., 1998). The matrix approach is relatively inexpensive, and it overcomes the shortage of experts qualified in a wide range of occupational exposures (Bouyer and Hemon, 1993). However, it does not take into account exposure variability within jobs due to variation from worker to worker in specific tasks, processes, and technology, and variations across industries and countries (Gérin et al., 1985; Benke et al., 2001).

Expert Evaluation. The approach of occupational exposure assessment by trained experts using general and job specific questionnaires (i.e., interview/expert method) was first described by Siemiatycki and coworkers (Siemiatycki et al., 1982; Gérin et al., 1985). Briefly, it consists of an extensive interview (including general and job-specific questionnaires) aimed at providing detailed information on each job study subjects have held. The questions elicit details on the work environment and exposure determinants. Subsequently, trained experts (e.g., industrial hygienists, chemists, occupational physiccians) examine each questionnaire in order to estimate levels (e.g., none, low, medium, high) of intensity, probability or frequency, and the expert's confidence in these estimates (Siemiatycki et al., 1997; Stewart and Stewart, 1994). Additional interviews with study subjects may follow this review to clarify reported information. Measures of intra- and inter-rater reliability may be assigned (e.g., kappa and correlation coefficients) to describe the validity and reliability of the exposure assessment and to understand how measures of exposure can be improved (Stewart and Stewart, 1994). Although expert assessment of exposure may not be completely reproducible as exposure assessment judgments may vary among individual researchers (Katz et al., 1994), results of previous studies indicate that occupational data derived from expert assessment may yield considerable improvements upon the use of self-assessed exposure, job titles and generic job exposure matrices (Siemiatycki et al., 1989; Tielemans et al., 1999; Louik et al., 2000).

REVIEW OF EPIDEMIOLOGIC STUDIES

In this section we will summarize the published literature on paternal occupational exposures and childhood cancer. We will focus on leukemia (and lymphoma) and brain tumors, the two most commonly studied childhood cancer types. We will build upon two previous reviews by Savitz and Chen (1990) and Colt and Blair (1998). We will summarize the previous evidence presented up to 1997 by Colt and Blair and present the findings for studies published subsequently.

Leukemia and Lymphoma

There have been over 20 studies that have examined paternal occupation in relation to the risk of leukemia or lymphoma in the offspring (Table 1). Previous reviews have suggested associations for paternal exposure to solvents, paints, ionizing radiation, electromagnetic fields, exhaust hydrocarbons, and pesticides. Although the following sections have been divided into different exposures, it should be noted that a given exposure may overlap different occupations and other exposure groups. For example, painters are exposed to both paints and solvents and pesticides also contain solvents.

Table 1. Key findings from recent studies of childhood lymphohematopoietic cancers and paternal occupations - Not included in Colt and Blair (1998) or Zahm and Ward (1998)

Reference	Histology	Industry or occupation	Exposure	Time frame	Relative risk	95% CI	# exposed cases	Comment
Draper et al. (1997)	Leukemia and NHL	Radiation worker	Radiation	Preconception	1.8	1.1-3.0	40	
Fear et al. (1998)	Leukemia		Pesticides	At time of death	0.9	0.8-1.0	180	
Fear et al. (1999)	Leukemia		Social contact	Childhood:				
				Low	1.0	1.0-1.0	4062	
				Medium	1.0	1.0-1.1	1180	
				High	0.9	0.9-1.0	648	
Infante-Rivard and Sinnett (1999)	ALL		Pesticides	Preconception	1.6	1.0-2.4	66	
			Fungicides		5.1	1.5-17.8	15	
			Insecticides		1.4	0.9-2.2	50	
			Herbicides		2.1	0.9-4.6	19	
			Fertilizers		3.0	1.3-6.6	25	
Meinert et al. (1999)	Leukemia	Occupational exposure to ionizing radiation	Radiation	Preconception	1.8	0.7-4.6	16	
Roman et al. (1999)	Leukemia	Nuclear workers	Radiation	Preconception: Before				
				employment	1.0		8	
				<50 mSv	1.9	0.7-5.7	8	
				50-100 mSv	1.7	0.2-14.2	1	
				100+ mSv	5.8	1.3-24.8	3	
Shu et al. (1999)	ALL		Plastic materials	Preconception	1.4	1.0-1.9	119	
			Solvents	Any time	1.0	0.9-1.2	602	
			Paints or thinners	Any time	0.9	0.8-1.1	619	

Reference	Histology	Industry or occupation	Exposure	Time frame	Relative risk	95% CI	# exposed cases	Comment
Smulevich et al. (1999)	Leukemia		Ionizing radiation	Preconception	6.7	2.8-15.8	21	
			EMF		4.6	1.8-11.9	14	
			VDU		2.4	1.0-5.8	12	
Buckley et al. (2000)	NHL		Pesticides	Around pregnancy	1.7	0.8-3.7	21	For 'parent'
Heacock et al. (2000)	Leukemia	Sawmill workers	Chloro-phenate fungicides	Until diagnosis: High ($\geq$3560)	0.8	0.2-3.6	5	
Feychting et al. (2000)	Leukemia		EMF	Preconception:				
				$\leq$0.12 μT	1.0	-	25	
				0.13-0.21 μT	1.6	1.0-2.5	80	
				$\geq$0.22 μT	1.3	0.8-2.1	36	
				0.13-0.29 μT	1.4	0.9-2.2	92	
				$\geq$0.30 μT	2.0	1.1-3.5	24	
Meinert et al. (2000)	Leukemia		Herbicides, insecticides, and fungicides	Preconception	1.5	1.1-2.2	62	
				During pregnancy	1.6	1.1-2.3	57	
				Postnatal	1.3	0.9-1.9	49	
				Ever	1.6	1.1-2.3	68	
Schüz et al. (2000)	Leukemia		Solvents	Any time	1.0	0.8-1.3	151	
			Painters or lacquers		1.1	0.9-1.4	157	
Wen et al. (2000)	AML	Military service in Vietnam or Cambodia		Any time before diagnosis	1.7	1.0-2.9	40	
	Leukemia		Herbicides		1.3	0.9-2.0	49	
Feychting et al. (2001)	Leukemia	Sheet metal workers	Metal	Preconception	4.1	1.9-8.8	7	
		Wood work			2.2	1.3-3.8	14	
		Agricultural, Horticultural and Forestry mangement	Pesticides		1.1	0.5-2.7	5	
			Solvents		1.3	0.8-2.0	23	
			Benzene		1.2	0.4-3.9	3	
			Pesticides		0.9	0.4-2.2	5	

NHL: non Hodgkin's lymphoma; ALL: acute lymphocytic leukemia; AML: acute myeloid leukemia; HCs: hydrocarbons;
EMF: electromagnetic fields; VDU: visual display units.

Solvents. Paternal occupational exposure to solvents, in general, or specific solvents such as benzene, xylene, toluene, carbon tetrachloride, trichloroethylene (TCE), and chlorinated solvents have been implicated as risk factors for leukemia (acute lymphocytic leukemia - ALL and acute non-lymphocytic leukemia ANLL) and non Hodgkin's leukemia (NHL). Colt and Blair (1998) had noted that the five studies examining these exposures had all reported odds ratios greater than 2. Recent studies have not fully supported the consistency of the solvent association. A notable study is the North American case control study of ALL (Shu et al., 1999). This large (1,842 cases) study conducted telephone interviews with fathers about specific exposures in time windows around and during pregnancy and estimated duration of exposure. None of the 12 solvents analyzed had an odds ratio (OR) above 1.2; the odds ratio for any solvent exposure was 1.0 [95% confidence interval (CI)=0.9-1.2]. There was no trend of increasing risk with increasing duration of exposure. Exposure data obtained from a self-administered questionnaire used in three German case-control studies of childhood ALL were pooled (1138 cases) and no association with any solvent exposure was found (OR=1.0; CI=0.8-1.3; Schuz et al., 2000). A Swedish linkage study of 235,635 children (161 cases of leukemia) found a small increase in risk (OR=1.3; CI=0.8-1.9) for solvent exposure, as determined from job title and industry, obtained from linked census data (Feychting et al., 2001).

The pattern of elevated risk associated with solvent exposure present in the earlier studies is generally not apparent in more recent studies. Most notable is the Shu et al. (1999) study that was very large, obtained self-reported exposure data on specific solvents, and examined duration of exposure. As discussed in the previous section, self-reported exposure data using simple checklists is not considered an ideal exposure source in the absence of additional exposure determinant information and can lead to an underestimate of the true risk. Nonetheless, this study and the other recent studies suggest that solvents may not be as promising a lead as previously suspected.

Paints and Pigments. Several studies of childhood leukemia and lymphoma have reported associations with father's employment as a painter (Savitz and Chen, 1990; Colt and Blair, 1998). Few earlier studies further explored the painter association by considering paint and pigment exposure more broadly, although one study reported elevated risks of ALL for spray paints, dyes, pigments (Lowengart et al., 1987). The North American ALL study (Shu et al., 1999) and the pooled German analysis (Schuz et al., 2000) reported odds ratios around 1.0 for paint exposure. As with solvent exposure, the new studies, particularly Shu et al., cast some doubt on the paint and pigment relationship. However, the consistency of the earlier reports of the painter association indicates that additional study of paints and other painter exposures such as solvents warrant consideration.

Exhaust Hydrocarbons. More than a dozen studies have examined the risk of childhood leukemia or lymphoma among fathers with exposure to potential exhaust hydrocarbons. Vehicle mechanics are exposed to exhaust hydrocarbons as well as other exposures such as solvents. The Shu et al., (1999) study reported a decreased risk of ALL for exhaust exposure any time during pregnancy (OR=0.6; CI=0.4-0.9). Interestingly, fuels exposure was associated with an elevated odds ratio (OR=1.5; CI=1.0-2.4). The Swedish study found an odds ratio of 0.8 (CI=0.5-1.6) for combustion products (Feychting et al., 2001). The preponderance of evidence still suggests an increased risk associated with paternal employment in vehicle-related occupations. Potential exposure to exhaust hydrocarbons may explain some of this risk, but the more recent studies cast some doubt on this explanation.

Pesticides. Recent reviews by Daniels et al. (1997) and Zahm and Ward (1998) have noted in twelve studies an elevated risk of childhood leukemia associated with potential paternal occupational exposure to pesticides. Most of these studies defined exposure simply as employment as a farmer or in the agricultural industry. A few studies did ask specifically about pesticides and examined duration and/or intensity of exposure. Those studies tended to report larger effect estimates with longer duration of exposure. In a smaller number of studies of non-Hodgkin's lymphoma (NHL), associations with farming were reported (Kristensen et al., 1996). Recent studies have also added support to the association between paternal occupational pesticide exposure and risk of leukemia or NHL. Buckley et al., analyzed data from the parents of 268 NHL cases and matched controls and found an elevated odds ratio (1.7) for parental occupational exposure to pesticides, although the father's exposure was not analyzed separately (2000). The Swedish cohort study did not report an increase in risk (RR=0.9; CI=0.4-2.2) based on only 5 exposed cases (Feychting et al., 2001). The German case-control study of leukemia (1,184 cases) and NHL (234 cases) found odds ratios of 1.6 (CI=1.1-2.3) for leukemia and 1.9 (CI=0.9-3.7) for lymphoma with paternal occupational pesticide exposure (Meinert et al., 2000). Presumed pesticide exposure was determined from job title and industry and a pesticide checklist (Olshan and Daniels, 2000). Other recent studies have also reported increased risks for leukemia after paternal exposure to different pesticides (Infante-Rivard and Sinnett, 1999; Heacock et al., 2000).

Radiation. Much of the interest in the possible relationship between childhood cancer and paternal occupational exposure to ionizing radiation stems from the 1990 report by Gardner (Gardner et al., 1990). His study suggested an increased risk of leukemia among the offspring of men employed at a nuclear fuel reprocessing plant in England. Subsequent case-control studies in Great Britain and Canada did not support this association (McLaughlin et al., 1993; Anderson et al., 2000). Several investigators also proposed alternative hypotheses involving population mixture and an infectious origin to explain the nuclear plant clusters (Kinlen, 1988). A record linkage study of United Kingdom child-hood cancer registration records and data on workers included in the National Registry for Radiation Workers found an odds ratio of 1.8 (CI=1.1-3.0) for leukemia and Non-Hodgkin's lymphoma combined (Draper et al., 1997). A recent cohort study (Roman et al., 1999) of children of nuclear industry employees reported an increased rate of leukemia among fathers with estimated cumulative external doses of > 100 mSv in the six months prior to conception (RR=7.7; CI=1.9-31.0). Interpretation is limited because of the small number of exposed cases (3) and the overlap with the Gardner study population. Recent case-control studies of paternal occupational exposure to radiation and childhood leukemia from Germany and Russia have also reported elevated effect estimates (Meinert et al., 1999; Smulevich et al., 1999). The German study found an odds ratio of 1.8 (CI=0.7-4.6) for jobs where fathers reported dosimetric monitoring (Meinert et al., 1999). There have been few studies of paternal occupational exposure to magnetic field exposures and risk of childhood leukemia or lymphoma (Feychting et al., 2000).

Brain Cancer

There have been nearly 30 studies that have examined paternal occupation in relation to the risk of brain cancer in the offspring (Table 2). Previous reviews found suggestive evidence for paternal exposure to paints and pigments, pesticides, (exhaust) hydrocarbons, and electro-magnetic fields, but considered the evidence strongest for exposure to paints and pesticides (Colt and Blair, 1998; Zahm and Ward, 1998).

Table 2. Key findings from recent studies of childhood brain cancer and paternal occupations - Not included in Colt and Blair (1998) or Zahm and Ward (1998)

Reference	Histology	Industry or occupation	Exposure	Time frame	Relative risk	95% CI	# exposed cases
Cordier et al. (1997)	All Brain	Agriculture	Pesticides	Before birth	2.2	1.0-4.7	16
		Printing	HCs		1.8	0.7-4.4	9
		Motor-vehicle related	HCs		1.6	1.0-2.8	29
		Painter	HCs		1.3	0.5-3.4	8
		Electrical work	EMF		0.8	0.4-1.6	37
	PNET		Solvents		1.2	0.7-1.9	16
		Motor-vehicle related			2.7	1.1-6.6	NR
Fear et al. (1998)	Brain		Pesticides	At time of death	0.8	0.7-1.0	109
McKean-Cowdin et al. (1998)	All Brain	Chemicals-petroleum	HCs	Before birth	1.7	0.9-2.9	28
		Electrical workers	EMF		2.3	1.3-4.0	34
		Agriculture-farming	Pesticides		1.2	0.7-2.1	22
Smulevich et al. (1999)	Brain		Oil products	Preconception	2.5	1.2-5.1	33
Heacock et al. (2000)	Brain	Sawmill workers	Chloro-phenate fungicides	Until diagnosis: High (≥3560)	1.5	0.4-6.9	5
Feychting et al. (2000)	Brain		EMF	Preconception:			
				≤0.12 µT	1.0	-	41
				0.13-0.21 µT	0.8	0.5-1.2	67
				≥0.22 µT	0.7	0.5-1.1	33
				0.13-0.29 µT	0.8	0.6-1.2	90
				≥0.30 µT	0.5	0.3-1.0	10
Feychting et al. (2001)	Nervous system		Pesticides	Preconception	2.4	1.3-4.4	11
		Painting work	Solvents		3.7	1.7-7.8	7
		Mechanical engineers and technicians			1.9	1.0-3.6	11
		Agricultural, horticultural, forestry, livestock work	Pesticides		2.1	1.1-4.2	9
		Firefighters			5.9	1.9-18.5	3
			Solvents		1.2	0.7-1.9	19

PNET: primitive neuroectodermal tumors; HCs: hydrocarbons; EMF: electromagnetic fields.

Paints and Pigments. Several studies of childhood brain cancer have reported associations with paternal occupations involving paints and inks, such as painters and printing workers (Savitz and Chen, 1990; Colt and Blair, 1998). Four out of five studies reviewed by Colt and Blair had reported relative risks greater than 2 (Peters et al., 1981; Hemminki et al., 1981; Johnson et al., 1987; Kuijten et al., 1992). Recent studies, however, have been inconsistent. The Swedish cohort study reported a relative risk of 3.7 (CI=1.7-7.8) with painting work before birth (Feychting et al.,2001). On the other hand, a three-center case-control study of childhood brain cancer in Europe (251 cases) did not find an association with paternal work as a painter (OR=1.3; CI=0.5-3.4) based on in-person interviews (Cordier et al., 1997). Nonetheless, an elevated risk of 1.8 (CI=0.7-4.4) was found for printing work.

The pattern of elevated risk associated with exposure to paints and pigments present in the earlier studies is supported to some extent by the more recent studies. Although the Swedish study found an elevated risk for painters, they did not observe an association with exposure to solvents (RR=1.2; CI=0.7-1.9). This observation is consistent with the Cordier et al. (1997) study which did not find solvent exposure associated with brain cancer risk (OR=1.2; CI=0.7-1.9), although an elevated risk was found for printing workers. In conclusion, recent studies were of similar or better quality as compared to the earlier studies and provide additional suggestive evidence for an elevated risk associated with exposure to paints and pigments. However, the inconsistencies with the results observed for solvents should be explored.

Pesticides. Paternal occupational exposure to pesticides as a risk factor for childhood nervous system tumors has been examined in 11 studies. Zahm and Ward (1998) included seven studies of paternal exposure in their review, of which four reported elevated risks associated with work in the agricultural sector (Wilkins and Koutras, 1988, Wilkins and Sinks, 1990; Kuijten et al., 1992; Kristensen et al., 1996). Additional evidence was provided by the Swedish study (Feychting et al., 2001), which reported increased risks for pesticide exposure (RR=2.4; CI=1.3-4.4) and agricultural, horticultural and forestry management (RR=2.1; CI=1.1-4.2). An increased risk in relation to employment in agriculture (RR=2.2; CI=1.0-4.7) was observed in another European study (Cordier et al., 1997). A cohort study of male British Columbian sawmill workers found an increased but imprecise risk (OR=1.5; CI=0.4-6.9) in relation to high chlorophenate fungicide exposure (Heacock et al., 2000). In contrast, a United States childhood brain cancer case-control study (540 cases) collected exposure information based on self-reported industry of employment and job tasks and found no association with employment in agriculture (OR=1.2; CI=0.7-2.1; McKean-Cowdin et al., 1998). In addition, a proportionate mortality study (109 cases) in the United Kingdom found a slightly decreased risk of brain cancer (PMR=0.8; CI=0.7-1.0) in relation to potential exposure to pesticides based on occupations reported on death certificates (Fear et al., 1998).

The evidence for an association between occupational pesticide exposure and childhood brain cancer remains inconclusive after consideration of the more recent studies. However, most notable is the Swedish study which used industrial hygiene expert assessment to estimate the probability of pesticide exposure. This study may be considered of better quality compared to previous studies that relied mostly on job title or industry, and provides suggestive leads for further research on pesticides and brain cancer.

Hydrocarbons. Colt and Blair (1998) reviewed eight studies on paternal hydrocarbon exposure and childhood nervous system tumors, and found some suggestive evidence for an association in half of these studies (Fabia and Thuy, 1974; Gold et al., 1982; Nasca et al., 1988). Industries or occupations with potential exposure to hydrocarbons include motor vehicle mechanics, painters, printers, machinists, and aircraft industry workers.

Some additional support for an association was found by recent studies in the United States (McKean-Cowdin et al., 1998), Europe (Cordier et al., 1997), and Russia (Smulevich, 1999). McKean-Cowdin and coworkers (1998) found the highest risk of brain cancer with paternal employment in the chemical-petroleum industry during pregnancy (OR=2.7; CI=1.3-5.8). Cordier and colleagues (1997) found an increased risk associated with motor vehicle-related work, either defined by industrial activity (OR=2.0; CI=0.7-5.5) or occupational codes (OR=1.6; CI=1.0-2.8). As mentioned previously, an elevated risk for printing work was also observed. Smulevich et al. reported an increased odds ratio of 2.5 (CI=1.2-5.1) for paternal exposure to oil products.

In conclusion, recent studies seem to support the association between childhood brain cancer and father's hydrocarbon related work. However, most of these studies were based on job title or industry that could only be considered a crude proxy for hydrocarbon exposure. In this context, it is interesting to note that the Swedish study did not find an increased risk with benzene exposure. More work needs to be done to fully characterize individual exposures and work practices among hydrocarbon related occupations, and evaluate specific hydrocarbon exposures (e.g., toluene, xylene, benzene) across a wide spectrum of occupations.

Electromagnetic Fields. Several associations between childhood brain cancer and paternal occupations involving electromagnetic field (EMF) exposure have been reported. Colt and Blair (1998) had noted that the six childhood brain cancer studies examining EMF exposures had all elevated relative risks in relation to "electrical work" (including welding) (Wilkins and Koutras, 1988; Nasca et al., 1988; Johnson and Spitz, 1989; Wilkins et al., 1991; Kuijten et al., 1992, Wilkins and Wellage, 1996). However, two of the three recently conducted studies could not replicate these findings. Feychting and coworkers (2000), using the Swedish linkage study, obtained quantitative measures of EMF exposure based on a measurement survey in a large number of occupations held by a sample of the general male population (Floderus et al., 1996). A decreased risk of brain cancer was found for paternal EMF exposure above 0.3 µT (RR = 0.5; 95% CI 0.3-1.0). Cordier et al. (1997) did not find elevated risks associated with electrical work (OR=0.8; CI=0.4-1.6). However, McKean-Cowdin et al. (1998) found an association with father's occupation as an electrical worker, with an odds ratio of 2.3 (CI=1.3-4.0) for all histology types combined.

These results indicate that the pattern of elevated risk associated with EMF exposure present in the earlier studies is generally not apparent in more recent studies. Most importantly, the Swedish cohort study used high quality exposure data and evaluated dose-response patterns (Feychting et al., 2000). This study suggests that EMF exposure may not be as important a risk factor as had been previously suspected.

DISCUSSION

Although some recent epidemiologic evidence has generated additional inconsistency, there remain suggestive findings for paternal occupational exposure to solvents, paints, pesticides, and radiation and an increased risk of childhood leukemia and brain cancer among offspring. The assessment of this evidence poses a significant challenge. There are several threats to study validity that can complicate the casual assessment of epidemiologic evidence. These include: selection bias, confounding, and misclassification of exposure. Furthermore, low study power can result in imprecise effect estimates adding to the difficulty. Many of the earlier studies of paternal occupation and childhood cancer were small, did not collect data on potentially confounding factors, and were not population-based. However, some recent studies of leukemia (Shu et al., 1999) and brain tumors (McKean-Cowdin et al., 1998) have overcome these limitations. For example, the

Children's Cancer Group study of acute lymphocytic leukemia included 1,842 cases and extensive interviews with the mother and father (Shu et al., 1999). Unfortunately, there have not been a large number of multiple studies with these superior design features, but these studies remain influential and instructive.

As discussed in a previous section, exposure misclassification is a serious concern. The vast majority of studies published to date have not employed the most advanced methods to collect and assess occupational exposures in case-control studies. Most have relied upon simple job title and industry and/or broad self-reported exposure checklists. These exposure sources are clearly subject to serious error in exposure classification. The central issue is the prediction of the direction of the error, that is, does the misclassification bias the effect estimates away from or towards the null (odds ratio or risk ratio of 1.0)? For some exposure surrogates, such as job title and for well-designed structured interview instruments, one would typically expect a non-differential bias, that is, the reported effect estimates may be *underestimates* of the actual risk. Nonetheless, there is a legitimate concern about recall bias (differential misclassification), whereby parents of affected children may be more motivated than control parents and either over-report exposures or report more accurately. This differential recall can lead to overestimates of the effect. Theoretical work has suggested that strong recall bias would typically produce only weak to moderate spurious effects (Little, 1999). A recent study of potential recall bias in studies of childhood leukemia reported that, for exposures such as proximity to power lines that result from special public concern, parental recall can be differential but otherwise is most often nondifferential (Infante-Rivard et al., 2000).

FUTURE STUDIES

There are a number of directions that future research can take to address the issues raised in the previous section. Major improvements in exposure assessment are needed. Several recommendations for improvements in data collection and assessment methods were previously described. The isolation of etiologic agents with greater specificity and confidence should be a major goal of future epidemiologic studies. Molecular epidemiologic approaches may help clarify associations. Investigation of interactions between exposures and genetic factors such as polymorphisms of carcinogen metabolizing enzymes and DNA repair genes may identify subgroups of individuals that may be more susceptible to the effects of occupational exposures. This approach has recently provided interesting results on the association between pesticides and childhood leukemia (Infante-Rivard et al., 1999). Another approach to identifying etiologic pathways is to define subgroups of cases based upon molecular characteristics of their tumors, such as alterations of oncogenes and tumor suppressor genes, and examine the relative effects of exposures within these subgroups (for example, MLL-rearranged leukemia).

There also needs to be a greater emphasis on experimental models. A wider array of substances encountered in the occupational situation needs to be evaluated in animal test systems for paternal transgenerational carcinogenesis. Epidemiologic studies can inform the animal research by providing some prioritization to the chemical testing inventory. One example is the epidemiologic findings related to paternal occupational metal exposure and the resulting evaluation in a mouse system (Anderson et al., 1994). In addition to assessing the potential for a given agent to induce a male-mediated effect, such studies can also provide additional information on mechanisms; the aforementioned study of metals pointed out the potential for epigenetic mechanisms of transgenerational carcinogenesis.

CONCLUSIONS

So, how far are we from coming to a consensus regarding the causal relationship between paternal occupational exposures and childhood cancer?   The cumulative epidemiologic evidence has provided some useful suggestive leads, especially for leukemia and brain tumors.   However, for the reasons noted previously, the evidence is far from consistent and compelling.   It cannot be claimed that there is a "smoking gun".   Years of additional epidemiologic study may not be warranted.   However, targeted epidemiologic studies that incorporate markers to define biologic subgroups and improved exposure assessment may provide a degree of clarity.   Nonetheless, the standard epidemiologic approach is not likely to yield final answers.   Animal studies, although not directly analogous with respect to the development of childhood tumors may provide critical evidence for describing the potential toxicity of a given agent, the synergism between agents, and mechanisms of a paternal pathway.   These data will help to assess biologic plausibility and provide a framework to consider results of epidemiologic studies.

REFERENCES

Anderson, L.M., Kasprzak, K.S., Rice, J.M., 1994, Preconception exposure of males and neoplasia in their progeny: Effects of metals and consideration of mechanisms. In: Male-Mediated Developmental Toxicity (Olshan AF, Mattinson DR, eds). New York:Plenum Press, 129-140.

Anderson, L.M., Diwan, D.A., Fear, N.T., Roman, E., 2000, Critical windows of exposure for children's health: cancer in human epidemiological studies and neoplasms in experimental animal models. *Environ Health Perspect.* 108 Suppl 3:573-594.

Benke, G., Sim, M., Fritschi, L., Aldred, G., Forbes, A., Kauppinen, T., 2001, Comparison of occupational exposure using three different methods: Hygiene panel, job exposure matrix (JEM), and self reports. *Appl Occup Environ Hyg.* 16:84-91.

Bouyer, J., Hémon, D., 1993, Retrospective evaluation of occupational exposures in population-based case-control studies: General overview with special attention to job exposure matrices. *Int J Epidemiol.* 22(Suppl. 2):S57-S64.

Buckley, J., Meadows, A., Kadin, M., Le Beau, M., Siegel, S., Robinson, L., 2000, Pesticide Exposures in Children with Non-Hodgkin Lymphoma. *Cancer.* 89: 2315-2321.

Checkoway, H., Seixas, N.S., Demers, P.A., 1996, The influence of occupational exposure assessment on epidemiological inferences. *Occup Hyg.* 3:7-21.

Colt, J.S., Blair, A., 1998, Parental occupational exposures and risk of childhood cancer. *Environ Health Perspect.* 106(Suppl 3):909-925.

Cordier, S., Lefeuvre, B., Filippini, G., Peris-Bonet, R., Farinotti, M., Lovicu, G., Mandereau, L., 1997, Parental occupation, occupational exposure to solvents and polycyclic aromatic hydrocarbons and risk of childhood brain tumors (Italy, France, Spain). *Cancer Causes Control.* 8:688-697.

Daniels, J.L., Olshan, A.F., Savitz, D.A., 1997, Pesticides and childhood cancers. *Environ Health Perspect.* 105:1068-1077.

Draper, G., Little, M., Sorahan, T., Kinlen, L., Bunch, K., Conquest, A., Kendall, G., Kneale, G., Lancashire, R., Muirhead, C., O'Connor, C., Vincent, T., 1997, Cancer in the offspring of radiation workers: a record linkage study. *BMJ.* 315:1181-1188.

Fabia, J., Thuy, T.D., 1974, Occupation of father at time of birth of children dying of malignant diseases. *Br J Prev Soc Med.* 28:98-100.

Fear, N.T., Roman, E., Reeves, G., Pannett, B., 1998, Childhood cancer and paternal employment in agriculture: the role of pesticides. *Br J Cancer.* 77:825-829.

Feychting, M., Floderus, B., Ahlbom, A., 2000, Parental occupational exposure to magnetic fields and childhood cancer (Sweden). *Cancer Causes Control.* 11:151-156.

Feychting, M., Plato, N., Nise, G., Ahlbom, A., 2001, Paternal occupational exposures and childhood cancer. *Environ Health Perspect.* 109:193-196.

Floderus, B., Persson, T., Stenlund, C., 1996, Magnetic-field exposures in the workplace: reference distribution and exposures in occupational groups. *Int J Occup Environ Health.* 2:226-238.

Gardner, M., Snee, M., Hall, A., Powell, C., Downes, S., Terrell, J., 1990, Results of case-control study of leukaemia and lymphoma among young people near Sellafield nuclear plant in West Cumbria. *BMJ.* 300: 423-429.

Gérin, M., Siemiatycki, J., Kemper, H., Bégin, D., 1985, Obtaining occupational exposure histories in epidemiologic case-control studies. *J Occup Med.* 27:420-426.

Gold, E.B., Diener, M.D., Szklo, M., 1982, Parental occupations and cancer in children. *J Occup Med.* 24:578-584.

Hales, B.F., Smith, S., Robaire, B., 1986, Cyclophosphamide in the seminal fluid of treated males: transmission to females by mating and effect on pregnancy outcome. *Toxicol Appl Pharmacol.* 84:423-430.

Heacock, H., Hertzman, C., Demers, P.A., Gallagher, R., Hogg, R.S., Teschke, K., Hershler, R., Bajdik, C.D., Dimich-Ward, H., Marion, S.A., Ostry, A., Kelly, S., 2000, Childhood cancer in the offspring of male sawmill workers occupationally exposed to chlorophenate fungicides. *Environ Health Perspect.* 108:499-503.

Hemminki, K., Saloniemi, S., Salonen, T., Partanen, T., Vainio, H., 1981, Childhood cancer and parental occupation in Finland. *J Epidemiol Comm Health.* 35:11-15.

Hoar, S.K., Morrison, A.S., Cole, P., Silverman, D.T., 1980, On occupation and exposure linkage system for the study of occupational carcinogenesis. *J Occup Med.* 22:722-726.

Infante-Rivard, C., Jacques, L., 2000, Empirical study of parental recall bias. *Am J Epidemiol.* 152:480-486.

Infante-Rivard, C., Labuda, D., Krajinovic, M., Sinnett, D., 1999, Risk of childhood leukemia associated with exposure to pesticides and with gene polymorphisms. *Epidemiology.* 10:481-487.

Johnson, C.C., Annegers, J.F., Frankowski, R.F., Spitz, M.R., Buffler, P.A., 1987, Childhood nervous system tumors – an evaluation of the association with paternal occupational exposure to hydrocarbons. *Am J Epidemiol.* 126:605-613.

Katz, E.A., Shaw, G.M., Schaffer, D.M., 1994, Exposure assessment in epidemiologic studies of birth defects by industrial hygiene review of maternal interviews. *Am J Ind Med.* 26:1-11.

Kauppinen, T., Toikkanen, J., Pukkala, E., 1998, From cross-tabulations to multipurpose exposure information systems: A new job-exposure matrix. *Am J Ind Med.* 33:409-417.

Kinlen, L., 1988, Evidence for an infective cause of childhood leukaemia: Comparison of a Scottish new town with nuclear reprocessing sites in Britain. *Lancet.* 2:1323-1327.

Knudson, A.G., Jr, Strong, L.C., 1972, Mutation and cancer: a model for Wilms' tumor of the kidney. *J Natl Cancer Inst.* 48:313-324.

Knudson, A.G., Jr, Hethcote, H.W., Brown, B.W., 1975, Mutation and childhood cancer: a probabilistic model for the incidence of retinoblastoma. *Proc Natl Acad Sci USA.* 72:5116-5120.

Kristensen, P., Andersen, A., Irgens, L.M., Bye, A.S., Sundheim, L., 1996, Cancer in offspring of parents engaged in agricultural activities in Norway: incidence and risk factors in the farm environment. *Int J Cancer.* 65:39-50.

Kuijten, R.R., Bunin, G.R., Nass, C.C., Meadows, A.T., 1992, Paternal occupation and childhood astrocytoma: results of a case-control study. *Cancer Res.* 52:782-786.

Linet, M.S., Ries, L.A., Smith, M.A., Tarone, R.E., Devesa, S.S., 1999, Cancer surveillance series: recent trends in childhood cancer incidence and mortality in the United States. *J Natl Cancer Inst.* 91:1051-1058.

Little, J., 1999, Epidemiology of childhood cancer. IARC Scientific Publications No. 149. International Agency for Research on Cancer, Lyon, France.

Louik, C., Frumkin, H., Ellenbecker, M.J., Goldman, R.H., Werler, M.M., Mitchell, A.A., 2000, Use of a job-exposure matrix to assess occupational exposures in relation to birth defects. *J Occup Environ Med.* 42:693-703.

Lowengart, R.A., Peters, J.M., Cicioni, C., Buckley, J., Bernstein, L., Preston-Martin, S.,

Rappaport, E., 1987, Childhood leukemia and parents' occupational and home exposures. *J Natl Cancer Inst.* 79:39-46.

McGuire, V., Nelson, L.M., Koepsell, T.D., Checkoway, H., Longstreth, W.T., Jr., 1998, Assessment of occupational exposures in community-based case-control studies. *Annu Rev Public Health.* 19:35-53.

McKean-Cowdin, R., Preston-Martin, S., Pogoda, J.M., Holly, E.A., Mueller, B.A., Davis, R.L., 1998, Parental occupation and childhood brain tumors: Astroglial and primitive neuroectodermal tumors. *J Occup Environ Med.* 4:332-340.

McLaughlin, J., King, W., Anderson, T., Clarke, E., Ashmore, J., 1993, Paternal radiation exposure and leukaemia in offspring: the Ontario case-control study. *BMJ.* 307:959-966.

Meinert, R., Kaletsch, U., Kaatsch, P., Schuz, J., Michaelis, J., 1999, Associations between Childhood Cancer and Ionizing Radiation: Results of a Population-based Case-Control Study in Germany. *Cancer Epidemiology, Biomarkers & Prevention.* 8:793-799.

Meinert, R., Schuz, J., Kaletsch, U., Kaatsch, P., Michaelis, J., 2000, Leukemia and Non-Hodgkin's Lymphoma in Childhood and Exposure to Pesticides: Results of a Register-based Case-Control Study in Germany. *Am J Epidemiol.* 151:639-646.

Nasca, P.C., Baptiste, M.S., MacCubbin, P.A., Metzger, B.B., Carlton, K., Greenwald, P., Armbrustmacher, V.W., Earle, K.M., Waldman, J., 1988, An epidemiologic case-control study of central nervous system tumors in children and parental occupational exposures. *Am J Epidemiol.* 128:1256-1265.

Olshan, A., Daniels, J., 2000, Invited Commentary: Pesticides and Childhood Cancer. *AJE.* 151(7):647-649.

Olshan, A.F., Faustman, E.M., 1993, Male-mediated developmental toxicity. *Annu Rev Public Health.* 14:159-181.

Pannett, B., Coggon, D., Acheson, E.D., 1985, A job-exposure matrix for use in population based studies in England and Wales. *Br J Ind Med.* 42:777-783.

Parkin, D.M., Kramarova, E., Draper, G.J., Masuyer, E., Michaelis, J., Neglia, J., Qureshi, S., Stiller, C.A., 1998, International Incidence of Childhood Cancer, Vol. II. IARC Scientific Publications No. 144. International Agency for Research on Cancer, Lyon, France.

Peters, J.M., Preston-Martin, S., Yu, M.C., 1981, Brain tumors in children and occupational exposure of parents. *Science.* 213:235-247.

Ries, L.A.G., Smith, M.A., Gurney, J.G., Linet, M., Tamra, T., Young, J.L., Bunin, G.R., Bernstein, L., Key, C.R., Lynch, C.F., Simone, J., Stevens, J., 1999, Cancer Incidence and Survival Among Children and Adolescents: United States SEER Program 1975-1995. Bethesda, MD: National Cancer Institute, SEER Program, NIH Pub. No. 99-4649; 51-63.

Roman, E., Doyle, P., Maconochie, N., Davies, G., Smith, P., Beral, V., 1999, Cancer in children of nuclear industry employees: report on children aged under 25 years from nuclear industry family study. *BMJ.* 318: 1443-1450.

Savitz, D.A., Chen, J., 1990, Parental occupation and childhood cancer: review of epidemiologic studies. *Environ Health Perspect.* 88:325-337.

Schuz, J., Kaletsch, U., Meinert, R., Kaatsch, P., Michaelis, J., 2000, Risk of Childhood Leukemia and Parental Self-Reported Occupational Exposure to Chemicals, Dusts, and Fumes: Results from Pooled Analyses of German Population-based Case-Control Studies, *Cancer Epidemiology. Biomarkers & Prevention.* 9:835-838.

Shu, X., Stewart, P., Wen, W., Han, D., Potter, J., Buckley, J., Heineman, E., Robison, L., 1999, Parental Occupational Exposure to Hydrocarbons and Risk of Acute Lymphocytic Leukemia in Offspring. *Cancer Epidemiology, Biomarkers & Prevention.* 8:783-791.

Sieber, W.K., Sundin, D.S., Frazier, T.M., Robinson, C.F., 1991, Development, use, and availability of a job exposure matrix based on national occupational hazard survey data. *Am J Ind Med.* 20:163-174.

Siemiatycki, J., Dewar, R., Richardson, L., 1989, Costs and statistical power associated with five methods of collecting occupational exposure information for population-based case-control studies. *Am J Epidemiol.* 130:1236-1246.

Siemiatycki, J., Fritschi, L., Nadon, L., Gérin, M., 1997, Reliability of an expert rating procedure for retrospective assessment of occupational exposures in community-based case-control studies. *Am J Ind Med.* 31:280-286.

Siemiatycki, J., Gérin, M., Richardson, L., Hubert, J., Kemper, H., 1982, Preliminary report of an exposure-based, case-control monitoring system for discovering occupational carcinogens. *Teratogenesis Carcinog Mutagen.* 2:169-177.

Smulevich, V., Solionova, L., Belyakova, S., 1999, Parental occupation and other factors and cancer risk in children: II. Occupational Factors. *Int J Cancer.* 83:718-722.

Stewart, P.A., Herrick, R.F., 1991, Issues in performing retrospective exposure assessment. *Appl Occup Environ Hyg.* 6:421-427.

Stewart, P.A., Stewart, W.F., 1994, Occupational case-control studies: II. Recommendations for exposure assessment. *Am J Ind Med.* 26:313-326.

Teschke, K., Kennedy, S.M., Olshan, A.F., 1994, Effect of different questionnaire formats on reporting of occupational exposures. *Am J Ind Med.* 26:327-337.

Tielemans, E., Heederik, D., Burdorf, A., Vermeulen, R., Veulemans, H., Kromhout, H., Hartog, K., 1999, Assessment of occupational exposures in a general populations: comparison of different methods. *Occup Environ Med.* 56:145-151.

Tomatis, L., Narod, S., Yamasaki, H., 1992, Transgeneration transmission of carcinogenic risk. *Carcinogenesis.* 13:145-151.

Wen, W.Q., Shu, X.O., Steinbuch, M., Severson, R.K., Reaman, G.H., Buckley, J.D., Robison, L.L., 2000, Paternal military service and risk for childhood leukemia in offspring. *Am J Epidemiol.* 151:231-240.

Wilkins, J.R., Koutras, R.A., 1988, Paternal occupation and brain cancer in offspring: a mortality-based case-control study. *Am J Ind Med.* 14:299-318.

Wilkins, J.R., McLaughlin, J.A., Sinks, T.H., Kosnik, E.J., 1991, Parental occupation and intracranial neoplasms of childhood: anecdotal evidence from a unique occupational cancer cluster. *Am J Ind Med.* 19:643-653.

Wilkins, J.R., Sinks, T., 1990, Parental occupation and intracranial neoplasms of childhood: results of a case-control study. *Am J Epidemiol.* 132:275-292.

Wilkins, J.R., Wellage, L.C., 1996, Brain tumor risk in offspring of men occupationally exposed to electric and magnetic fields. Scan*d J Work Environ Health.* 22:339-345.

Zahm, and Ward, 1998, Pesticides and childhood cancer. *Environ Health Perspect.* 106(Suppl 3):893-908.

RADIATION AND MALFORMATIONS IN A MURINE MODEL

Wolfgang-U. Müller

Institut für Medizinische Strahlenbiologie
Universitätsklinikum Essen
D45122 Essen, Germany

INTRODUCTION

Many textbooks claim that ionizing radiation induces malformations only when acting during organogenesis. Meanwhile, quite a number of authors have shown that some mouse strains respond with malformations also after radiation exposure during the preimplantation stage and, most notably, even when the zygote stage is affected (Pampfer and Streffer, 1988; Generoso et al., 1988; Jacquet et al., 1995; Gu et al., 1997). Because the zygote is just one single cell, this result rules out any cell killing mechanism, which is most common in the case of radiation exposure during organogenesis.

Thus, one is inclined to assume an effect on the genomic material. This, however, is not necessarily so, because there could be some indirect effect on the mother which is inevitably also radiation-exposed, when oocytes or preimplantation stages are irradiated. The situation is, of course, quite different, when male mice are exposed to ionizing radiation. Most probably, modifications of the genomic material will be responsible for effects seen in the offspring.

Therefore, we initiated some experiments to determine, whether radiation exposure of different stages of spermatogenesis result in malformed fetuses in our mouse strain.

THE MURINE MODEL SYSTEM

The 'Heiligenberger' mouse strain was originally established at the University of Freiburg, Germany, in 1939. The genetic traits are similar to those of NMRI mice. Since 1975, the strain was colony-bred in Essen, Germany; after more than 50 inbred generations it was registered as 'HLG/Zte' in 1994. A major characteristic of the strain is the spontaneous occurrence of gastroschisis, i.e. some of the fetuses are born with an open belly and protruding guts (Fig. 1). At the time of colony-breeding the frequency of spontaneous gastroschises was around 1%, whereas this frequency increased to 3-4% after inbreeding.

Advances in Male Mediated Developmental Toxicity, edited by Bernard Robaire and Barbara F. Hales.
Kluwer Academic/Plenum Publishers, 2003.

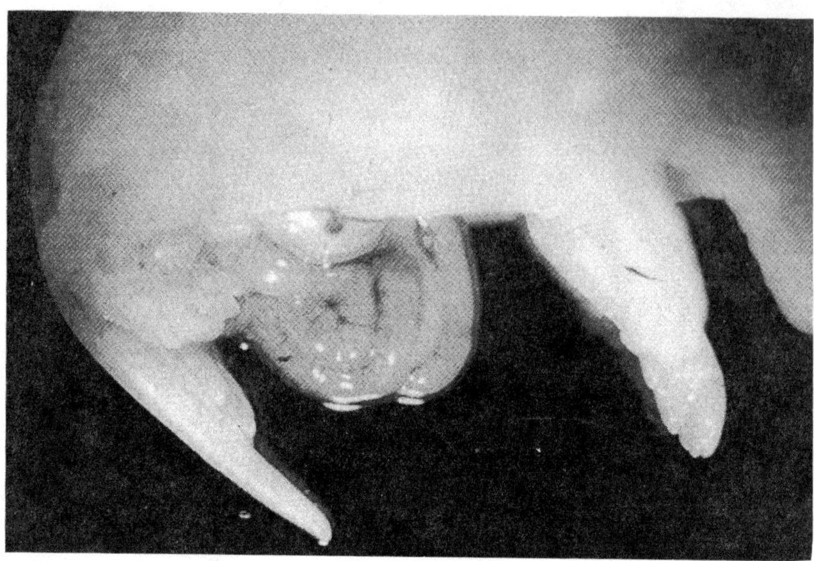

Figure 1. Gastroschisis of a Heiligenberger fetus on day 19 of gestation

A description of the mating procedure was published by Müller and Streffer (1990). A detailed overview of the most important results obtained after exposure of preimplantation stages of this strain was given by Streffer and Müller (1996).

SUMMARY OF RESULTS OBTAINED AFTER RADIATION EXPOSURE OF OOGENESIS AND PREIMPLANTATION STAGES

The topic of the paper presented here is the question whether radiation exposure of male germ cells results in teratogenic effects in the offspring. The malformation studies in the murine model system used in our Institute started in the middle 1980s with radiation exposure of the zygote (Pampfer, and Streffer, 1988); they were followed by experiments with other preimplantation stages (Müller, and Streffer, 1990; Müller et al., 1994) and with stages of oogenesis (Müller, and Schotten, 1995). Prior to describing the experiments carried out with male mice, I would like to summarize the results of the other experiments, because a lot can be learned from these results for the interpretation of the results obtained in the spermatogenesis analyses. As mentioned above, a more detailed summary is given by Streffer and Müller (1996), but the most important aspects will be described in the following:

- ➢ It is possible to induce malformations by radiation exposure of all stages of oogenesis and the preimplantation period, but there are different sensitivities (Pampfer, and Streffer, 1988; Müller, and Streffer, 1990; Müller, and Schotten, 1995).
- ➢ This effect is mouse strain dependent, i.e. C57Bl mice do not show an increase in malformed fetuses after radiation exposure of zygotes (Müller et al., 1996).
- ➢ A significant increase is observed only for that malformation that is already quite frequent in unexposed mice (i.e. gastroschisis with 1 to 4%).
- ➢ The effect is transmittable to the next generation (Pils et al., 1999).

➤ An increase in the number of chromosomal aberrations was observed in the skin fibroblasts of gastroschisis fetuses after radiation exposure of zygotes (Pampfer, and Streffer, 1989).

➤ Protein patterns were modified in gastroschisis fetuses induced by radiation exposure (Hillebrandt, and Streffer, 1994).

➤ Presumably 3 genes are involved in the process of radiation-induced gastroschisis formation (Hillebrandt et al., 1996; Hillebrandt et al., 1998).

We suggested 'genetic predisposition' and induction of 'genomic instability' as most likely explanations for the outcome of the experiments carried out with oocytes and preimplantation stages (Streffer, and Müller, 1996). A major point to support these mechanisms is, of course, the demonstration that damage to the genomic material of the oocyte or the early embryo is responsible for the induction of malformations and not some indirect effect (e.g. radiation sickness) of the mother. Evidence for the involvement of the genomic material was indeed obtained in experiments with male mice (Müller et al., 1999), as described in the following section. (In the Müller et al. publication (1999) you will also find a full data set of all results.)

RADIATION EXPOSURE OF STAGES OF SPERMATOGENESIS

The inbred version of the Heiligenberger mice (HLG/Zte) was used in these experiments. Male mice were exposed to a total dose of 2.8 Gy cesium gamma rays at a dose rate of 0.28 Gy/h (this rather low dose rate was used, because we were interested in a comparison with similar experiments done with female mice). The interval between radiation exposure and successful mating gives the information on the exposed stage of spermatogenesis: mating during the first week after radiation exposure means that spermatozoa were exposed during weeks 2 and 3 as spermatids, during weeks 4 and 5 as spermatocytes, and during weeks 6 and 8 as spermatogonia (Oakberg, 1956). Thus, pre-meiotic stages were affected, when the interval between radiation exposure and mating was during weeks 6 and 8, meiotic stages during weeks 4 and 5, and post-meiotic stages during weeks 1 and 3. There is, of course, some overlap, but at least a rough allocation is possible in this way.

Figure 2 gives the information on the killing effects during pregnancy induced by radiation exposure of male germ cells. As we did a lot of in vitro culturing of preimplantation embryos in the past, we know that on average 9.2 zygotes are observed per female in our mouse strain. Thus, using the information of uterus inspection on day 19 of gestation (detection of the vaginal plug = day 1), we are able to calculate the number of zygotes that did not develop successfully during gestation. In the controls, this number amounts to about 30%. With the exception of immature spermatogonia (interval between exposure and conception = 8 weeks), all stages of spermatogenesis showed pronounced effects with regard to survival of embryos and fetuses.

It is, of course, interesting to look for the point of time at which the embryo or fetus died. Figure 3 shows the results. The decrease in survival (Fig. 2) is entirely due to preimplantation loss and early resorptions. That is, the time around implantation seems to be particularly critical. Neither total litter loss (i.e. no sign of an implantation on day

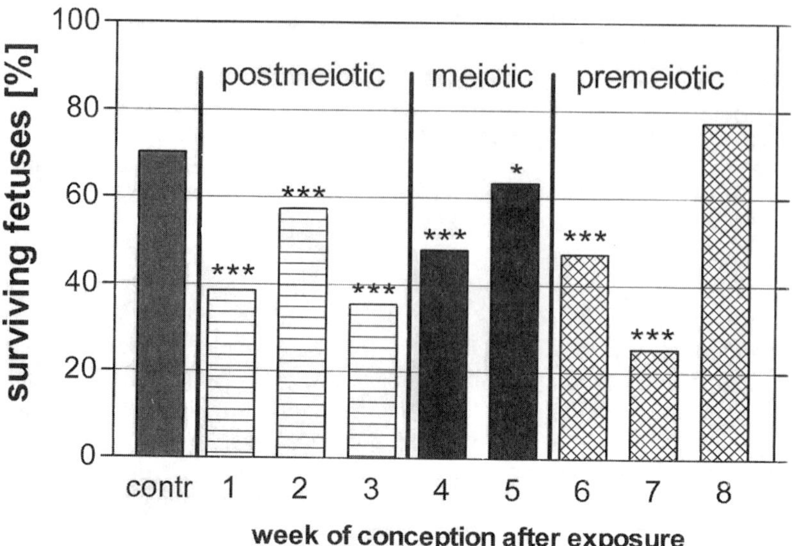

Figure 2. Percentage of survival of embryos or fetuses after a 2.8 Gy radiation exposure of various stages of spermatogenesis. (*,*** Significantly different from the control at P<0.1 or P<0.01, respectively); significance was tested using the control of the same week, not the average control shown in the figure.)

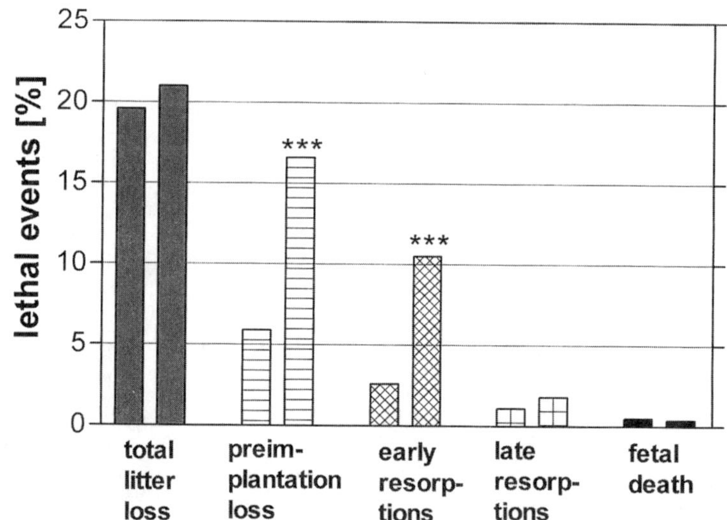

Figure 3. Types of lethal events after radiation exposure of various stages of spermatogenesis. (*** Significantly different from control at P<0.01; left bars are controls, right bars are exposed.)

19, although a plug on day 1 has indicated successful mating), nor late resorptions (placenta and fetus are clearly discernable), nor fetal death (eyelids are visible) are higher after radiation exposure of male mice.

The most interesting results are shown in Figure 4, which summarizes the effect of radiation exposure of male mice on the malformation frequency in the offspring. If one pools the results of all stages, one observes a statistically significant increase in the number of malformed fetuses (53 malformed out of 2308 living fetuses in the controls, i.e. 2.3%, and 72 malformed out of 1610 living fetuses in the exposed group, i.e. 4.5%). This increase is not dramatic, but significant at P<0.01.

Figure 4 reveals another aspect worth mentioning: when one looks at the affected stages of spermatogenesis, there is a strong indication that the meiotic stages are most sensitive with regard to the induction of malformations.

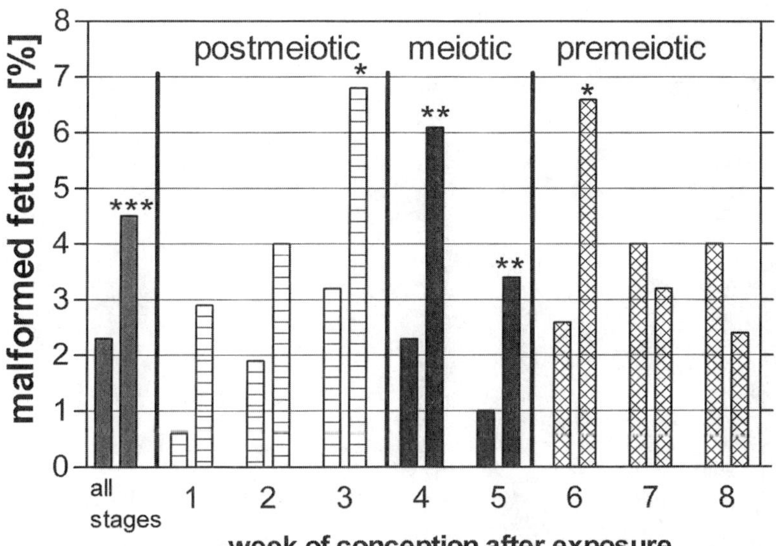

Figure 4. Malformation frequency on day 19 fetuses sired by male mice exposed at various stages of spermatogenesis to 2.8 Gy gamma rays. (*, **, *** Significantly different at P<0.1, P<0.5, and P<0.01, respectively. Left bars are controls, right bars are exposed. The seemingly pronounced difference of week 1 was not significant, because the bars are based on only 1 malformed out of 160 living fetuses in the control, and on 3 malformed out of 103 living fetuses in the exposed group.)

SUMMARY AND CONCLUSIONS

Thus, the results of radiation exposure of the different stages of spermatogenesis with regard to the induction of malformations in the offspring of these exposed male mice can be summarized as follows:

> - It is possible to induce malformations in the offspring by radiation exposure of <u>male</u> mice of the HLG strain.
> - This means that <u>direct DNA damage</u> is responsible for the induction of the observed malformations.
> - The highest sensitivity is observed during meiosis.

The fact that direct DNA damage is involved in the induction of radiation induced malformations underlines the hypothesis mentioned above, that two processes may have a major impact on malformation induction in germ cells and preimplantation stages: 'genetic predisposition' and 'genomic instability'.

REFERENCES

Generoso, W.M., Rutledge, J.C., Cain, K.T., Hughes, L.A., and Downing, J.J., 1988, Mutagen-induced fetal anomalies and death following treatment of females within hours after mating. *Mutat Res.* 199:175-181.

Gu, Y., Kai, M., and Kusama, T., 1997, The embryonic and fetal effects in ICR mice irradiated in the various stages of the preimplantation period. *Radiat Res.* 147:735-740.

Hillebrandt, S., Streffer, C., and Müller, W.-U., 1996, Genetic analysis of the cause of gastroschisis in the HLG mouse strain. *Mutat Res.* 372:43-51.

Hillebrandt, S., Streffer, C., Montagutelli, X., and Balling, A., 1998, A locus for radiation-induced gastroschisis on mouse chromosome 7. *Mamm Genome.* 9:995-997.

Hillebrandt, S., and Streffer, C., 1994, Protein patterns in tissues of fetuses with radiation-induced gastroschisis. *Mutat Res.* 308:11-22.

Jacquet, P., de Saint Georges, L., Vankerkom, J., and Baugnet Mahieu, L., 1995, Embryonic death, dwarfism and fetal malformations after irradiation of embryos at the zygote stage: studies on two mouse strains. *Mutat Res.* 332:73-87.

Müller, W.-U., Streffer, C., and Pampfer, S., 1994, The question of threshold doses for radiation damage: malformations induced by radiation exposure of unicellular or multicellular preimplantation stages of the mouse. *Radiat Environ Biophys.* 33:63-68.

Müller, W.-U., Streffer, C., and Knoelker, M., 1996, A genetic characterization of differences in the sensitivity to radiation-induced malformation frequencies in the mouse strains Heiligenberger, C57Bl, and HeiligenbergerxC57Bl. *Radiat Environ Biophys.* 35:37-40.

Müller, W.-U., Streffer, C., Wojcik, A., and Niedereichholz, F., 1999, Radiation-induced malformations after exposure of murine germ cells in various stages of spermatogenesis. *Mutat Res.* 425:99-106.

Müller, W.-U., and Schotten, H., 1995, Induction of malformations by X-ray exposure of various stages of the oogenesis of mice. *Mutat Res.* 331:119-125.

Müller, W.-U., and Streffer, C., 1990, Lethal and teratogenic effects after exposure to X-rays at various times of early murine gestation. *Teratology.* 42:643-650.

Oakberg, E.F., 1956, Duration of spermatogenesis in the mouse and timing of stages of the cycle of the seminiferous epithelium. *Am J Anat.* 99:507-516.

Pampfer, S., and Streffer, C., 1988, Prenatal death and malformations after irradiation of mouse zygotes with neutrons or X-rays. *Teratology.* 37:599-607.

Pampfer, S., and Streffer, C., 1989, Increased chromosome aberration levels in cells from mouse fetuses after zygote X-irradiation. *Int J Radiat Biol.* 55:85-92.

Pils, S., Müller, W.-U., and Streffer, C., 1999, Lethal and teratogenic effects in two successive generations of the HLG mouse strain after radiation exposure of zygotes - association with genomic instability?. *Mutat Res.* 429:85-92.

Streffer, C., and Müller, W.-U., 1996, Malformations after radiation exposure of preimplantation stages, *Int J Dev Biol.* 40:355-360.

MECHANISMS OF ACTION OF CYCLOPHOSPHAMIDE AS A MALE-MEDIATED DEVELOPMENTAL TOXICANT

Bernard Robaire[1,2] and Barbara F. Hales[1]

Departments of Pharmacology and Therapeutics[1]
and of Obstetrics and Gynecology[2]
McGill University, Montréal, QC, H3G 1Y6, Canada

INTRODUCTION

The consequences of exposure to drugs, radiation, and environmental toxicants on reproduction and development are a growing concern. The extent to which paternal exposures contribute to human infertility and pregnancy loss is unknown. Several epidemiological studies presented at this conference and in the literature show that certain paternal occupations, e.g. as a welder, painter, auto mechanic, or fireman, involving exposure to metals, combustion products, solvents, and pesticides, are associated with an increase in spontaneous abortions, birth defects, and childhood cancer (Savitz et al., 1994; Ries et al., 1999; Olshan and van Wijngaarden, this volume; Bonde et al, this volume). In most instances, investigators have not identified specific chemical or exposure paradigms as causative for male-mediated adverse effects on progeny outcome. First, the exposures are usually to a plethora of chemicals; with rare exceptions, there are no data on dose, duration of exposure, or potential chemical interactions. Second, in humans, infertility and pregnancy loss are very frequent events; to demonstrate a significant increase would require both a large study population and a marked augmentation in risk. Consequently, studies of the mechanisms by which paternal exposures adversely affect progeny outcome are essential for us to understand how alterations in the male affect fertility and progeny and to provide the scientific basis for the development of biomarkers for risk assessment.

Unlike for man, there is extensive evidence from animal studies that exposure of males to specific drugs, environmental chemicals, or radiation results in adverse progeny outcome. There are three major mechanisms by which exposure of males to these agents may affect their progeny adversely. The **first** is by direct exposure of the embryo or developing fetus during mating to a chemical present in the seminal fluid. Methadone, morphine, thalidomide, and cyclophosphamide have been demonstrated to have adverse effects on progeny outcome by this mechanism (Soyka et al., 1978; Lutwak-Mann, 1974; Hales et al., 1986). Studies from our laboratories have shown that cyclophosphamide (and/or its metabolites), administered to the male rat, was found in seminal fluid, was transmitted to the female partner during mating, and caused a dose-dependent increase in

Advances in Male Mediated Developmental Toxicity, edited by Bernard Robaire and Barbara F. Hales.
Kluwer Academic/Plenum Publishers, 2003.

preimplantation loss (Hales et al., 1986). The **second** is by modulating the hypothalamic pituitary testicular axis. Endocrine disruptors may interfere with the ability of the hypothalamic pituitary complex to communicate with the testis; p,p'-DDE, a metabolite of the pesticide DDT, has been identified as an endocrine disruptor (Kelce et al., 1995). Leydig cells and epididymal principal cells are targeted by ethane dimethane sulphonate (EDS) (Klinefelter et al., 1991), while Sertoli cells may be targeted by compounds such as 2,3,7,8-tetrachlorodibenzo-p-dioxin (TCDD) or hexanedione (Kleeman et al., 1990; Hall et al., 1995); disruption of the functions of these cells may lead to alterations in germ cell quantity or quality. The **third** is by a direct effect of the toxicant on the male germ cell, either during sperm maturation in the epididymis or spermatogenesis in the testis. While previous studies from our labs and from others have shown that each of these mechanisms occurs, we will focus our attention on the third, on the male germ cell itself as a target for toxicant action. We will first discuss the effects of drug treatment on progeny outcome, next, identify when during embryonic development it becomes possible to detect effects of paternal drug exposure, then how the male germ cells can be affected while it is maturing in the epididymis or being made in the testis, and finally how specific gene expression in germ cells can be modulated by paternal drug treatment.

EFFECTS OF CYCLOPHOSPHAMIDE ON PROGENY OUTCOME

From the timing of the effect of a toxicant exposure, one can deduce the stage specificity of the susceptibility of the germ cells during spermatogenesis (Trasler et al., 1985; Russell, 1994). Germ cells at the spermatogonial stage are very susceptible to drugs such as procarbazine or to X-ray exposure, resulting in reduced sperm numbers and an increased percentage of morphologically abnormal surviving sperm (Russell, 1994; Kangasniemi et al., 1995). Following exposure of mice to anticancer drugs such as chlorambucil, a peak in mutation yield is observed when offspring are conceived from germ cells exposed as spermatids. In contrast, spermatozoa are the germ cells which are maximally sensitive to specific locus mutations induced by acrylamide and dominant lethality induced by ethylnitrosourea (Russell, 1994).

Depending on the window of exposure (Figure 1), we and others have shown that cyclophosphamide, a commonly used anticancer drug, has marked dose-dependent and time-specific effects on progeny outcome (Trasler et al, 1985; 1986; 1987; Qiu et al, 1992; Anderson et al, 1995). An increase in post-implantation loss was found after 2 weeks of chronic low dose cyclophosphamide treatment of male rats. This post-implantation loss rose dramatically to plateau at a level that was dependent on drug dose by 4 weeks of treatment and was reversed within 4 weeks of the termination of drug treatment (Hales and Robaire, 1990); thus, cyclophosphamide induced post-implantation loss was associated primarily with germ cell exposure during spermiogenesis. Post-meiotic germ cells were also most susceptible to the induction of learning abnormalities in the progeny after paternal exposure to cyclophosphamide (Fabricant et al., 1983). Heritable translocations have been found in mice after exposure of spermatids and spermatozoa to a single high dose of cyclophosphamide (Generoso et al., 1981). Exposure of rat spermatocytes to cyclophosphamide results in increased pre-implantation loss, as well as in synaptic failure, fragmentation of the synaptonemal complex and altered centromeric DNA sequences (Backer et al., 1988).

An increase in malformed and growth retarded fetuses was observed after 7-9 weeks treatment of male rats with a chronic low dose of cyclophosphamide; the malformations observed were principally hydrocephaly, edema and micrognathia (Trasler et al., 1986). Therefore, external malformations and growth retardation were produced in progeny sired by germ cells first exposed to cyclophosphamide as spermatogonia. Spermatogonia have

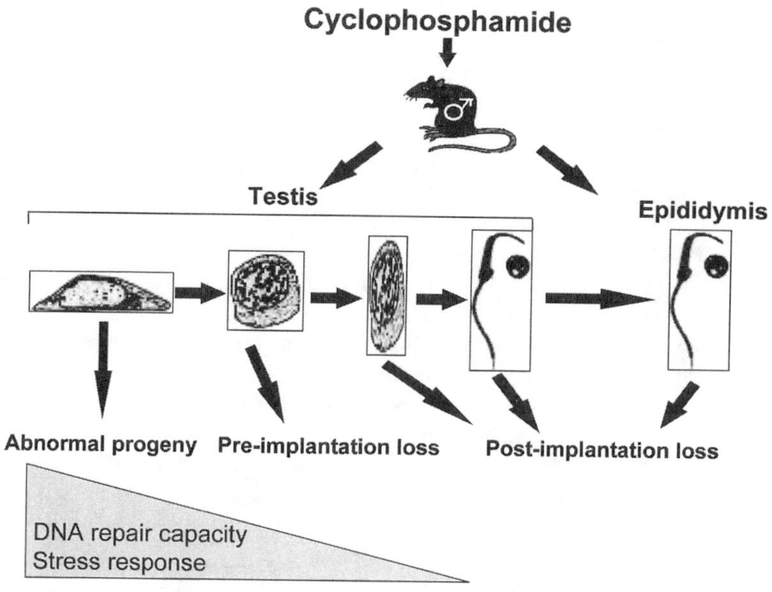

Figure 1. The progeny outcome resulting from exposure of male germ cells to cyclophosphamide is dependent on when germ cells are first exposed during spermatogenesis and sperm maturation (adapted from Robaire and Hales, 1999).

been reported to be at low risk for the induction of heritable translocations by cyclophosphamide (Generoso et al., 1981; Sotomeyer and Cumming, 1975). Significantly, an increase in post-implantation loss and malformations persisted to the F_2 generation (Hales et al., 1992); the malformations observed in the offspring of rats whose fathers were exposed to cyclophosphamide included open eyes, omphalocele, generalized edema, syndactyly, gigantism and dwarfism; the malformations observed in the F_2 generation were similar to those found in the F_1 (Trasler et al., 1986). Some of the behavioural abnormalities caused by paternal cyclophosphamide treatment also persisted in subsequent generations (Auroux et al., 1988, 1990).

Thus, adverse effects on progeny outcome after exposure of males to model drugs such as cyclophosphamide range from pre- and post-implantation death, to growth retardation, malformations, and behavioural abnormalities. Furthermore, outcome is influenced by the stage at which germ cells are first exposed to the drug.

EARLY EMBRYONIC DEVELOPMENT IN EMBRYOS SIRED AFTER PATERNAL
EXPOSURE TO CYCLOPHOSPHAMIDE

An increase in post-implantation loss may reflect either the inability of a "normal" embryo to develop further after implantation or the ability of an "abnormal" embryo to trigger an implantation reaction. Examination of the implantation sites of embryos sired by cyclophosphamide-treated male rats revealed that cell death occurs selectively in tissues derived from the inner cell mass, while the trophoblast-derived trophectoderm cells appear morphologically normal (Kelly et al., 1992). However, blastocyts sired by cyclophospha-mide-treated males have about half the cell number of their control counterparts; interestingly, the trophectoderm and inner cell mass cell lineages are decreased proportionally. Thus, the inner cell mass and trophoblast-derived cells are not equally able to handle insult, as only the trophoblast-derived cells survived until day 7. We found a decrease in cell number, in the rate of cell proliferation, and in thymidine incorporation among the embryos sired by cyclophosphamide treated males as early as day 3 of gestation (4-8 cell embryos) (Austin et al., 1994). Decreased cell numbers in these embryos result in a decrease in the cell-cell contacts that are essential for successful embryonic development (Harrouk et al., 2000a).

The ability of zona-free mouse embryos to develop to term depends on the number of total points of contact among blastomeres; more contacts result in a higher number of inner cell mass cells and, subsequently, an increased number of live offspring (Suzuki et al., 1995). In control embryos, steady state concentrations of the mRNAs for cell adhesion molecules (cadherins and connexin 43) and structural proteins (ß-actin, collagen, and vimentin) are low in 2- and 4-cell control embryos, and expression increases dramatically by the 8-cell stage (Harrouk et al., 2000a). In contrast, embryos sired by cyclophosphamide-treated males display the highest expression of most transcripts at the 2-cell stage. In parallel with the mRNA profiles, E-cadherin immmunoreactivity is nearly absent in 2-cell control embryos and strong by the 8-cell stage; immunoreactivity in embryos sired by drug treated fathers is strong at the 2-cell stage, but absent at later stages. Embryos from litters sired by cyclophosphamide-treated males have lower cell numbers and decreased cell-cell contacts. Thus, drug exposure of the paternal genome leads to decreased cell interactions during early embryo development

The alkaline comet assay was used to analyse DNA strand breaks in 1-cell embryos fertilized by control and drug-exposed spermatozoa in order to explore the presence and role of damaged DNA in embryos after exposure of the male genome to cyclophos-phamide. Comet formation, as assessed by measuring the tail/head ratio and tail length of electrophoresed, stained DNA, reveals that cyclophosphamide-exposed spermatozoa impart a significantly greater DNA damage to the newly fertilized egg at the 1-cell stage than controls (Harrouk et al., 2000b).

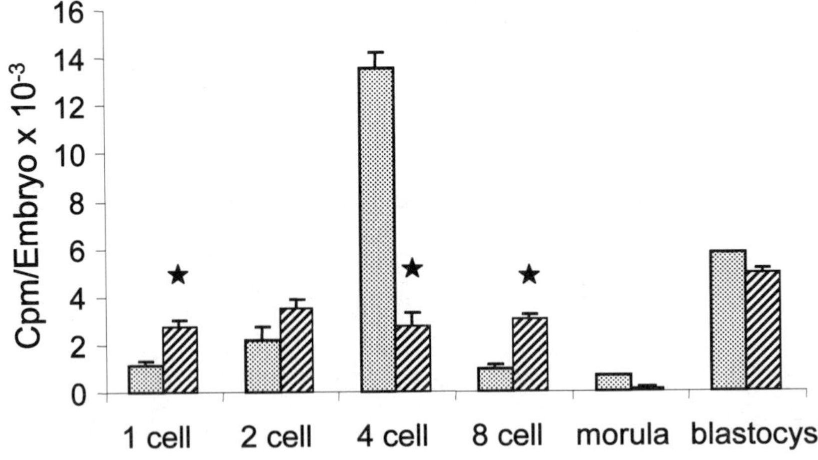

Figure 2. RNA synthesis in rat pre-implantation embryos. Embryos (n=30-50 per stage per treatment) were collected on days 0-2 p.c., permeabilised and incubated in the presence of ^{32}P-UTP for 4 hours. Total RNA synthesis per embryo is represented as the total counts per minute (cpm) obtained. To assess the difference in the capacity of embryos sired by control (dotted bars) and cyclophosphamide (hatched bars) exposed males to synthesise RNA, a t-test analysis was conducted followed by Tukey's post-hoc test with a threshold of significance value of $P \leq 0.05$ (*). Each data point was repeated 3-5 times with embryos sired by 6 different males for each treatment group. Error bars represent the standard error of the mean (adapted from Harrouk et al., 2000b).

172

Transcription is turned on earlier in the male pronucleus than in the female pronucleus (Worrad et al., 1994); transcriptional activity can be detected in the one-cell murine embryo by G_2 when the male pronucleus, but not when the female, is injected with a Sp1 dependent luciferase reporter gene (Ram and Schultz, 1993). Therefore, if the damaged male genome does not have the same ability to serve as a template for gene transcription as in the control, the impact on transcription may occur as early as the one cell stage. The assessment of RNA synthesis in embryos sired by control and cyclophosphamide-treated males revealed that total RNA synthesis ($[^{32}P]$-UTP incorporation) was constant in 1-8 cell embryos sired by drug-treated fathers, while in control embryos RNA synthesis increased four-fold to peak at the 4-cell stage (Figure 2). Moreover, both BrUTP incorporation into RNA and Sp1 transcription factor immunostaining were increased and spread over the cytoplasmic and nuclear compartments in 2-cell embryos sired by cyclophosphamide-treated males (Harrouk et al., 2000c). In contrast, in embryos fertilized by control spermatozoa, both BrUTP incorporation and Sp1 immunostaining were nuclear.

We have discovered, using the antisense RNA approach, that the profile of expression of specific genes is altered in embryos sired by drug-treated males even as early as the one cell stage (Harrouk et al., 2000a,b,c). By the 2-cell stage, the relative abundance of transcripts for candidate imprinted, growth factor, and cell adhesion genes was elevated significantly above control in embryos sired by cyclophosphamide-treated males; a peak in the expression of many of these genes was not observed until the 8-cell stage in control embryos (Harrouk et al., 2000c). Thus, paternal drug exposure temporally and spatially dysregulates rat zygotic gene activation, altering the developmental clock.

THE EFFECTS OF A DAMAGED MALE GENOME ON CHROMATIN REMODELLING AND PRONUCLEAR FORMATION.

After the spermatozoon enters the oocyte, chromatin decondenses to restore the paternal genome to an active conformation (Naish et al., 1987). The complete decondensation process is characterized by reduction of the disulfide bonds of the protamines, followed by their degradation and replacement with histones (Perreault etal., 1987; Perreault and Zirkin, 1982). Sperm nuclei with a low disulfide bond content will decondense more quickly in a hamster oocyte, forming a male pronucleus sooner (Perreault et al., 1988). Spermatozoa can be decondensed *in vitro* by incubation with a reducing agent (dithiothreitol) and proteinase K. Using this "test tube" approach, we showed that exposure to cyclophosphamide alters spermatozoal decondensation in vitro (Qiu et al., 1995a). The first phase of decondensation of spermatozoa from rats exposed to cyclophosphamide for 6 weeks was similar to control; however, in the second phase, spermatozoa from drug treated males remained quite compacted, whereas chromatin from control spermatozoa dispersed completely. Using a more "in vivo" model, the denuded hamster oocyte, we have also demonstrated that the decondensation of spermatozoa from cyclophosphamide treated males was more rapid than that of control spermatozoa (Harrouk et al., 2000c). Indeed, we have established that male pronucleus formation is earlier in rat oocytes sired by drug treated males. One possibility is that cyclophosphamide-induced chromatin damage prevents "normal" condensation during spermiogenesis. It is interesting to speculate that that a disturbance in male germ cell chromatin decondensation, remodelling, and pronucleus formation results in the dysregulation of zygotic gene activation and the adverse effects on embryo development observed.

Fertilization with damaged sperm may lead to aberrant pronuclear development; a significantly higher incidence of mosaicism was found in human embryos that exhibited two pronuclei differing in size by at least 4 μm (Sadowy et al., 1998). It is likely that chromatin remodelling is required to make the male genome accessible to DNA polymerases for DNA replication. The first round of DNA replication is important to reprogram gene expression in the zygote. Restriction of sperm template function is released in vitro in sperm that are decondensed by treatment with disulfide-reducing agents. Our labs have shown that the in vitro template function of spermatozoal DNA was markedly affected after six weeks of treatment with cyclophosphamide (Qiu et al., 1995b); we found that at zero time, with no decondensation, there was a significant amount of [^{3}H] thymidine incorporated into nuclei from cyclophosphamide-treated spermatozoa, while incorporation was similar to control after 30 minutes of decondensation. Neuber et al. (1999) showed that human sperm undergo nuclear envelope breakdown and chromatin reorganization in an extract from *Xenopus* eggs arrested in meiotic metaphase II, and then acquire envelopes, form pronuclei, and undergo DNA replication upon dilution into an interphase extract (eggs activated by calcium ionophore treatment). Interestingly, two of the eight samples from subfertile men displayed rates of DNA replication that were higher than any of those for the proven fertile donors.

THE MALE GENOME AS A TARGET FOR TOXICANT ACTION

Germ Cells in the Epididymis

As spermatozoa transit through the epididymis, they acquire the ability to fertilize an egg. In the nucleus, cysteine-rich protamines become progressively more tightly compacted by the formation of disulfide bonds. There is evidence that spermatozoa in transit in the epididymis are susceptible to insult with male-mediated developmental toxicants. A 2-fold increase in mutation frequency was found in bone marrow cells from the F_1 generation of male mice irradiated 1-7 days prior to mating (Luke et al., 1997). Methyl chloride induced an increase in dominant lethal mutations that was thought to be due to a selective inflammatory action of this drug on the epididymis; the increase in embryo loss was reversed by an anti-inflammatory agent (Chellman et al., 1986a,b). A dose-dependent increase in post-implantation loss was observed four days after treatment with cyclophosphamide. In addition, cyclophosphamide treatment of animals whose efferent ducts had been ligated for seven days revealed that there was no effect on progeny outcome, but ligation followed by administration of cyclophosphamide produced the same percentage of post-implantation loss as that produced by cyclophosphamide treatment alone. Together these results demonstrate that short-term paternal treatment with cyclophosphamide exerts an adverse effect on progeny outcome; this effect occurs during spermatozoal transit through the caput and corpus epididymides (Qiu et al., 1992).

Germ Cells in the Testis

Most studies on the mechanisms by which drugs administered to the male have adverse effects on progeny outcome have emphasized the male germ cell in the testis as the target. It is not surprising that many agents can affect spermatogenesis given that this process is highly ordered and remarkably complex. First, stem cells (spermatogonia) must decide to either repopulate their cell pool or progress toward making spermatozoa. Spermatogonia then undergo five mitotic divisions (rat) to become spermatocytes, two meiotic divisions to form spermatids, and a complex differentiation process (condensing nuclear elements, developing a propulsion mechanism, and shedding most of their cytoplasm) to become spermatozoa (Clermont, 1972). As a consequence of the removal of most histones and their replacement with protamines, the DNA in spermatozoa is packaged very tightly.

The overall DNA damage in male germ cells after exposure of rats to cyclophosphamide has been assessed using alkaline elution (Qiu et al, 1995b). Under alkaline conditions DNA unwinds and is eluted through a filter at a rate reflecting the extent of DNA single stand breaks or cross links (Kohn et al, 1979). Treatment with cyclophosphamide for one week caused DNA single strand breaks which could be detected only with proteinase K in the lysis solution; no DNA cross-links were observed (Qiu et al., 1995b). In contrast, six weeks of treatment with cyclophosphamide induced a significant increase in DNA single strand breaks and cross-links in sperm chromatin; the cross-links were due primarily to DNA-DNA and not DNA-protein linkages (Qiu et al, 1995b).

Although alkaline elution provides a powerful test of the interaction of a drug with chromatin in a large population of cells, it cannot be used on an individual cell basis (Fairbain et al., 1995). With the development of the comet assay, the in situ nick translation assay, and techniques to assess chromosome structure in male germ cells, including banding patterns in hamster eggs, and fluorescent in situ hybridization (FISH), these goals became attainable. In the comet assay, individual cells, lysed and decondensed, are electrophoresed in agar either after treatment with alkali, to assess double strand breaks, or under neutral conditions, to identify single strand breaks (Fairbain et al., 1995; Monteith and Vanstone, 1995; Irvine et al., 2000). Chromosomal structures in the male pronucleus can be analysed by allowing denuded hamster eggs to be fertilized by spermatozoa. This is a time consuming approach, but it has been used successfully to demonstrate that age, X-irradiation, and drugs affect the chromosomal banding pattern of human sperm (Martin et al., 1989). More recently, FISH has been developed for the analysis of aneuploidy in the male genome (Wyrobek et al., 1994; Martin et al., 1995; Lowe et al., 1998) not only of men but also of mice and rats (see chapters in this volume by Marchetti and Wyrobek, Martin, and Robbins).

The overall consequence of germ cell exposure to a toxicant will depend on both the extent and specificity of DNA damage and on the DNA repair processes that are active at various stages of spermatogenesis (Figure 1). The repair of chemically induced DNA damage in post-mitotic male germ cells has been assessed with the unscheduled DNA synthesis (UDS) assay (Sotomayor et al., 1978). The few studies that have assessed the presence of components of DNA repair processes in male germ cells during spermatogenesis indicate that the level of expression of DNA repair genes decreases dramatically during spermatogenesis. With the advent of gene expression profiling technology, it has become possible to determine which genes, or families of genes, are expressed in specific spermatogenic cell types and how this expression is altered after exposure to drugs or environmental chemicals.

TREATMENT WITH CYCLOPHOSPHAMIDE ALTERS GENE EXPRESSION
DURING SPERMATOGENESIS

The effects of cyclophosphamide on progeny outcome depend on the stage of spermatogenesis at which germ cells are first exposed (Trasler et al., 1986). Furthermore, the germ cell stages that can be separated by unit gravity sedimentation (spermatocytes, round spermatids and elongated spermatids/spermatozoa) exhibit different gene expression profiles (Aguilar-Mahecha et al., 2001a). Thus, we hypothesized that investigating the effects of cyclophosphamide treatment on gene expression in germ cells would provide new insight into mechanisms underlying the male-mediated developmental toxicity of this drug. Gene expression profiling for each germ cell type was done using Atlas StressTox membranes (Clontech, Palo Alto, CA); data were analyzed using GeneSpring software (Silicon Genetics, Redwood, CA). Sixteen hours after a single acute treatment of cyclophosphamide (70 mg/kg), the germ cell type that was primarily affected was the round spermatid, with 30 of 74 detected genes increasing in expression by at least 1.5 fold, while only 10 decreased; far fewer genes were affected in pachytene spermatocytes (8 of 87 detected, with 7 increasing in expression and only 1 decreasing). The genes that were most dramatically affected in the round spermatids were the heat shock proteins and their co-chaperones, and genes associated with DNA repair (Aguilar-Mahecha et al., 2001b).

In contrast to the effects of a single acute treatment with cyclophosphamide, chronic low dose drug treatment (6 mg/kg/day for up to six weeks) resulted in a dramatically decreased gene expression in pachytene spermatocytes and round spermatids. For both of these cell types, more than 50% of the genes expressed were down-regulated by at least 1.5 fold; no genes were up-regulated by this treatment (Figure 3). Unlike these cells, in elongated spermatids far fewer genes were affected; only about 10% of the expressed genes were down-regulated and about 20% were up-regulated in these cells by chronic cyclophosphamide treatment (Aguilar-Mahecha et al., 2002). The three major families of genes that were down-regulated in pachytene spermatocytes and round spermatids were oxidative stress related genes, heat shock protein genes and DNA repair genes, while the expression of most other genes remained unchanged.

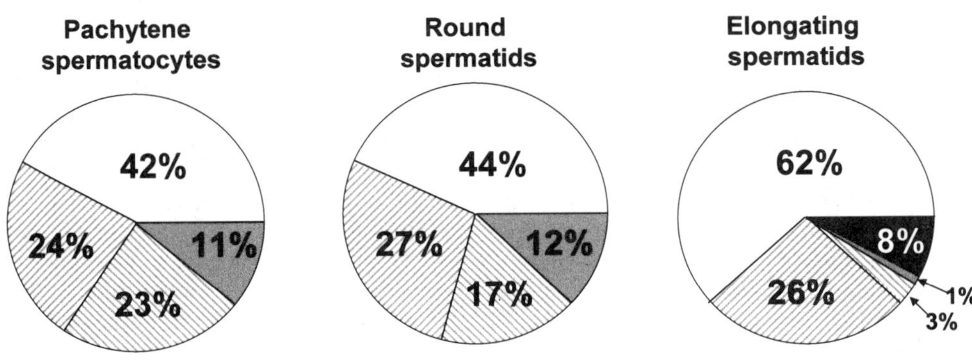

Figure 3. Effect of chronic cyclophosphamide treatment (6 mg/kg/day for 5-6 weeks) on gene expression in testicular germ cells. The numbers depict the percent of genes affected from the total number of genes studied (216) in pachytene spermatocytes, round spermatids, and elongating spermatids. White wedges: not detected; hatched wedges: unchanged; cross-hatched wedges: expression decreased by 33-50%; grey wedges: decreased > 50%; black wedge: increased gene expression by 50% or greater; n=5 (adapted from Aguilar-Mahecha et al., 2002).

Reduced expression of these important defense mechanisms in male germ cells after cyclophosphamide treatment may allow damage to accumulate and result in altered sperm function and infertility. The mechanism by which chronic cyclophosphamide leads to abnormal male germ cell gene expression and the impact on fetal loss and abnormal development remains to be established.

CONCLUSIONS

A clear picture is emerging from the data generated in studies on male germ cells with cyclophosphamide and other drugs. Depending on the point in spermatogenesis when a germ cell is first exposed to a damaging agent, and on the extent of damage caused by the agent, it will respond in one of several ways. If the insult occurs when the appropriate

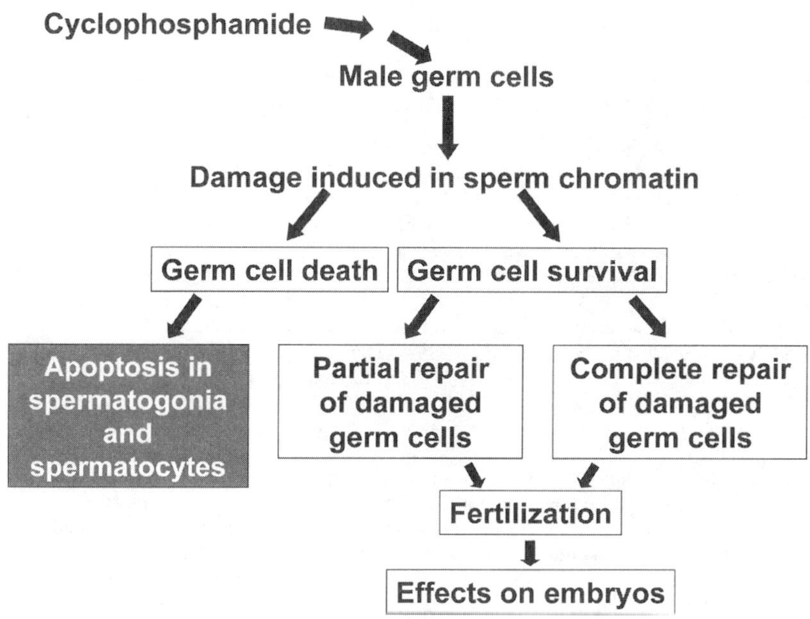

Figure 4. Potential consequences of male germ cell exposure to drugs such as cyclophosphamide.

DNA repair systems are present, one of three potential scenarios may ensue (Figure 4). The first is complete repair of the damage, and hence there are no adverse consequences to the progeny. The second is partial repair of the damage; in this case, spermatogenesis may continue and damaged chromatin may enter the oocyte. DNA repair in the oocyte may rescue the zygote or abnormal progeny may result. The third is for the insult to overwhelm the repair processes and the damaged male germ cell to undergo apoptosis. A low spontaneous incidence of apoptosis was observed in the seminiferous tubules of testes from control rats; in cyclophosphamide-exposed rats, the incidence of apoptosis increased to a level 3.5 fold above control by 12 hours after treatment (Cai et al., 1997). We and others have observed that drug-induced apoptosis was most pronounced in pre-meiotic germ cells (spermatogonia and spermatocytes) (Hikim et al., 1995; Brinkworth and Nieschlag, 2000). Apoptosis of damaged pre-meiotic germ cells may serve a critical role in protecting subsequent generations from the diverse effects of male mediated developmental toxicants. If the insult occurs when DNA repair systems have been lost (post-meiotic germ cells), the ability to mount an apoptotic response is likely to have been lost. Hence, germ cells can neither rectify the inflicted DNA damage nor trigger the suicidal apoptotic program; the net result is that sperm retain their ability to fertilize, but their chromatin is damaged. We have hypothesized (Robaire and Hales, 1999) that cyclophosphamide exerts its maximal effects on elongating spermatids and spermatozoa (post-meiotic germ cells) because these cells have lost both the ability to repair DNA and to undergo apoptosis.

ACKNOWLEDGEMENTS

Studies from our labs that are discussed in this chapter were supported by a grant from the Canadian Institutes of Health Research (CIHR).

REFERENCES

Aguilar-Mahecha, A., Hales, B.F., Robaire, B., 2001a, Expression of stress response genes in germ cells during spermatogenesis. *Biol Reprod.* 65:119-127.

Aguilar-Mahecha, A., Hales, B.F., Robaire, B., 2001b, Acute cyclophosphamide exposure has germ cell specific effects on the expression of stress response genes during rat spermatogenesis. *Molec Repro Dev.* 60:302-311.

Aguilar-Mahecha, A., Hales, B.F., Robaire, B., 2002, Chronic cyclophosphamide treatment alters the expression of stress response genes in rat male germ cells. *Biol Reprod.* 66:1024-1032.

Anderson, D., Bishop, J.B., Garner, R.C., Ostrosky-Wegman, P., Selby, P.B., 1995, Cyclophosphamide: review of its mutagenicity for an assessment of potential germ cell risks. *Mutat Res.* 330:115-181.

Auroux, M., Dulioust, E.J.B., Nawar, N.N.Y., Yacoub, S.G., Mayaux, M.J., Schwartz, D., David, G., 1988, Antimitotic drugs in male rat. Behavioral abnormalities in the second generation, *J Androl.* 9:153-159.

Auroux, M., Dulioust, E., Selva, J., Rince, P., 1990, Cyclophosphamide in the F_0 male rat: physical and behavioral changes in three successive adult generations. *Mutat Res.* 229:189-200.

Austin (Kelly), S.M., Robaire, B., Hales, B.F., 1994, Paternal cyclophosphamide exposure causes decreased cell proliferation in cleavage-stage embryos. *Biol Reprod.* 50:55-64.

Backer, L.C., Gibson. M.J., Moses, M.J., Allen, J.W., 1988, Synaptonemal complex damage in relation to meiotic chromosome aberration after exposure of male mice to cyclophosphamide. *Mutat Res.* 203:317-330.

Brinkworth, M.H., Nieschlag, E., 2000, Association of cyclophosphamide-induced male-mediated, foetal abnormalities with reduced paternal germ-cell apoptosis. *Mutat Res.* 447:149-154.

Cai, L., Hales, B.F., Robaire, B., 1997, Induction of apoptosis in the germ cells of adult male rats after exposure to cyclophosphamide. *Biol Reprod.* 56:1490-1497.

Chellman, G.J., Bus, J.S., Working, P.K., 1986a, Role of epididymal inflammation in the induction of dominant lethal mutations in Fischer 344 rat sperm by methyl chloride. *Proc Natl Acad Sci. USA* 83:8087-8091.

Chellman, G.J., Morgan, K.T., Bus, J.S., Working, P.G., 1986b, Inhibition of methyl chloride toxicity in male F-344 rats by the anti-inflammatory agent BW755C. *Toxicol. Appl. Pharmacol.* 85:367-379.

Clermont, Y., 1972, Kinetics of spermatogenesis in mammals: seminiferous epithelium cycle and spermatogenic renewal, *Physiol Rev.* 52:198-236.

Fabricant, J.D., Legator, M.S., Adams, P.M., 1983, Post-meiotic cell mediation of behavior in progeny of male rats treated with cyclophosphamide, *Mutat Res.* 119:185-190,.

Fairbain, D. W., Olive, P. L., and O'Neill, K. L. The comet assay: a comprehensive review. Mutat. Res. 339:37-59, 1995.

Generoso, W.M., Cattanach, B., Malashenko, A.M., 1981, Mutagenicity of selected chemicals in mammals; the heritable translocation test, in: F. J. de Serres and M. D. Shelby, eds., Comparative Chemical Mutagenesis, Plenum Press, New York, pp. 681-707.

Hales, B.F., Crosman,K., Robaire, B., 1992, Increased postimplantation loss and malformations among the F_2 progeny of male rats chronically treated with cyclophosphamide, *Teratology* 45:671-678.

Hales, B.F., Robaire, B., 1990, Reversibility of the effects of chronic paternal exposure to cyclophosphamide on pregnancy outcome in rats. *Mutat Res.* 229:129-134.

Hales, B.F., Smith, S., Robaire, B., 1986, Cyclophosphamide in the seminal fluid of treated males: transmission to females by mating and effects on progeny outcome, *Toxicol Appl Pharmacol.* 84:423-430.

Hall, E.S., Hall, S.J., Boekelheide, K., 1995, 2,5 Hexanedione exposure alters microtubule motor distribution in adult rat testis. *Fund Appl Toxicol.* 24:173-182.

Irvine, D.S., Twigg, J.P., Gordon, E.L., Fulton, N., Milne, P.A., Aitken, R.J., 2000, DNA integrity in human spermatozoa: relationships with semen quality. *J Androl* 21:33-44.

Harrouk, W., Codrington, A., Vinson, R., Robaire, B., Hales, B.F., 2000b, Paternal exposure to cyclophosphamide induces DNA damage and alters the expression of DNA repair genes in the rat preimplantation embryo. *Mutat Res.* 461:229-241.

Harrouk, W., Khatabaksh, S., Robaire, B., Hales, B.F., 2000c, Paternal exposure to cyclophosphamide dysregulates the gene activation program in rat preimplantation embryos. *Molec Reprod Dev.* 57:214-223.

Harrouk, W., Robaire, B. Hales, B. F., 2000a, Paternal exposure to cyclophosphamide alters cell-cell contacts and activation of embryonic transcription in the pre-implantation rat embryo. *Biol Reprod.* 63:74-81.

Hikim, A.P., Wang, C., Leung, A., Swerdloff, R.S., 1995, Involvement of apoptosis in the induction of germ cell degeneration in adult rats after gonadotropin-releasing hormone antagonist treatment. *Endocrinology.* 136:2770-2775.

Kangasniemi, M., Wilson, G., Parchuri, N., Huhtaniemi, I., Meistrich, M. L., 1995, Rapid protection of rat spermatogenic stem cells against procarbazine by treatment with gonadotropin-releasing hormone antagonist (Nal-Glu) and an antiandrogen (flutamide). *Endocrinology* 136:2881-2888.

Kelce, W.R., Stone, C.R., Laws, S.C., Gray, L.E., Kemppainen, J.A., Wilson, E.M., 1995, Persistent DDT metabolite p,p'-DDE is a potent androgen receptor antagonist, *Nature* 375:581-585.

Kelly, S.M., Robaire, B. and Hales, B.F. (1992) Paternal cyclophosphamide treatment causes post-implantation loss via inner cell mass-specific cell death. *Teratology* 45:313-318.

Kleeman, J.M., Moore, R.W., Peterson, R.E., 1990, Inhibition of testicular steroidogenesis in 2,3,7,8-tetrachloro-p-dioxin treated rats: evidence that the key lesion occurs prior to or during pregnenolone formation, *Toxicol Appl Pharmacol.* 106:112-115.

Klinefelter, G.R., Laskey, J.W., Roberts, N.L., 1991, In vitro/in vivo effects of ethane dimethanesulfonate on Leydig cells of adult rats. *Toxicol Appl Pharmacol.* 107:460-471.

Kohn, K.W., 1979, DNA as a target in cancer chemotherapy: Measurement of macromolecular and damage produced in mammalian cells by anticancer agents and carcinogens. *Methods Cancer Res.* 16:291-345.

Lowe, X.R., de Stoppelaar, J.M., Bishop, J., Cassel, M., Hoebee, B., Moore II, D., Wyrobek, A.J., 1998, Epididymal sperm aneuploidies in three strains of rats detected by multicolor fluorescence in situ hybridization. *Env Mol Mutagen.* 31:125-132.

Luke, G.A., Ricches, A.C., Bryant, P.E., 1997, Genomic instability in haematopoietic cells of F_1 generation mice of irradiated male parents. *Mutagenesis* 12:147-152.

Lutwak-Mann, C., 1964, Observations on the progeny of thalidomide-treated male rabbits. *Br Med J.* 1:1090-1091.

Martin, R.H., Rademaker, A., Hildebrand, K., Barnes, M., Arthur, K., Ringrose, T., Brown, I.S., Douglas, G., 1989, A comparison of chromosomal aberrations induced by in vivo radiotherapy in human sperm and lymphocytes. *Mutat Res.* 226:21-30.

Martin, R.H., Spriggs, E., Ko, E., Rademaker, A.W., 1995, The relationship between paternal age, sex ratios, and aneuploidy frequencies in human sperm, as assessed by multicolor FISH. *Am J Hum Genet.* 57:1395-1399.

Monteith, D.K. Vanstone, J., 1995, Comparison of the microgel electrophoresis assay and other assays for genotoxicity in the detection of DNA damage. *Mutat Res.* 345:97-103.

Naish, S.J., Perreault, S.D., Zirkin, B.R., 1987, DNA synthesis in the fertilizing hamster sperm nucleus: sperm template availability and egg cytoplasmic control. *Biol Reprod.* 36:245-253.

Neuber, E., Havari, E., Sanchez, J.A., Powers, R.D., Wangh, L.J., 1999, Efficient human sperm pronucleus formation and replication in Xenopus egg extracts. *Biol Reprod.* 61:912-920.

Perreault, S.D., Barbee, R.R., Elstein, K.H., Zucker, R.M., Keefer, C.L., 1988, Interspecies differences in the stability of mammalian sperm nuclei assessed in vivo by sperm microinjection and in vitro by flow cytometry. *Biol Reprod.* 39:157-167.

Perreault, S.D., Naish, S.J., Zirkin, B.R., 1987, The timing of hamster sperm nuclear decondensation and male pronucleus formation is related to sperm nuclear disulfide bond content. *Biol Reprod.* 36:239-244.

Perreault, S.D., Zirkin, B. R., 1982, Sperm nuclear decondensation in mammals: role of sperm-associated proteinase in vivo. *J Exp Zool.* 224:253-257.

Qiu, J., Hales, B.F., Robaire, B., 1992, Adverse effects of cyclophosphamide on progeny outcome can be mediated through post-testicular mechanisms in the rat. *Biol Reprod.* 46:926-931.

Qiu, J., Hales, B.F., Robaire, B., 1995a, Effects of chronic low dose cyclophosphamide exposure on the nuclei of rat spermatozoa. *Biol Reprod.* 52:33-40.

Qiu, J., Hales, B.F., Robaire, B., 1995b, Damage to rat spermatozoal DNA after chronic cyclophosphamide exposure, *Biol Reprod.* 53:1465-1473.

Ram, P.T., Schultz, R.M., 1993, Reporter gene expression in G2 of the 1-cell embryo. *Dev Biol.* 156:552-556.

Ries, L.A.G., Smith, M.A., Gurney, J.G., Linet, M., Tamra, T., Young, J.L., Bunin, G.R., Bernstein, L., Key, C.R., Lynch, C.F., Simone, J., Stevens, J., 1999, Cancer Incidence and Survival Among Children and Adolescents: United States SEER Program 1975-1995. Bethesda, MD: National Cancer Institute, SEER Program, NIH Pub. No. 99-4649; 51-63.

Robaire, B., Hales, B.F., 1999, The Male Germ Cell as a Target for Drug and Toxicant Action. In: The Male Gamete: From Basic Science to Clinical Applications, Gagnon, C. Ed., Cache River Press, FL, pp 469-474.

Russell, L.B., 1994, Effects of spermatogenic cell type on quantity and quality of mutations. In Male-Mediated Developmental Toxicity, Mattison, D. R. and Olshan, A. F., Eds., Plenum Press, New York, pp. 37-48.

Sadowy, S., Tomkin, G., Munné, S., Ferrara-Congedo, T., Cohen, J., 1998, Impaired development of zygotes with uneven pronuclear size. *Zygote* 6:137-141.

Savitz, D.A., Sonnenfeld, N.L., Olshan, A.F., 1994, Review of epidemiologic studies of paternal occupational exposure and spontaneous abortion. *Am J Ind Med.* 25:361-383.

Sotomayor, R.E., Cumming, R.B., 1975, Induction of translocations by cyclophosphamide in different germ cell stages of male mice: cytological characterization and transmission. *Mutat Res.* 27:375-388.

Sotomayor, R.E., Sega, G.A., Cumming, R.B., 1978, Unscheduled DNA synthesis in spermatogenic cells of mice treated in vivo with the indirect alkylating agents cyclophosphamide and mitomen. *Mutat Res.* 50:229-240.

Soyka, L.F., Peterson, J.M., Joffe, J.M., 1978, Lethal and sublethal effects on the progeny of male rats treated with methadone. *Toxicol Appl Pharmacol.* 45:797-807.

Suzuki H, Togashi M, Adachi J, Toyoda Y., 1995, Developmental ability of mouse embryos is influenced by cell association at the 4-cell stage. *Biol Reprod* 53:78-83.

Trasler, J.M., Hales, B.F., Robaire, B., 1985, Paternal cyclophosphamide treatment of rats causes fetal loss and malformations without affecting male fertility. *Nature* 316:144-146.

Trasler, J.M., Hales, B.F., Robaire, B., 1986, Chronic low dose cyclophosphamide treatment of adult male rats: effect on fertility, pregnancy outcome and progeny. *Biol Reprod.* 34:275-283.

Trasler, J.M., Hales, B.F., Robaire, B., 1987, A time course study of chronic paternal cyclophosphamide treatment in rats: effects on pregnancy outcome and the male reproductive and hematologic systems. *Biol Reprod.* 37:317-326.

Worrad, D.M., Ram, P.T., Schultz, R.M., 1993, Regulation of gene expression in the mouse oocyte and early preimplantation embryo: developmental changes in Sp1 and TATA box-binding protein, TBP. *Development.* 120:2347-2357.

Wyrobek, A.J., Robbins, W.A., Mehraein, Y., Pinkel, D., Weier, H.U., 1994, Detection of sex chromosomal aneuploidies X-X, Y-Y, and X-Y in human sperm using two-chromosome fluorescence in situ hybridization. *Am J Med Genet.* 53:1-7.

CHROMOSOME ABNORMALITIES IN HUMAN SPERM

Renée H. Martin

Department of Medical Genetics
University of Calgary
Calgary, Alberta, Canada

INTRODUCTION

Chromosome abnormalities are common in humans, with approximately 0.6% present in newborns, 6% in stillbirths and 60% in spontaneous abortions. In comparison to other species, humans appear to have the highest incidence of chromosome anomalies with up to 50% of conceptions being abnormal (Bond and Chandley 1983; Boué et al., 1975). Most chromosome abnormalities are lethal and are lost early in embryologic development, manifesting as infertility or spontaneous abortions. Chromosomally abnormal concepti that survive to term include trisomy 13, 18, 21 and aneuploidy for the sex chromosomes as well as some structural alterations such as translocations and inversions (Jacobs 1992). Children with these chromosome abnormalities often have physical disabilities, reduced intellectual capabilities, behavioural problems, and/or impaired sexual development.

To study the etiology of these chromosome abnormalities, it is best to study human gametes, since the great majority of errors occur during meiosis and analysis of newborns or spontaneous abortions yields information on only those concepti that have survived. It has been difficult to study human oocytes since they are not available in large numbers. However, there have been significant developments of techniques for studying chromosome abnormalities in human sperm during the past two decades. In 1978, Rudak et al. published a technique that permits the visualization of human sperm chromosomes. This provided the first information on the frequency and type of chromosome abnormalities in normal men (Martin et al., 1983, 1987; Brandriff et al., 1985; Kamiguchi and Mikamo, 1986) as well as research on men at increased risk of anomalies such as cancer patients (Martin et al., 1986; Genesca et al., 1990; Jenderny et al., 1992). Despite the great excitement generated by the sperm karyotyping technique, there are significant disadvantages: it is extremely time consuming, provides low yields of karyotypes and has been successfully applied in only a dozen laboratories worldwide. The past decade has seen the development of fluorescence in situ hybridization (FISH) for analysis of human sperm chromosomes as a rapid, simple technique that can yield data on very large numbers of sperm. However, it must be remembered that FISH is an indirect method, more prone to error than direct karyotyping of human sperm. To date, over five million sperm from more

Advances in Male Mediated Developmental Toxicity, edited by Bernard Robaire and Barbara F. Hales.
Kluwer Academic/Plenum Publishers, 2003.

than 500 normal men have been analyzed by a number of laboratories from around the world for sperm disomy frequencies by FISH analysis (Shi and Martin, 2000).

Several studies have analyzed sperm chromosome abnormalities in normal men (McInnes et al., 1998b; Scarpato et al., 1998; Spriggs et al., 1996) and men with an increased risk of sperm chromosome abnormalities because of cancer treatment (Martin et al., 1986; Robbins et al., 1997a; Martin et al., 1997a,b), constitutional chromosomal abnormalities (Martin and Spriggs, 1995; Escudero et al., 2000; Shi and Martin, 2000b) environmental factors (Robbins et al., 1997b; Shi and Martin, 2000a), and infertility (McInnes et al., 1998a; Pang et al., 1999; Shi and Martin, 2001). In this paper, we will review information from sperm chromosome studies on cancer patients and men exposed to pesticides.

THE FREQUENCY AND DISTRIBUTION OF SPERM CHROMOSOME ABNORMALITIES IN NORMAL MEN

Sperm karyotyping using the hamster oocyte/human sperm fusion technique has demonstrated that the frequency of numerical chromosome abnormalities in normal men is 1-2% and varies from 7-14% for structural chromosome abnormalities (reviewed in Martin, 1993). On average, sperm karyotyping studies have shown that approximately 10% of human sperm from normal men have abnormalities (reviewed in Guttenbach et al., 1997). As well as looking at the frequency of chromosomal abnormalities, it is also of interest to study the distribution of numerical abnormalities (aneuploidy) to determine if some chromosomes are more susceptible to nondisjunction. Since all chromosomes have been observed as trisomies in human embryos, it appears that all chromosomes are susceptible to nondisjunction. However, it is not clear whether all chromosomes display the same frequency of nondisjunction at meiosis. Our studies on sperm karyotypes have demonstrated that all chromosomes undergo nondisjunction during spermatogenesis but the G group chromosomes (21 and 22) and the sex chromosomes have a significantly increased frequency of aneuploidy (Martin et al., 1991). The technique of human sperm karyotyping provided our first glimpse at the frequency, type, and distribution of chromosome abnormalities in human sperm. This technique has a number of distinct advantages: precise information is obtained on all chromosomes, numerical as well as structural abnormalities can be assessed, and information on individual bands of chromosomes can be studied (e.g., in deletions or inversions). However, there are significant disadvantages to this technique: the hamster oocyte/human sperm analysis is extremely time-consuming, requires the sperm to have the ability to fertilize an egg, yields are low, and requires considerable expertise, as has been demonstrated by the fact that it has only been successful in a dozen laboratories worldwide. This led to the search for easier, cheaper and more productive methods.

Sperm chromosome analysis by fluorescence in situ hybridization (FISH) has been embraced in the past 10 years as this technique can be performed on interphase sperm nuclei without fertilization of an egg, is very rapid, inexpensive and allows information on very large samples of sperm. However, there are a number of disadvantages: results are indirect and a fluorescent signal is inferred to indicate the presence of a chromosome, scoring criteria must be adhered to strictly, and information is only available for those chromosomes being tested. In general, only numerical abnormalities are studied by FISH analysis, as analysis for structural abnormalities requires a number of centromeric and telomeric probes for one chromosome and is still an indirect assessment of the structural chromosome complement (Van Hummelen et al., 1996; McInnes et al., 1998b).

An approximate frequency of chromosome abnormalities in human sperm is more difficult to estimate with the use of FISH analysis as frequencies for individual chromosomes vary tremendously. A few studies, including ours, have found frequencies of aneuploidy that are very similar to those observed in the hamster oocyte system (Spriggs et al., 1996; Williams et al., 1993; Wyrobek et al., 1994) whereas others have found much higher frequencies (Pang et al., 1999; Rousseaux et al., 1998). This highlights the fact that FISH is a subjective technique and stringent criteria must be maintained for acceptance of a sperm with 2 signals (designated as a disomic sperm, representing 2 chromosomes). We have determined that most chromosomes have a disomy frequency of approximately 0.1% or 1/1000 using the FISH technique (Spriggs et al., 1996). However, chromosomes 21, 22 (Martin and Rademaker, 1999) and the sex chromosomes have a significantly increased frequency of aneuploidy (Spriggs et al., 1996). These data corroborate our results from human sperm karyotypes (Martin et al., 1991) and suggest that the sex chromosome bivalent and the G group chromosomes are more susceptible to nondisjunction during spermatogenesis. Other FISH studies on human sperm also suggest an increase in the frequency of aneuploidy for the sex chromosomes compared with the autosomes (Williams et al., 1993; Bischoff et al., 1994; Scarpato et al., 1998) and also for chromosome 21 disomy compared to other autosomes (Blanco et al., 1996). The increased proportion of paternal meiotic errors in sex chromosomal aneuploidies (MacDonald et al., 1994) concurs with the elevated frequency of sex chromosomal disomy in human sperm. Both the XY bivalent and chromosome 21 normally only have one chiasma (Laurie and Hulten 1985). If the frequency of recombination is reduced or absent for these chromosomes, they may be particularly susceptible to nondisjunction. Indeed, it has been shown that 47,XXY of paternal origin is associated with a decreased recombination frequency (Hassold et al., 1991) and trisomy 21 of paternal origin has an altered recombination pattern (Savage et al., 1998). We have recently demonstrated, by single-sperm PCR, that aneuploid 24,XY sperm have a significantly reduced frequency of recombination in the pseudoautosomal region compared to normal sperm (Shi et al., 2001). Thus this susceptibility to nondisjunction for these chromosomes may be due to the reduced fidelity of pairing of homologues in non-recombinant chromosomes.

CANCER PATIENTS

Both sperm karyotyping and FISH analysis have been used to study cancer patients exposed to mutagenic, aneugenic and clastogenic agents. We have studied 13 cancer patients before radiotherapy (RT) and at regular intervals after RT to determine the effect of RT on chromosomal abnormalities in sperm (Martin et al., 1986). The majority of the men had seminoma of the testis and had undergone unilateral orchidectomy. The total body radiation doses were 30 to 40 Gy and the testicular radiation doses ranged from 0.4 to 5.0 Gy. Traditional semen analysis and sperm karyotyping were performed before and after RT at 1, 3, 12, 24, and 36 months. Radiation treatment had a dramatic effect on sperm count. The majority of men became azoospermic in the first year and some men never regained sperm production throughout the study. Most of the men were producing sperm 2 to 3 years after RT, but low sperm counts were common. By 36 months post-RT, 8 men regained sperm production and had an average sperm chromosome abnormality rate of 20.9%. This was significantly higher than control donors (8.5%). The range in the frequencies of abnormal sperm chromosome complements in individuals at 36 months post-RT was 6% to 67% and these frequencies were significantly correlated with the testicular radiation dose. The results demonstrate a significant dose-dependent increase in the frequency of sperm chromosomal abnormalities after RT.

We studied the induction of chromosome abnormalities in lymphocytes of the cancer patients, as well as sperm, to determine if there was any correlation in the frequency of chromosomal abnormalities in these two cell types (Martin et al., 1989). The frequency of abnormalities before RT was 0% and increased to 42% 1 month post-RT. From this level it gradually decreased to 9% by 36 months and 4% by 60 months. There was no correlation or constant ratio between chromosome abnormalities in lymphocytes and sperm, thus lymphocytes cannot be used as surrogates to estimate germ cell damage.

Other studies of cancer patients using the human sperm/hamster oocyte system have had both positive and negative results. For example, Jenderny and Röhrborn (1987) studied one cancer patient 18 to 20 months after exposure to 30 Gy. They did not find a significantly increased frequency of chromosome abnormalities in this man. Genesca and colleagues (1990) studied two patients with Wilm's tumour who received 22 and 40 Gy, respectively, 11-18 years previously. They found a significant increase in the frequency of structural chromosome abnormalities in the sperm of these two patients. It is not surprising that various studies on both RT and chemotherapy (CT) patients have produced divergent results since the studies have included small sample sizes of sperm from small numbers of men with different types of cancer, different treatment regimens, analyzed at different time periods.

Recent studies using FISH analysis have attempted to examine more homogeneous groups of men with the same type of cancer and treatment. We studied four testicular cancer patients (all with embryonal carcinoma) before and 2 to 13 years after CT (Martin et al., 1997 a & b). All men had the same treatment with BEP therapy (bleomycin, etoposide, cisplatin). FISH analysis with chromosome-specific DNA probes was utilized to allow analysis of a large sample of sperm in each patient as well as sperm karyotype analysis to provide precise information on both structural and numerical chromosomal abnormalities. Multicolour FISH analysis for assessment of aneuploidy for chromosomes 1, 12, X, and Y was performed on a total of 161,097 sperm, 80,445 before and 80,642 after treatment (Martin et al., 1997b). There was no significant difference in the frequency of disomy pre-CT vs post-CT for any chromosome. The sex ratios and frequency of diploid sperm were also not significantly different in pre- and post-CT samples. None of the values in the cancer patients differed significantly from those of ten normal control donors. Using the human sperm/hamster oocyte system, a total of 788 sperm chromosome complements was studied, 236 before CT and 552 post-CT. There was no significant difference in the total frequency of sperm chromosomal abnormalities pre-CT (10.2%) compared with post-CT (10.7%). Similarly, there were no significant differences in the frequencies of numerical abnormalities or structural abnormalities. The percentage of X-bearing sperm was also not significantly different before and after CT. The results in cancer patients were not significantly different from those in control donors (Martin et al., 1997a).

Together, these two types of analyses suggest that men treated with BEP CT do not have an increased risk of chromosomal aneuploidy in their sperm after treatment. These studies are reassuring but they only provide information on sperm chromosome complements obtained 2 or more years after treatment. Any chromosome abnormality observed after this time period would have occurred because of damage to the spermatogonial stem cells. Animal studies have suggested that meiotic (spermatocytes) and post-meiotic (spermatids, spermatozoa) cells are more sensitive to the induction of mutations than are pre-meiotic cells (stem cells) (Adler and Tarras 1990; Qiu et al., 1992). Thus it is important to assay the effect on spermatocytes, spermatids, and spermatogonia during CT as well as post-CT.

In an elegant study on FISH analysis in Hodgkin's patients treated with NOVP therapy (novantrone, oncovin, vinblastine, prednisone), Robbins et al., 1997a demonstrated an elevation in aneuploidy frequencies during treatment. We have also studied testicular cancer patients before, during and 12 months after BEP CT (Martin et al., 1999). A total of 60,400 sperm were analyzed: 20,004 before CT, 20,005 during CT (60 days after the start of treatment) and 20,391 after CT (1 year after the start of CT). Disomy frequencies for chromosome 1, 12, XX and YY were not significantly different pre-, during or post-CT. There was a significant increase in the frequency of 24,XY sperm during CT (0.33%) and post-CT (0.34%) compared with pre-CT (0.14%). There was also a significant increase in the frequency of diploid sperm post-CT (0.29%) compared with pre- (0.10%) and during (0.14%) CT values. Thus, it is possible that there may be a significantly increased risk of chromosomal abnormalities in sperm of CT patients during and immediately post-CT and consequently it would be unwise to bank sperm samples at this time. Animal studies have also demonstrated that this time period is the most sensitive for induction of chromosome abnormalities (Marchetti et al., 2001). A recent study by De Mas et al. (2001) has confirmed that this time period is sensitive for the induction of sperm chromosome abnormalities in humans. They studied 5 testicular cancer patients 6 to 18 months after BEP CT and found an increased frequency of disomy 16, 18 and XY as well as diploidy compared to control donors. Clearly more studies must be conducted, but the evidence to date does suggest an increased risk of induced sperm chromosome abnormalities during chemotherapy and up to 18 months following treatment.

MEN EXPOSED TO PESTICIDES

There is increasing concern about the consequences of the burden of chemicals in our environment. Pesticides are used liberally all over the world, and there may be adverse consequences to reproduction and fertility. A few studies utilizing FISH analysis have attempted to assess the risk of chromosome abnormalities in men exposed to high levels of pesticides. We have studied 20 pesticide appliers exposed to a variety of herbicides, insecticides and fungicides. Multicolour FISH analysis performed blindly with 20 controls from the same region in over 800,000 sperm failed to demonstrate a significant difference for disomy 13, 21, XX, XY, YY, or for diploidy. The only significant difference observed was an increased frequency of X-bearing sperm in the exposed men (unpublished results). Harkonen et al., (1999) studied sperm from 30 farmers exposed to fungicides; they found no significant increase in disomy frequencies for chromosome 1 or 7. In contrast, Padungtod et al., (1999) studied aneuploidy in sperm of 32 Chinese pesticide factory workers and found an increase in YY aneuploidy. These differing results may reflect different pesticides used or the level of exposure. Certainly there are too few studies to date to draw any conclusions. One recommendation for future studies would be to include data on aneuploidy for chromosomes 21 and the sex chromosomes as these chromosomes have been shown to be most susceptible to nondisjunction in normal (Martin et al., 1991; Spriggs et al., 1996; Scarpato et al., 1998; Blanco et al., 1996) and infertile men (Moosani et al., 1995; McInnes et al., 1998a; Pfeffer et al., 1999).

INFERTILE MEN

Infertile men have an increased risk of chromosome abnormalities in their sperm; these abnormalities can be passed on to their offspring. This has become more clinically relevant with the advent of intracytoplasmic sperm injection (ICSI) since even men with extremely poor sperm parameters can successfully father a pregnancy. The chromosome abnormalities in sperm can occur because of constitutional chromosome abnormalities in the somatic cells of the infertile men (e.g., translocations or sex chromosomal aneuploidy) or because of a susceptibility to nondisjunction in chromosomally normal infertile men. Many studies employing FISH analysis have been published in this area, and this subject has been reviewed recently (Shi and Martin 2001).

ACKNOWLEDGEMENTS

Renée Martin holds a Canada Research Chair in Genetics and her research is supported by the Canadian Institute of Health Research. Special thanks to Linda Macaulay for preparing the manuscript.

REFERENCES

Adler, I., El Tarras, A., 1990, Clastogenic effects of cis-diamminedichloroplatinum. II. Induction of chromosomal aberrations in primary spermatocytes and spermatogonia stem cells of mice. *Mutat Res.* 243:173-178.

Bischoff, F., Nguyen, D., Burt, K., Shaffer, L., 1994, Estimates of aneuploidy using multicolor fluorescence in situ hybridization on human sperm. *Cytogenet Cell Genet.* 66:237-243.

Blanco, J., Egozcue, J., Vidal, F., 1996, Incidence of chromosome 21 disomy in human spermatozoa as determined by fluorescent in-situ hybridization. *Hum Reprod.* 11:722-726.

Bond, D.J., Chandley, A.C., 1983. Aneuploidy. In: *Oxford Monographs on Medical* Genetics, (J.A.F. Roberts, C.O. Carter, A.G. Motulsky, eds.) Oxford University Press, Oxford, p198.

Boue, J., Boue, A., Lazar, P., 1975, Retrospective and prospective epidemiological studies of 1500 karyotyped spontaneous human abortions. *Teratology.* 12:11-26.

Brandriff, B., Gordon, L., Ashworth, L., Watchmaker, G., Moore, D. 2d, Wyrobek, A., Carrano, A., 1985, Chromosomes of human sperm: variability among normal individuals. *Hum Genet.* 70:18-24.

De Mas, P., Daudin, M.-C., Bourrouillou, G., Calvas, P., Mieusset, R., Bujan, L., 2001, Increased aneuploidy in spermatozon from testicular tumour patients after chemotherapy with cisplatin, etoposide and bleomycin. *Hum repod.* 16:1204-1208.

Escudero, T., Lee, M., Carrel, D., Blanco, J., Munne, S., 2000, Analysis of chromosome abnormalities in sperm and embryos from two 45,XY,t(13;14)(q10;q10) carriers. *Prenatal Diagnosis.* 20:599-602.

Genesca, A., Caballin, M.R., Miro, R., Benet, J., Bonfill, X., Egozcue, J., 1990, Human sperm chromosomes. Long-term effect of cancer treatment. *Cancer Genet Cytogent.* 46:251-260.

Guttenbach,M,. Engel, W., Schmid, M., 1997, Analysis of structural and numerical chromosome abnormalities in sperm of normal men and carriers of constitutional chromosome aberrations: A review. *Hum Genet.* 100:1-21.

Harkonen, K., Viitanen, T., Larsen, S., Bonde, J., Lahdetie, J., 1999, Aneuploidy in sperm and exposure to fungicides and lifestyle factors. ASCLEPIOS. A European concerted action on occupational hazards to male reproductive capability. *Environ Mol Mutagen.* 34:39-46.

Hassold, T., Sherman, S., Pettay, D., Page, D., Jacobs, P., 1991, XY Chromosome nondisjunction in man is associated with diminished recombination in the pseudoautosomal region. *Am J Hum Genet.* 49:253-260.

Jacobs, P., 1992, The chromosome complement of human gametes. In *Oxford Review of Reproductive Biology*, (S.R. Milligan, ed.). Oxford University Press, Oxford, pp 47.

Jenderny, J., Jacobi, M., Ruger, A., Rohrborn, G.,1992, Chromosome aberrations in 450 sperm complements from eight controls and lack of increase after chemotherapy in two patients. *Hum Genet.* 90:151-154 .

Jenderny, J., Rohrborn, G., 1987, Chromosome analysis of human sperm. 1. First results with a modified method. *Hum Genet.* 76:385-388.

Kamiguchi, Y., Mikamo, K., 1986, An improved, efficient method for analyzing human sperm chromosomes using zona-free hamster ova. *Am J Hum Genet.* 38:724-740.

Laurie, D., Hulten, M., 1985, Further studies on chiasma distribution and interference in the human male. *Ann Hum Genet.* 49:203-214.

MacDonald, M., Hassold, T., Harvey, J., Wang, L.H., Morton, N.E., Jacobs, P., 1994, The origin of 47,XXY and 47,XXX aneuploidy: heterogeneous mechanisms and role of aberrant recombination. *Hum Mol Genet.* 3:1365-1371.

Marchetti, F., Bishop, J.B., Lowe, X., Generoso, W.M., Hozier, J., Wyrobek, A.J., 2001, Etoposide induces heritable chromosomal aberrations and aneuploidy during male meiosis in the mouse. *Proc Natl Acad Sci USA.* 98:3952-3957.

Martin, R.H., 1993, Detection of genetic damage in human sperm. *Reprod Tox.* 7:47-52.

Martin, R.H., Balkan, W., Burns, K., Rademaker, A.W., Lin, C.C., Rudd, N.L., 1983, The chromosome constitution of 1000 human spermatozoa. *Hum Genet.* 63:305-309.

Martin, R.H., Ernst, S., Rademaker, A., Barclay, L., Ko, E., Summers, N., 1997a, Analysis of human sperm karyotypes in testicular cancer patients before and after chemotherapy. *Cytogenet Cell Genet* 78:120-123.

Martin, R.H., Ernst, S., Rademaker, A., Barclay, L., Ko, E., Summers, N., 1997b, Chromosomal abnormalities in sperm from testicular cancer patients before and after chemotherapy. *Hum Genet.* 99:214-218.

Martin, R.H., Ernst, S., Rademaker, A., Barclay, L., Ko, E., Summers, N., 1999, Analysis of sperm chromosome complements before, during and after chemotherapy. *Cancer Genet Cytogenet.* 108:133-136.

Martin, R.H., Hildebrand, K., Yamamoto, J., Rademaker, A., Barnes, M., Douglas, G., Arthur, K., Ringrose, T., Brown, I.S., 1986, An increased frequency of human sperm chromosomal abnormalities after radiotherapy. *Mutat Res.* 174:219-225.

Martin, R.H., Ko, E., Rademaker, A., 1991, Distribution of aneuploidy in human gametes: comparison between human sperm and oocytes. *Am J Med Genet.* 39:321-331.

Martin, R.H., Rademaker, A.W., 1999, Nondisjunction in human sperm: Comparison of frequencies in acrocentric chromosomes. *Cytogenet Cell Genet.* 86:43-45.

Martin, R.H., Rademaker, A., Hildebrand, K., Barnes, M., Arthur, K., Ringrose, T., Brown, I.S., Douglas, G., 1989, A comparison of chromosomal aberrations induced by in-vivo radiotherapy in human sperm and lymphocytes. *Mutat Res.* 226:21-30.

Martin, R.H., Rademaker, A.W., Hildebrand, K., Long-Simpson, L., Peterson, D., Yamamoto, J., 1987, Variation in the frequency and type of sperm chromosomal abnormalities among normal men. *Hum Genet.* 77:108-114.

Martin, R.H., Spriggs, E.L., 1995, Sperm chromosome complements in a man heterozygous for a reciprocal translocation 46,XY,t(9;13)(q21.1;q21.2) and a review of the literature. *Clin Genet.* 47:42-46.

McInnes, B., Rademaker, A., Greene, C., Ko, E., Barclay, L., Martin, R.H., 1998a, Abnormalities for chromosomes 13 and 21 detected in spermatozoa from infertile men. *Hum Reprod.* 13:2787-2790.

McInnes, B., Rademaker, A., Martin, R.H., 1998b, Donor age and the frequency of disomy for chromosomes 1, 13, 21 and structural abnormalities in human spermatozoa using multicolour fluorescence in-situ hybridization. *Hum Reprod.* 13:2489-2494.

Moosani, N., Pattinson, H A , Carter, M D , Cox, D.M., Rademaker, A.W., Martin, R.H., 1995, Chromosomal analysis of sperm from men with idiopathic infertility using sperm karyotyping and fluorescence in situ hybridization. *Fertil Steril.* 64:811-817.

Padungtod, C., Hassold, T.J., Millie, E., Ryan, L.M., Savitz, D.A., Christiani, D.C., Xu, X., 1999, Sperm aneuploidy among Chinese pesticide factory workers: scoring by the FISH method. *Am J Ind Med.* 36:230-238.

Pang, M., Hoegerman, S., Cuticchia, A., Moon, S., Doncel, G., Acosta, A., and Kearns, W., 1999, Detection of aneuploidy for chromosomes 4,6,7,8,9,10,11,12,13,17,18,21,X and Y by fluorescence in situ hybridization in spermatozoa from nine patients with oligoasthenoteratozoospermia undergoing intracytoplasmic sperm injection. *Hum Reprod.* 14:1266-1273.

Pfeffer, J., Pang, M., Hoegerman, S., Osgood, C., Stacey, M., Mayer, J., Oehninger, S., Kearns, W., 1999, Aneuploidy frequencies in semen fractions from ten oligoasthenoteratozoospermic patients donating sperm for intracytoplasmic sperm injection. *Fertil Steril.* 72:472-478.

Qiu, J., Hales, B., Robaire, B., 1992, Adverse effects of cyclophosphamide on progeny outcome can be mediated through post-testicular mechanisms in the rat. *Biol Reprod.* 46:926-931.

Robbins, W, Meistrich, M., Moore, D., Wyrobek, A., Hagemeister, F., Heinz-Ulrich, W., Cassel, M. et al., 1997a, Aneuploidy in sperm of Hodgkin's disease patients receiving NOVP chemotherapy. *Nature Genetics.* 16:74-78.

Robbins, W., Vine, M., Truong, K., Everson, R., 1997b, Use of fluorescence in situ hybridization (FISH) to assess effects of smoking, caffeine, and alcohol on aneuploidy load in sperm of healthy men. *Environ Mol Mutagen.* 20:175-183.

Rousseaux, S., Hazzouri , M., Pelletier, R., Monteil, M., Usson, Y., Sele, B., 1998, Disomy rates for chromosomes 14 and 21 studied by fluorescent in-situ hybridization in spermatozoa from three men over 60 years of age. *Mol Hum Reprod.* 4:695-699.

Rudak, E., Jacobs, P., Yanagimachi, R., 1978, Direct analysis of the chromosome constitution of human spermatozoa. *Nature.* 274:911-913.

Savage, A.R., Petersen, M.B., Pettay, D., Taft, L., Allran, K., Freeman, S.B., Karadima, G., Avramopoulos, D., Torfs, C., Mikkelsen, M., Hassold, T.J., Sherman, S.L., 1998, Elucidating the mechanisms of paternal non-disjunction of chromosome 21 in humans. *Hum Mol Genet.* 7:1221-1227.

Scarpato, R., Naccarati, A., Mariani, M., Migliore, L.,1998. Aneuploidy and diploidy rates in sperm of five men after three-colour hybridization: indication of X chromosome-associated autosome 2 aneuploidy. *Mutat Res.* 412:227-233.

Shi, Q., Martin, R.H., 2000a, Aneuploidy in human sperm: a review of the frequency and distribution of aneuploidy, effects of donor ages and lifestyle factors. *Cytogenet Cell Genet.* 90: 219-226.

Shi, Q., Martin, R.H., 2000b, Multicolour fluorescence in situ hybridization analysis of meiotic chromosome segregation in a 47,XYY male and a review of the literature. *Amer J Med Genet.* 93:40-66.

Shi, Q., Martin, R.H., 2001, Aneuploidy in human sperm: a review of FISH analysis in men with constitutional chromosome abnormalities, and in infertile men. *Reproduction.* 121:655-666.

Shi, Q., Spriggs, E., Field, L.L., Ko, E., Barclay, L., Martin, R.H., 2001, Single sperm typing demonstrates that reduced recombination is associated with the production of aneuploid 24, XY human sperm. *Amer J Med Genet.* 99:34-38.

Spriggs, E., Rademaker, A., Martin, R., 1996, Aneuploidy in human sperm: the use of multicolor FISH to test various theories of nondisjunction. *Am J Hum Genet.* 58:356-362.

Van Hummelen, P., Lowe, X.R., Wyrobek, A.J., 1996, Simultaneous detection of structural and numerical chromosome abnormalities in sperm of healthy men by multicolor fluorescence in situ hybridization. *Hum Genet.* 98:608-615.

Williams, B., Ballenger, C., Malter, H., Bishop, F., Tucker, M., Swingman, T., Hassold, T., 1993, Non-disjunction in human sperm: results of fluorescence in situ hybridization studies using two and three probes. *Hum Mol Genet.* 2:1929-1936.

Wyrobek, A.J., Robbins, W.A., Metvalin, Y., Pinkel, D., Weier, H.-U.,1994, Detection of sex chromosomal aneuploidies X-X,Y-Y and X-Y in human sperm using two chromosome fluorescence in situ hybridization. *Am J Med Genet.* 53:1-7.

DISTINGUISHING BETWEEN FERTILIZATION FAILURE AND EARLY PREGNANCY LOSS WHEN IDENTIFYING MALE-MEDIATED ADVERSE PREGNANCY OUTCOMES

Sally D. Perreault

Reproductive Toxicology Division
U.S. EPA, ORD, NHEERL
Research Triangle Park, NC 27711

INTRODUCTION

Successful reproduction depends upon the precise orchestration of many physiological processes. With respect to male reproductive performance, normal copulatory behavior and ejaculatory function are required to insure that semen is deposited in the female tract. Then, a sufficient number of sperm must reach the site of fertilization at the optimal time for the oocyte to be fertilized, and those sperm that reach the oocyte must be capable of fertilizing the egg (binding and penetrating the zona pellucida and fusing with the oolemma). Finally, the fused spermatozoon must be intact genetically in order to support normal embryonic and postnatal development.

Adverse male-mediated developmental outcomes can occur if the fertilizing spermatozoon is genetically abnormal. Genetic abnormalities in sperm can be due to numerical or structural chromosomal abnormalities, or DNA damage such as breaks or alkylations that may give rise to mutations in the zygote. Depending on the severity of the genetic damage and the ability of the oocyte to repair it, the embryo might fail at any stage of pregnancy or might develop to term with abnormalities. When embryos are lost before implantation, it is difficult, if not impossible, at least in humans, to distinguish between infertility due to very early embryo death and that due to failed fertilization.

Why does this distinction matter? For infertile couples and their physicians, this distinction may influence the choice of therapeutic strategies. For example, if the sperm appear normal but are unable to fertilize an oocyte, then intracytoplasmic sperm injection may be appropriate. However, if the defect is genetic in nature, then the couple may decide to turn to donor insemination or adoption. For companies developing male contraceptives, it is imperative to determine that a sperm-directed strategy (whether designed to kill sperm or block their production or maturation) prevents conception without damaging the sperm DNA. Users may not want a post-conception approach for ethical reasons, and there should be no risk of producing an abnormal pregnancy, should fertilization occur despite the use of the contraceptive. On the other hand, toxicologists testing various types of

Advances in Male Mediated Developmental Toxicity, edited by Bernard Robaire and Barbara F. Hales.
Kluwer Academic/Plenum Publishers, 2003.

therapeutic drugs need to identify potential side effects on reproduction, including any deficits in spermatozoa that could affect either fertility or early embryonic development. Finally, reproductive and developmental toxicology risk assessors in Federal Agencies such as the US EPA, FDA and USDA, need information about the modes of action of reproductive toxicants identified in animal test species in order to extrapolate potential risks to humans (U.S. Environmental Protection Agency, 1996; Clegg et al., 2001). For example, if a chemical alters fertility in rats, it is important to know the affected sex, and to characterize the mode and mechanism of action to determine if that chemical is likely to produce the same effect in humans. Thus, there are many reasons why it is important to distinguish between fertilization failure and early pregnancy loss. However, it is not easy to do this, especially in humans.

ANIMAL TEST PROTOCOLS FOR IDENTIFYING MALE-MEDIATED REPRODUCTIVE RISKS

When screening new drugs and chemicals for their potential to disrupt reproduction, Federal Agencies use specific test guidelines for reproductive and fertility effects (U.S. Environmental Protection Agency, 1998). These guidelines specify a multigenerational test protocol in which rats of both sexes are dosed continuously from early adulthood through two subsequent generations (Figure 1). This comprehensive test is designed to detect adverse effects of exposures to test substances at any and all stages of the reproductive life cycle, including transgenerational effects (such as male-mediated adverse developmental outcomes). The test relies heavily on breeding success in each generation, and also includes histological examination of the reproductive organs, and measures of sperm production, motility and morphology that may help to localize adverse effects to the male germ cells (Clegg et al., 2001).

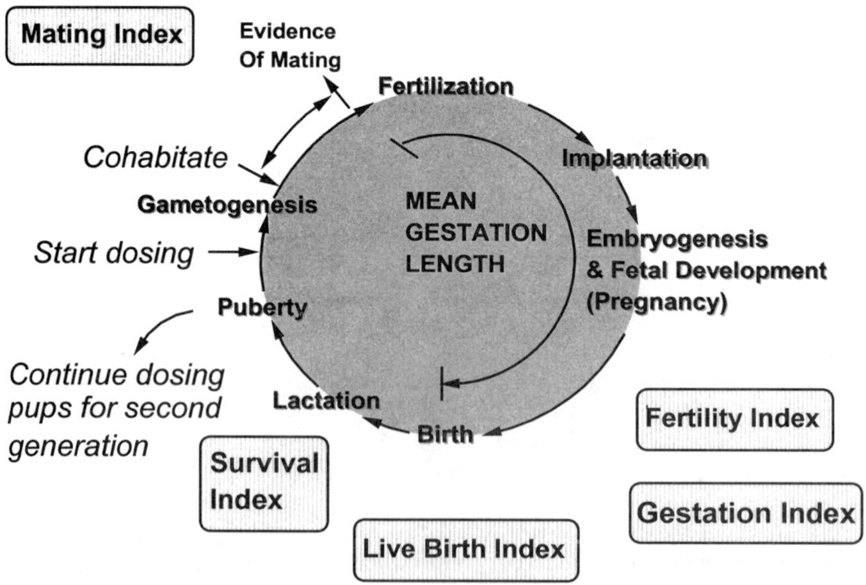

Figure 1. Multigeneration Study Diagram: Dosing of young adult rats (both sexes) begins after puberty and continues through gametogenesis, mating, pregnancy and lactation. After weaning, the first generation of pups is similarly dosed through a second generation. Various indices of reproductive performance are obtained for the parental, and subsequent two generations.

This test design has several limitations. Since both sexes are dosed, it may not be possible to identify the affected sex, although crossover breedings (with untreated animals) can be added to obviate this limitation (Chapin and Sloane, 1997). In addition, since reproductive success is measured at birth, the standard multigeneration test does not distinguish between reduced litter size (or infertility) occurring secondary to failed fertilization and the same outcomes arising from post-fertilization embryo or fetal loss. The latter shortcoming might be overcome to some extent by adding a test of sperm genetic integrity. As discussed in detail at this conference, several new tests for DNA and/or chromosome damage in sperm have been developed and are at differing stages of characterization and validation (Perreault et al., 2003).

A different study design was developed to test suspected mutagens and DNA-reactive chemicals for their ability to induce genetic damage in sperm. The "dominant lethal test" (Rohrborn, 1970; Green et al., 1985), involves acute dosing of the males only (typically for 5 days) and subsequently mating each male with one to two untreated females per week, in serial fashion, for about ten weeks, a period of time equivalent to the duration of spermatogenesis plus epididymal transit time (Figure 2). Females are examined at mid- to late-gestation to enumerate corpora lutea in the ovaries (an indication of the number of potential embryos), and implantations in the uterus. The implants are further classified as normal fetuses, dead fetuses, or early or late resorptions, and fetal weights are monitored. With this information both pre- and post-implantation embryonic loss can be calculated.

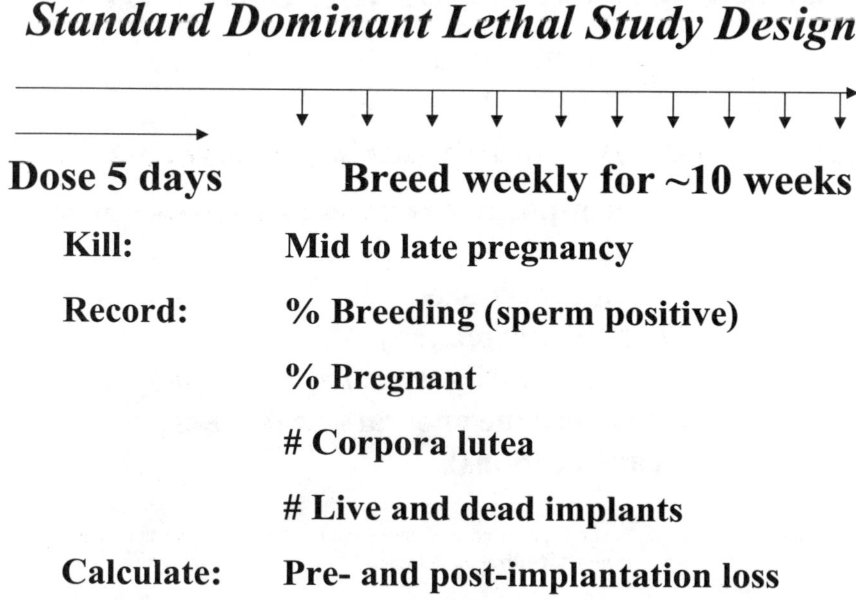

Figure 2. Diagram of standard dominant lethal test. Pre-implantation loss can be calculated with this test, but it cannot be attributed to fertilization failure vs. very early embryo loss.

Fetal loss detected by the dominant lethal test is assumed to be due to lethal gene mutations or chromosome defects derived from the fertilizing spermatozoon. This test is particularly useful for identifying the stage(s) of spermatogenesis that is most susceptible to the toxicant. In rats, for example, effects observed during the first week after exposure are indicative of damage to the spermatozoa during epididymal maturation. Effects evident 2-3 weeks after exposure are more likely due to post-meiotic damage in round or elongating spermatids, while those evident after longer periods of time are indicative of damage to meiotic spermatocytes, or to the spermatogonia. A variant of this test is to add it at the end of a subchronic or multigeneration test. In this case, when the males go off treatment, they are serially bred with untreated females as described above.

When pre-implantation loss is found, an *in vivo* fertilization assay can be added to the dominant lethal test to help distinguish fertilization failure from very early (pre-implantation) embryo loss. After using essentially the same treatment protocol, females with evidence of having mated the night before, i.e., with vaginal plugs and/or sperm in a vaginal lavage, can be killed that first day of pregnancy in order to recover the eggs and examine them microscopically for direct evidence of fertilization (Figure 3). Using phase contrast microscopy, with or without nuclear stains, the presence of two pronuclei and a sperm tail within the egg provides such evidence (Perreault and Mattson, 1993).

In addition to identifying fertilization failure (percentage of unfertilized eggs), details of the zygote phenotype can be revealed by using specific DNA dyes, and these may show post-fusion defects such as fragmented pronuclei or delayed male pronucleus formation (Perreault, 1998). However, a zygote can appear normal based on morphological criteria and still contain severe genetic damage. New molecular methods such as chromosome painting and fluorescence in situ hybridization (FISH) can also be applied to identify highly specific defects in the male pronucleus, such as chromosomal breaks and translocations, as described by Marchetti et al. (1997, see this volume).

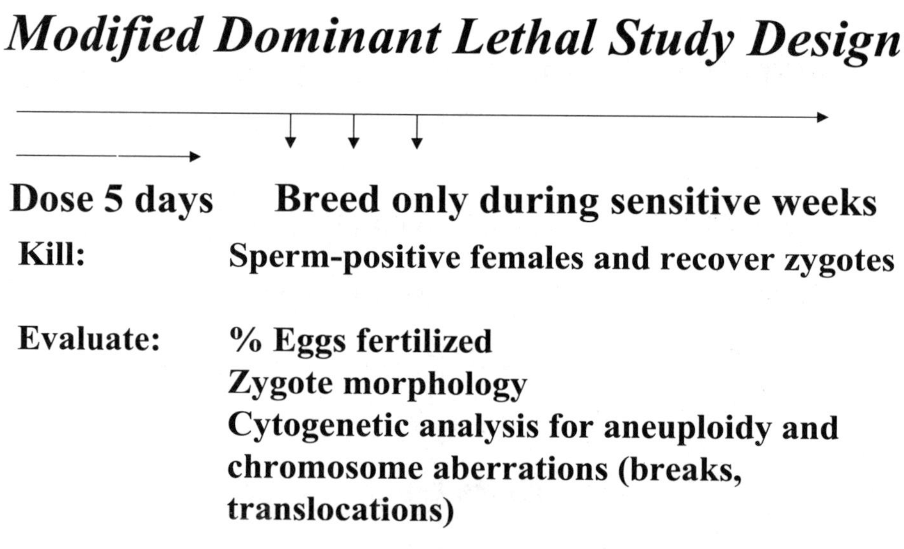

Modified Dominant Lethal Study Design

Dose 5 days   Breed only during sensitive weeks

Kill: Sperm-positive females and recover zygotes

Evaluate: % Eggs fertilized
Zygote morphology
Cytogenetic analysis for aneuploidy and chromosome aberrations (breaks, translocations)

Figure 3. Diagram of modified dominant lethal protocol with add-on fertilization assays at selected, sensitive times. Detailed information about fertilization failure and zygotic chromosome structure can be obtained with this protocol.

Alternatively, pre-implantation embryos can be recovered several days after breeding (e.g. day 4 after breeding in mice and rats) to determine whether they have developed to the morula or blastocyst stage. Enumeration of cell numbers can be used to detect altered development, as can staining for cell death and apoptosis. A one-cell embryo found at this time can be assumed to be unfertilized, but fragmentation of unfertilized eggs is not uncommon, making them difficult to distinguish from abnormal blastocysts. Since the flagellum persists, at least in rats, to the blastocyst stage, it is advisable to stain these embryos to identify the presence of a sperm tail and thereby confirm whether fertilization has occurred.

CASE STUDY: ACRYLAMIDE

Multiple Reactivities and Modes of Action of Acrylamide

Acrylamide is an example of a reactive chemical about which we have learned a lot from both multigenerational and dominant lethal tests, with and without the modifications mentioned above. It is discussed here to illustrate the important point that chemicals suspected of damaging sperm by virtue of their reactivity, may produce reproductive toxicity in a complex manner, targeting more than one cell type, acting by more than one mechanism, and impacting reproductive function in more than one way. Such chemicals may also impair more than a single aspect of sperm function, resulting in failed fertilization as well as impaired embryo development. Indeed, acrylamide can affect breeding performance and fertility through a neurotoxic mode of action, fertility through testicular toxicity and adverse effects on sperm function, and/or pregnancy outcome through a genotoxic mode of action (reviewed by Dearfield et al., 1988; Dearfield et al., 1995). The extent to which each of these modes acts to impair reproductive success depends on the dose and duration of the exposure.

Acrylamide, in its monomeric form, is an industrial chemical that is widely used to produce polymers and copolymers for many purposes, including: mining, paper strengthening, water purification (as a flocculent in waste water treatment plants and as a coagulant for water purification), oil recovery enhancement, experimental research (for chromatography and electrophoresis), and construction (in grouts and soil stabilizers) (reviewed by Dearfield et al., 1988). It is known to exhibit reproductive toxicity, genotoxicity and carcinogenicity. The National Toxicology Program (NTP) recently prioritized reproductive toxicants for potential human field studies based on their toxicity in animals and the likelihood of human exposure (Moorman et al., 2000). Acrylamide was given a high priority with respect to its known reproductive toxicity, and a medium priority with respect to estimates of population exposed. The presence of acrylamide in food, possibly arising as a consequence of heating, may mean that human exposures are higher than previously believed (Tareke et al., 2000; Harder, 2002).

Acrylamide can potentially affect sperm in several ways (reviewed by Dearfield et al., 1995). It can react directly with sperm nuclear protamines, particularly just after protamine has been added late during spermiogenesis and before its free sulfhydryls are oxidized to disulfides during epididymal maturation (Sega et al., 1989). It has been speculated that this could lead to chromosome breakage in sperm DNA, possibly during sperm chromatin remodeling *after* fertilization. Acrylamide can also directly bind glutathione (GSH) and deplete this important antioxidant in the testes, sperm and other tissues. Lack of GSH, in turn, may affect the rate of acrylamide elimination since this is accomplished by conjugation of GSH with reactive intermediates. During metabolism, a very reactive epoxide metabolite of acrylamide, glycidamide, can be formed. Glycidamide, in turn, can form adducts (especially with DNA but also with protein) and appears to be responsible for at least some of the dominant lethal effects (Generoso et al., 1996; Adler et al., 2000). While a detailed discussion of the complex modes of action of acrylamide is not the goal of this chapter, the point here is that the varied reactivities of acrylamide may account for its diverse effects in dominant lethal and other male reproduction studies (as well as its neurotoxicity). Indeed, acrylamide can affect fertility indirectly, by impairing copulatory behavior (Zenick et al., 1986). The present objective is to summarize the extent to which the dominant lethal effects of acrylamide can be attributed to fertilization failure (often detected as pre-implantation loss), as well as pre- and post-implantation loss due to genetic damage in sperm.

Antifertilization effects of Acrylamide Revealed in Dominant Lethal Studies after Acute Exposure

Acrylamide has been known for some time to induce dominant lethal effects, characterized by both pre- and post-implantation embryo loss, in mice and rats (Shelby et al., 1986; Smith et al., 1986; Working et al., 1987). These studies confirmed that late stage spermatids (those incapable of DNA repair) appear to be the most susceptible germ cell type, although delayed effects associated with exposure to meiotic and pre-meiotic germ cells may occur at higher doses. More recent research has been conducted in part to distinguish between the severity of and mechanisms involved in the reproductive vs. neurotoxicity of acrylamide, and in part to elucidate the nature of the genetic toxicity (e.g. clastogenicity, mutagenicity, protamine alkylation). This research has involved more specific, direct observations on zygotes and pre-implantation embryos, and thereby also provides information about fertilization failure.

Dominant Lethal Studies with Direct Assessments of Zygotes. Using the dominant lethal study design, Sublet et al. (1989) observed markedly decreased fertility in male rats during the first week of breeding after acute exposure to acrylamide (15, 30 or 45 mg/kg/day, p.o., for 5 days). By examining eggs recovered the day after breeding, they found that most of this loss was due to failed fertilization, and by flushing the female reproductive tract to recover sperm, they observed that this effect was associated with a lack of sperm transport from the vagina to the uterus in many of the females examined shortly after breeding. In contrast, during the second and third weeks after dosing, post-implantation loss was significantly elevated (again by these three doses), apparently due to genetic damage in the sperm exposed to acrylamide when they were differentiating spermatids. Only at the highest dose (45 mg/kg) was increased pre-implantation loss seen (together with the post-implantation loss) at these times, and this was again attributed to fertilization failure by examining zygotes. The authors also examined uterine sperm recovered from the bred females and found statistically significant, but modest, decrements in the percentage of motile sperm and their curvilinear velocity three weeks after exposure (45 mg/kg group). They concluded that the decrements in sperm motility could account, at least in part, for their reduced fertilizing ability three weeks after dosing, and that the genetic toxicity at this time was accompanied by impaired sperm function. Indeed, it has been suggested that a possible explanation of the decreased fertilization during the first week of breeding after acrylamide exposure in rats could be interference with sperm flagellar kinesin that could impact sperm motility and thereby affect the ability of the sperm to reach and/or penetrate the oocytes (Tyl et al., 2000b).

We capitalized upon the Sublet et al. (1989) study results in designing an experiment to test the hypothesis that GSH depletion could exacerbate the male reproductive effects of acrylamide. As mentioned above, lack of GSH could delay the elimination of acrylamide and its metabolites. Accordingly, male rats were exposed to acrylamide (25 or 50 mg/kg/day, i.p., for five days) with and without a GSH depleting agent given on the first exposure day only. These rats were bred with two untreated females during the second and third weeks after dosing (skipping the first week when neurotoxicologic effects would be expected to preclude fertilization). The first female that bred was killed at late gestation to evaluate pregnancy outcome, while the second female was killed the day after breeding to recover eggs and examine fertilization. Results showed that GSH depletion did exacerbate the dominant lethal effects of acrylamide, but the effects were due to increased pre-implantation loss, not post-implantation loss (Slott, Dyer, and Perreault, unpublished observations). Examination of oocytes the day after breeding confirmed that the enhanced effects were due to fertilization failure. We concluded that GSH depletion exacerbated adverse effects on sperm function, but not apparently on DNA damage in the sperm.

The genetic effects of acrylamide exposure (50 mg/kg/day, i.p, for five days) on mouse sperm were characterized in more detail by Pacchierotti et al. (1994). They recovered oocytes arrested at first-cleavage metaphases (the day after breeding) in order to examine the zygotic chromosomes. When the males were bred 7 days (but not 28 days) after dosing, they found significant increases in chromosome aberrations (fragments, dicentrics, rings and translocations), indicative of chromatin damage arising in late spermatids. Interestingly, the percentage of females that mated at this time was significantly reduced, and a significant number (32%) of the zygotes recovered from the mated females were found to be at the second meiotic metaphase, that is, they were unfertilized. They concluded that this relatively high dose of acrylamide produced three effects: reduced mating, reduced percentage of fertilized eggs in the females that did mate, and increased chromosome aberrations in the oocytes that did undergo fertilization.

More recently, Marchetti et al. (1997) used a similar approach to more thoroughly characterize the chromosomal aberrations in zygotes arising from acrylamide treatment of male mice (50 mg/kg/day for five days). Zygotes, arrested at first metaphase, were recovered the day after breeding and examined for stable and unstable chromosome rearrangements using a combination of multicolor chromosome painting and DAPI to visualize the chromosomes. This method was very successful for detecting an increased incidence of various forms of chromosome aberrations, especially during the first two weeks after exposure (corresponding to exposure of postmeiotic spermatids and sperm). During the same time period, they also noted a significant elevation in the percentage of eggs that were either unfertilized, or exhibited arrested or delayed development such that they did not progress to mitotic metaphase on schedule.

Dominant Lethal Studies with Direct Assessment of Pre-implantation Embryos. The consequences of acute paternal acrylamide exposure on pre-implantation embryo development were studied by Titenko-Holland et al. (1998) and Holland et al. (1999), again with the aim of characterizing the genetic effects of acrylamide, but also providing evidence for fertilization failure. Mouse embryos were recovered 4 days after breeding exposed males (50 mg/kg/day, i.p. for five days) with unexposed females, and stained to detect micronuclei as evidence of chromosomal damage (Titenko-Holland et al., 1998). At this time the embryos should be at the morula-blastocyst stage of development. However, when males were bred 5-17 days after treatment, a substantial percentage (38%) of the pre-implantation embryos remained at the egg/zygote stage, consistent with failed fertilization. Using the same protocol, Holland et al. (1999) found a high proportion of zygotes/eggs, 69%, 52% and 26% for males bred 1, 2 or 3 weeks after treatment, along with a variety of other abnormalities including arrested cleavage and embryo lysis. In parallel dominant lethal experiments, levels of pre-implantation loss were similar to the percentages of all classes of abnormal pre-implantation embryos, while post-implantation loss was lower (~13%).

Taken together, these acute exposure dominant lethal studies show that acrylamide treatments that induce genetic defects in zygotes and early embryos also produce a substantial antifertilization effect. These observations suggest that the spermatozoa that mature after being exposed during the later stages of spermatid differentiation and perhaps during epididymal maturation, sustain physiologic as well as genetic damage. Only by directly observing eggs and early embryos could this fertilization failure be distinguished from very early embryo demise.

Insights from Multigeneration Studies that include a Dominant Lethal Component

Two recent multigeneration studies were conducted to determine the relationship between the neurotoxicity and the reproductive toxicity of acrylamide (Chapin et al., 1995;

Tyl et al., 2000a). Of relevance to the present discussion, these two studies also added a dominant lethal test at the end of parental dosing. In addition to the longer duration of exposure, these studies differed in several ways from the acute exposure dominant lethal studies described above. First, the route of exposure was via drinking water (as opposed to p.o. or i.p.), a more relevant route for human risk assessment. Since acrylamide is cleared rapidly, and exposures in drinking water are delivered more gradually than bolus doses (oral or i.p.), the delivered dose to the testes may be lower. Second, lower doses of acrylamide (0.5-5mg/kg/day, or about an order of magnitude lower than used in the acute dominant lethal tests) were selected to better define the lowest dose responsible for each form of toxicity. Nevertheless, evidence of pre-implantation embryo loss was observed in rats at 5 mg/kg/day; the total number of implantations per litter was reduced by about 20% (Tyl et al., 2000a). However, oocytes were not examined to determine whether this was due to fertilization failure. In mice, fertility effects observed at the highest dose tested (30 ppm or about $4\mu g/kg$) could be attributed to post-implantation loss. Taken together, these studies confirm that late stage spermatids are vulnerable to damage by relatively low levels of acrylamide, and suggest that acrylamide may be slightly more potent as a genotoxicant than as a reproductive toxicant. Both acute and chronic exposure studies are important for evaluating reproductive risks of acrylamide since exposures in drinking water would be expected to be chronic and low dose, while consumption of certain foods high in cooking-generated acrylamide could add an acute (bolus) exposure. In such cases, an understanding of both the anti-fertility and genetic risks would be important (along with the neurotoxicity and potential carcinogenic risks, of course).

IMPLICATIONS FOR MALE-MEDIATED DEVELOMENTAL TOXICITY IN HUMANS

Sperm-mediated developmental toxicity can occur only when a sperm with damaged DNA is able to fertilize an oocyte. Thus, distinguishing between early pregnancy loss due to failed fertilization and that due to embryo death before implantation may be important for risk assessment. Although this can be difficult, if not impossible to do in humans, rodent models have proven useful in this regard.

Routine evaluation of human semen (sperm concentration, motility, morphology) can provide indirect evidence for poor fertilizing ability, and new tests for sperm chromosome integrity, aneuploidy and DNA damage are being developed in an attempt to predict risks of male-mediated developmental toxicity (Perreault et al., 2003; Evenson et al., 2002). Yet even when a significant association is seen between a measure of DNA damage, such as the sperm chromatin structure assay, and infertility (Evenson et al., 1999; Spano et al., 2000), the components of the infertility due to failed fertilization and those due to post-fertilization embryo failure, often remain obscure.

An intriguing question is whether there is a predictable relationship between poor semen quality and poor genetic integrity in sperm (Irvine et al, 2000). There is some evidence that men with poor semen quality (low motility and/or a high percentage of morphologically abnormal sperm) produce poor quality embryos, even when the sperm manage to fertilize eggs *in vitro* (Janny and Menezo, 1994). Such observations suggest that poor semen quality may be predictive not only of poor fertilizing ability but also of genetic defects in sperm. Indeed, sperm aneuploidy appears to be elevated in men with poor semen quality (Templado et al., 2002; Ryu et al., 2001).

The case study of acrylamide presented here suggests that reactive chemicals may damage sperm in more than one way, such that both reduced fertilization and reduced embryo survival may occur as a syndrome. Clearly, advances in reproductive risk assessment, including identification of the risks of male-mediated developmental toxicity, will depend upon an improved understanding of the factors that impair both aspects of sperm function.

This discussion is relevant to the recommendations of the breakout group charged with integrating tests of genetic integrity into routine semen analysis. Clearly, we need to know "the predictive value of individual sperm biomarkers for infertility, spontaneous abortions, malformations, birth with a chromosomal defect, and other transmitted genetic diseases." Once we have more specific test batteries, including bioindicators of both sperm fertilizing ability and genetic integrity, we can "apply them in large scale epidemiological studies to identify risk factors for paternally transmitted genetic damage (i.e., environmental or occupational exposures, inherent factors such as age, and lifestyle factors)...," and we can attempt to "identify genetically susceptible subpopulations of men at risk for infertility, spontaneous abortions, and birth defects" (Perreault et al., 2003).

REFERENCES

Adler, I.-D., Baumgartner, A., Gonda, H., Friedman, M.A. and Skerhut, M., 2000, 1-Aminobenotriazole inhibits acrylamide-induced dominant lethal effects in spermatids of male mice. *Mutagenesis.* 15:133-136.

Chapin, R.E. and Sloane, R.A., 1997, Reproductive assessment by continuous breeding: evolving study design and summaries of ninety studies. *Environ Health Perspect.* 105, Supple 1:199-205.

Chapin, R.E., Fail, P.A., George, J.D., Grizzle, T.B., Heindel, J.J., Harry, G.J., Collins, B.J. and Teague, J., 1995, The reproductive and neural toxicities of acrylamide and three analogues in Swiss mice, evaluated using the continuous breeding protocol. *Fund Appl Toxicol.* 27:9-24.

Clegg, E.D., Perreault, S.D. and Klinefelter, G.R., 2001, Assessment of Male Reproductive Toxicity, in: *Principles and Methods of Toxicology*, Fourth Edition, A. Wallace Hayes, ed., Taylor & Francis, Philadelphia, p. 1263.

Dearfield, K.L., Abernathy, C.O., Ottley, M.S., Brantner, J.H. and Hayes, P.F., 1988, Acrylamide: its metabolism, developmental and reproductive effects, genotoxicity, and carcinogenicity. *Mutat. Res.* 195:45-77.

Dearfield, K.L., Douglas, G.R., Ehling U.H., Moore, M.M., Sega, G.A. and Brusick, D.J., 1995, Acrylamide: a review of its genotoxicity and an assessment of heritable genetic risk. *Mutation Res.* 330:71-99.

Evenson, D.P., Jost, L.K., Marshall, D., Zinaman, M.J., Clegg, E., Purvis, K., de Angelis, P., and Claussen, O.P., 1999, Utility of the sperm chromatin structure assay as a diagnostic and prognostic tool in the human fertility clinic. *Human Reprod.* 14:1039-1049.

Evenson, D.P., Larson, K., and Jost, L.K., 2002, Sperm chromatin structure assay: its clinical use for detecting sperm DNA fragmentation related to male infertility and comparisons with other techniques. *J Androl.* 23:25-43.

Generoso, W.M., Sega, G.A., Lockhart, A.M., Hughes, L.A., Cain, K.T., Cacheiro, N.L.A., and Shelby, M.D., 1996, Dominant lethal mutations, heritable translocations, and unscheduled DNA synthesis induced in male mouse germ cells by glycidamide, a metabolite of acrylamide. *Mutat Res.* 371:175-183.

Green, S., Auletta, A., Fabricant, R., Kapp, M., Sheu, C., Springer, J., and Whitfield, B., 1985, Current status of bioassays in genetic toxicology: The dominant lethal test. *Mutation Res.* 154:49-67.

Harder, B., 2002, Cancer link cooks up doubt. Heating may form potential carcinogen in food. *Science News.* 161:277.

Holland, N., Ahlborn, T., Turteltaub, K., Markee, C., Moore II, D., Wyrobek, A.J. and Smith, M.T., 1999, Acrylamide causes preimplantation abnormalities in embryos and induces chromatin-adducts in male germ cells of mice. *Reprod Toxicol.* 13:167-178.

Irvine, D.S., Twigg, J.P., Gordon, E.L., Fulton, N., Milne, P.A., and Aitken R.J., 2000, DNA integrity in human spermatozoa: relationships with semen quality. *J Androl.* 21: 33-44.

Janny, L. and Menezo, Y.J.R., 1994, Evidence for a strong paternal effect on human preimplantation embryo development and blastocyst formation. *Mol Reprod Dev.* 38:36-42.

Marchetti, F., Lowe, X., Bishop, J. and Wyrobek, A.J., 1997, Induction of chromosomal aberrations in mouse zygotes by acrylamide treatment of male germ cells and their correlation with dominant lethality and heritable translocations. *Environ Mol Mutagen.* 30:410-417.

Moorman, W.J., Ahlers, H.W., Chapin, R.E., Daston, G.P., Foster, P.M.D., Kavlock, R.J., Morawetz, J.S., Schnorr, T.M. and Schrader, S.M., 2000, Prioritization of NTP reproductive toxicants for field studies. *Reprod Toxicol.* 14:293-301.

Pacchierotti, F., Tiveron, C., D'Archivio, M., Bassani, B., Cordelli, E., Leter, G., and Spano, M, 1994, Acrylamide-induced chromosomal damage in male mouse germ cells detected by cytogenetic analysis of one-cell zygotes. *Mutat Res.* 309:273-284.

Perreault, S.D. and Mattson, B.A., 1993, Recovery and morphological evaluation of oocytes, zygotes and preimplantation embryos, in: *Methods in Toxicology, Vol. 3, Part B, Female Reproductive Toxicology,* J.J. Heindel and R.E. Chapin (eds.), Academic Press, Inc., Orlando, FL, p. 110.

Perreault, S.D., 1998, Gamete Toxicology: The Impact of New Technologies, in: K. Korach, ed., *Reproductive and Developmental Toxicology*, Marcel Dekker, Inc., New York, p., 635.

Perreault, S.D., Aitken, R.J., Baker, H.W.G., Evenson, D.P., Huszar, G., Irvine, D.S., Morris, I.D., Morris, R.A., Robbins, W.A., Sakkas, D., Spano, M., and Wyrobek, A.J., 2003, Integrating new tests of sperm genetic integrity into semen analysis: Breakout group discussion, in: *Advances in Male-Mediated Developmental Toxicity,* ed. B. Robaire and B. Hales, Kluwer-Plenum Press, *in press.*

Rohrborn, G., 1970, The dominant lethals: Method and cytogenetic examination of early cleavage stages, in: *Chemical Mutagenesis in Mammals and Man*, F. Vogel and G. Rohrborn, eds., Springer-Verlag, Heidelberg, p.148.

Ryu, H.M., Lin, W.W., Lamb, D.J., Chuang, W., Lipshultz, L.I., and Bishoff, F.Z., 2001, Increased chromosome X, Y, and 18 nondisjunction in sperm from infertile patients that were identified as normal by strict morphology: implication for intracytoplasmic sperm injection. *Fertil Steril.* 76:879-883.

Sega, G.A., Alcota, R.P.V., Tancongco, C.P. and Brimer, P.A., 1986, Acrylamide binding to the DNA and protamine of spermiogenic stages in the mouse and its relationship to genetic damage. *Mutat Res.* 216:221-230.

Shelby, M.D., Cain, K.T., Hughes, L.A., Braden, P.W., and Generoso, W.M., 1986, Dominant lethal effects of acrylamide in male mice. *Mutat Res.* 173:35-40.

Smith, M.K., Zenick, H., Preston, R.J., George, E.L. and Long, R.E., 1986, Dominant lethal effects of subchronic acrylamide administration in the male Long-Evans rat, *Mutat Res.* 173:273-277.

Spano, M., Bonde, J.P., Hjollund, H.I., Kolstad, H.A., Cordelli, E, and Leter, G., 2000, Sperm chromatin damage impairs human fertility. The Danish Pregnancy Planner Study Team. *Fertil Steril.* 73:43-50.

Sublet, V.H., Zenick, H., and Smith, M.K., 1989, Factors associated with reduced fertility and implantation rates in females mated to acrylamide-treated rats. *Toxicology.* 55:53-67.

Tareke, E., Rydberg, P., Karlsson, P., Ericksson, S., and Tornqvist, M., 2000, Acrylamide: A cooking carcinogen?. *Chem Res Toxicol.* 13:517-522.

Templado, C., Hoang, T., Greene, C., Rademaker, Z., Chernos, J., and Martin, R., 2002, Aneuploid spermatozoa in infertile men: teratozoospermia. *Mol Reprod Dev.* 61:200-204.

Titenko-Holland, N., Ahlborn, T., Lowe, X., Shang, N., Smith, M.T. and Wyrobek, A.J., 1998, Micronuclei and developmental abnormalities in 4-day mouse embryos after paternal treatment with acrylamide, *Environ. Molec Mutagen.* 31:206-217.

Tyl, R.W., Friedman, M.A., Losco, P.E., Fisher, L.C., Johnson, K.A., Storther, D.E. and Wolf, C.H., 2000a, Rat two-generation reproduction and dominant lethal study of acrylamide in drinking water. *Reprod Toxicol.* 14:385-401.

Tyl, R.W., Marr, M.C., Myers, C.B., Ross, W.P., and Friedman, M.A., 2000b, Relationship between acrylamide reproductive and neurotoxicity in male rats. *Reprod Toxicol.* 14:147-157.

U.S. Environmental Protection Agency, 1998, *Health Effect Test Guidelines OPPTS 870.3800 Reproduction and Fertility Effects*, U.S. Government Printing Office, Washington, D.C.

U.S. Environmental Protection Agency, 1996, *Guidelines for reproductive toxicity risk assessment.* Fed. Reg., 61(212):56274.

Working, P.K., Bentley, K.S., Hurtt, M.E. and Mohr, K.L., 1987, Comparison of the dominant lethal effects of acrylonitrile and acrylamide in male Fischer 344 rats. *Mutagenesis.* 2:215-220.

Zenick, H., Hope, E. and Smith, M.K., 1986, Reproductive toxicity associated with acrylamide treatment in male and female rats. *J Toxicol Environ Health.* 17:457-472.

DISCLAIMER: This document has been reviewed in accordance with the U.S. Environmental Protection Agency policy and approved for publication. Mention of trade names or commercial products does not constitute endorsement or recommendation for use.

ICSI, MALE PRONUCLEAR REMODELING AND CELL CYCLE CHECKPOINTS

Laura Hewitson, Calvin R. Simerly and Gerald Schatten

Pittsburgh Development Center
Magee-Womens Research Institute
University of Pittsburgh,
Pittsburgh, PA 15213

INTRODUCTION

Intracytoplasmic sperm injection (ICSI; Palermo et al., 1992; Van Steirteghem et al., 1993) has heralded an era of enormous improvements in treating male infertility; it has resulted in the births of thousands of babies. In fact its success has led some clinicians to claim that male infertility is now "cured", while others raise concerns over the global application of ICSI for all cases of in vitro fertilization (Tucker et al., 1995). These concerns may be, in part, based on the lack of supporting pre-clinical studies in relevant animal models (Ng et al., 1995; Yanagimachi et al., 1995). Others are worried about the possible long-term effects of ICSI on the offspring (Hamberger et al., 1998; Moutel et al., 1999). While the current view of ICSI is one of cautious optimism, the development of a preclinical, non-human primate model has been established to address the safety of ICSI and other methods of assisted reproduction (Hewitson et al., 1998; 1999; 2001). Using advanced fluorescent imaging techniques, the cytoskeletal events that occur during rhesus fertilization have been examined after both in vitro fertilization (IVF) and ICSI. It was found that rhesus monkeys share many similarities with humans in terms of cytoskeletal and chromatin dynamics after fertilization by IVF (Simerly et al., 1995; Wu et al., 1996). However, rhesus oocytes fertilized by ICSI resulted in abnormal nuclear remodeling leading to asynchronous chromatin decondensation in the apical region of the sperm head, delaying the onset of DNA synthesis. The persistence of the acrosome and perinuclear theca on the apex of sperm introduced into the oocyte by ICSI may constrict the DNA in this region (Hewitson et al., 1999; Ramalho-Santos et al., 2000a). Despite these differences, normal rhesus ICSI embryos have been produced and, after transfer to surrogate females, have resulted in several births (Hewitson et al., 1999; 2001). However, the irregularities described raise concerns that the ICSI procedure differs from IVF in several fundamental ways perhaps leading to chromatin damage during DNA decondensation.

Advances in Male Mediated Developmental Toxicity, edited by Bernard Robaire and Barbara F. Hales.
Kluwer Academic/Plenum Publishers, 2003.

Furthermore, this highlights the need for devising improved preclinical assessment prior to the global acceptance of novel methods of assisted reproduction.

ASSISTED REPRODUCTION AND INFERTILITY

ICSI was first introduced to enable oligospermic men to achieve parenthood (Palermo et al., 1992). Its early success was soon followed by the injection of immature sperm obtained from the epididymis (Devroey et al., 1995; Schlegal et al., 1995) or testis (Bourne et al., 1995; Silber et al., 1995) and even sperm with abnormal morphology (Harari et al., 1995; Liu et al., 1995; Rybouchkin et al., 1997), even though ICSI using sub-optimal sperm may pose a long-term risk to the offspring (Silber et al., 1995; Kent-First et al., 1996; Vogt, 1999; Meschede et al., 2000; Nudell et al., 2000). In some extreme cases, spermatids from men exhibiting spermatogenic impairment have been used to achieve fertilization by injection (Fishel et al., 1995; Tesarik et al., 1995) but this leads to concerns that achieving pregnancies in couples with extreme male factor infertility may be perpetuating the same or similar fertility problems in the next generation (Cummins, 1997; Kamischke et al., 1999; Schatten et al., 1998; Silber, 2000).

The injection of a sperm into an egg demonstrates that several stages of fertilization, typically thought to be necessary for successful reproduction, can now be bypassed by ICSI. These include events such as capacitation, the acrosome reaction and sperm-egg membrane fusion events. More importantly, there is no natural sperm selection process. Rather than letting Nature take its course, the sperm is typically selected by the embryologist performing the technique; since no specific criteria have been developed to aid in this process, any sperm that looks apparently 'normal' could be selected. In fact many of the sperm function tests developed for IVF, could now be viewed as obsolete as far as ICSI is concerned. In severe cases of male infertility, there may only be a handful of sperm isolated from a biopsy resulting in no selection process at all since the number of oocytes may be higher than the number of sperm available.

Perhaps one of the most worrisome concerns regarding ICSI is the possibility that abnormal genes may be transferred to the offspring because many of the natural barriers to conception have been bypassed. Before ICSI was developed, couples with genetic abnormalities relating to infertility were unlikely to conceive. With the advent of ICSI, however, these couples are now suitable candidates, providing that a thorough genetic evaluation is undertaken followed by genetic counseling, if necessary. The three most common genetic factors related to male infertility are Y-chromosome microdeletions leading to spermatogenic impairment, cystic fibrosis gene mutations leading to congenital absence of the vas deferens (CABVD), and karyotype abnormalities. While outside the realm of this paper, there are many excellent reviews on this subject (Kim et al., 1998; Vogt, 1998; Kupker et al., 1999; Hargreave et al., 2000; Kent-First et al., 2000; Lamb and Lipshultz, 2000). Other concerns for ICSI offspring include developmental outcome (Bowen et al., 1998; Bonduelle et al., 1998a), cytoplasmic changes (St. John et al., 2000), influences on oocyte polarity (Pedersen, 2001), contamination of the sperm (Chan et al., 2000), or other damage resulting from the technique itself (Hewitson et al., 1999; Luetjens et al., 1999; Dumoulin et al., 2001). So far, most of the reports on the incidence of major and minor malformations among ICSI offspring have been reassuring (Bonduelle et al., 1998b). However, the true risks of adverse health outcomes for children conceived by ICSI cannot be determined without large scale, long-term, follow-up studies of those offspring conceived naturally and by conventional IVF. These concerns underscore the importance of appropriate animal research followed by extensive preclinical studies before new methods of assisted reproduction are adopted as routine for treating infertility. This is especially relevant in regards to the use of spermatids (Fishel et al., 1995; Tesarik et al., 1995) and even spermatocytes (Sofikitis et al., 1998) for fertilization.

THE CYTOSKELETAL EVENTS OF FERTILIZATION

In Vitro Fertilization

Understanding the events that lead to the successful completion of fertilization, the intermixing of the gametes at first mitosis, is paramount to understanding why fertilization sometimes fails and for devising new methods of assisted reproduction. However, due to the ethical and political constraints of using donated human material, animal models are generally used which may not accurately reflect the events seen in humans (Schatten et al., 1998). Fluorescence microscopy has been an invaluable tool for determining the fates of sperm components during fertilization. Recent studies in the rhesus monkey looking at the cytoskeletal events of fertilization have shown that rhesus and humans share many similarities in that both systems require a sperm-derived centrosome to direct microtubule-mediated motility, critical for pronuclear apposition (Wu et al., 1996; Hewitson et al., 1996; Sutovsky et al., 1996).

Prime, fertile breeding rhesus monkeys have been used to examine the microtubule and chromatin patterns of oocytes fertilized by in vitro fertilization (Wu *et al.* 1996). In the unfertilized, metaphase-arrested rhesus oocyte, the only microtubules present are those of the barrel-shaped, anastral meiotic spindle that is oriented radially to the cell cortex. Human oocytes display these similar meiotic spindle arrangements, in contrast to mice which have a much larger spindle that is oriented perpendicular to the cell cortex, as well as numerous small cytasters organized within the cytoplasm (Schatten et al., 1986). Shortly after insemination of a rhesus oocyte, the oocyte activates, separating the female chromosomes, whilst a small aster of microtubules emanates from the introduced sperm centrosome (Wu et al., 1996). During early development, the sperm aster enlarges as the sperm DNA decondenses and pronuclear migration is initiated. As the two pronuclei move together, they occupy an eccentric position within the cytoplasm, quite similar to human zygotes (Simerly et al., 1995). The introduced sperm centrosome must then duplicate and split, so that there are now two centrosomes; each one will serve as a mitotic spindle pole. By prophase, the interphase microtubule array disassembles, having united the male and female pronuclei, and is replaced by a dense monaster of microtubules during chromosome condensation within the cytoplasm (Wu et al., 1996). By about 18-20 hrs post-insemination (prometaphase), a bipolar spindle emerges and the parental genomes begin aligning along the equator of the mitotic spindle. Mitotic metaphase is marked by the eccentrically positioned, barrel-shaped spindle, upon which the chromosomes become intermixed in preparation for mitosis; thus completing the fertilization process. These observations closely parallel the events seen in humans, in particular demonstrating the requirement for a sperm-derived centrosome to direct microtubule-mediated motility during fertilization (Schatten, 1994; Simerly et al., 1995; Wu et al., 1996). This work also validates the use of rhesus monkey gametes as a valuable animal model for understanding the events of human fertilization.

Intracytoplasmic Sperm Injection

When the first clinical ICSI pregnancies were reported (Palermo et al., 1992; Van Steirteghem et al., 1993), the technique was heralded as a major breakthrough for treating male infertility. No experimental phase preceded its introduction, partly because animal models were thought to be unsuitable and partly because of its immediate and overwhelming success (te Velde et al., 1998). It wasn't until several years later that data obtained in the rhesus monkey were shown to be clinically relevant, and by this time, ICSI using immature, or even morphologically abnormal, sperm was commonplace. Some clinicians may still doubt the extrapolation of the rhesus data to humans, but the ongoing discussions about possible long-term consequences for the health of ICSI offspring (Kent-First et al., 1996; Dowsing et al., 1999; Kamischke et al., 1999; Nudell et al., 2000), strongly suggest otherwise.

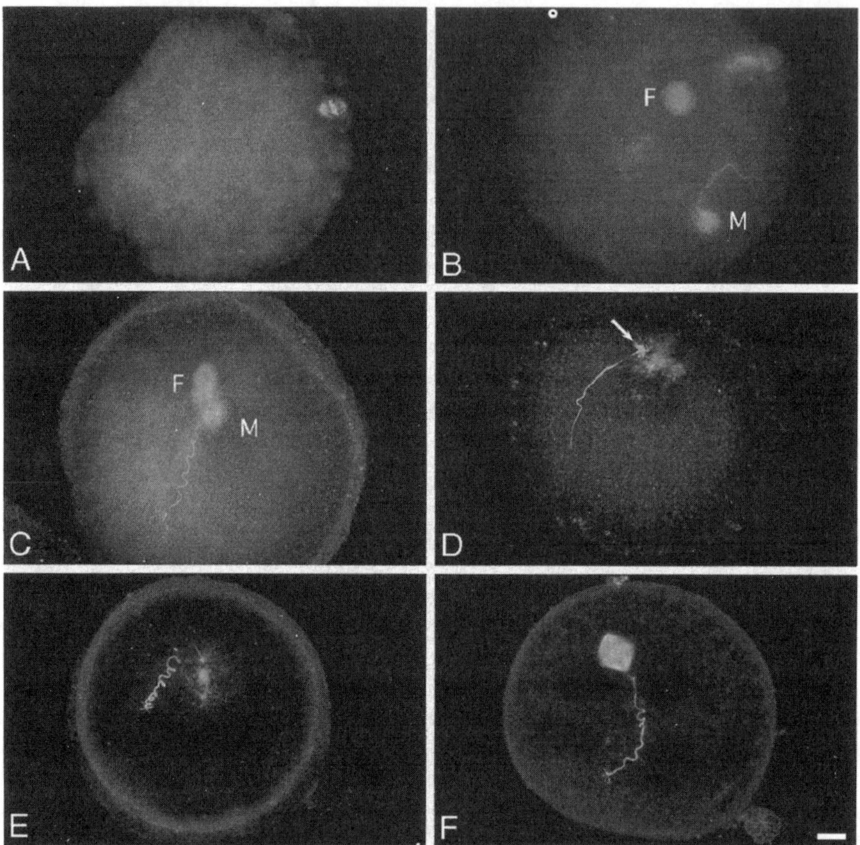

Figure 1. Microtubule and chromatin dynamics in rhesus ICSI zygotes obtained by conventional (A-B) and confocal (C-F), epifluorescence microscopy. The only microtubules present in the unfertilized rhesus oocyte are those found in the metaphase-arrested, meiotic spindle (A). After the injection of a sperm, the oocyte activates and the sperm and egg pronucleus begin to form (B). As the sperm decondenses an aster of microtubules emanates from the base of the sperm head. These microtubules elongate to fill the entire cytoplasm during pronuclear formation and apposition (C). By prophase, most of the cytoplasmic microtubules disassemble leaving a small tuft of microtubules associated with the introduced centrosome (arrow, D). The centrosome duplicates and separates to form a bipolar array (E). The chromosomes align across the metaphase spindle (F) in preparation for first mitosis. Magnification: A-B and D-F x100; C,x90. Reprinted with permission (Hewitson et al., 1996). M, male; F, female.

The first study examining the cytoskeletal dynamics during rhesus ICSI fertilization (Hewitson et al., 1996; Figure 1A-F) revealed that sperm microinjected into rhesus oocytes maintained the ability to activate the oocyte, organize a sperm aster and complete repositioning of the maternal and paternal pronuclei in a manner similar to that of rhesus

(Wu et al., 1996) and human (Simerly et al., 1995) oocytes fertilized by IVF. This is in contrast to rodents that do not rely on a paternally derived centrosome for pronuclear migrations (Schatten et al., 1986; Schatten, 1994). The similarities between rhesus and human IVF are also noted after ICSI (Hewitson et al., 1996; Rawe et al., 2000) and rhesus oocytes, like human, do not require an additional stimulus to induce oocyte activation, unlike oocytes from domestic species (Keefer et al., 1990; Kim et al., 1999).

After pronuclear apposition in rhesus ICSI zygotes, the introduced sperm centrosome normally duplicates and splits, establishing a bipolar microtubule array. The sperm tail remains attached to the centrosome that was introduced by the sperm during ICSI and constitutes one of the poles of the mitotic spindle. These observations validate the rhesus monkey as a model for studying the early events of human fertilization. Furthermore, since this study used *in vitro* matured rhesus oocytes obtained from older, unstimulated animals, a number of fertilization failures were also noted. These included: i) the inability to resume meiosis, shown by metaphase II arrest and premature chromosome condensation and ii) cytoskeletal defects characterized by microtubule nucleation arrest, premature detachment of the sperm axoneme and aster from the paternal pronucleus and sperm aster microtubule growth defects (Hewitson et al., 1996). This suggests that fertilization arrest may occur at several points in the cell cycle as a result of improper centrosomal functioning. Arrested human embryos inseminated by IVF and ICSI also demonstrated similar types of fertilization failures (Asch et al., 1995; Simerly et al., 1995; Rawe et al., 2000) suggesting that centrosomal defects may be a novel form of infertility since excess oocytes from some of the same patients were successfully fertilized with donor, but not spousal, sperm.

ABNORMAL MALE PRONUCLEAR REMODELING AND CELL CYCLE CHECKPOINTS

The similarities between IVF- and ICSI-derived rhesus oocytes in terms of their cytoskeletal dynamics suggest that an injected sperm behaves in much the same manner as a sperm that undergoes oolemma binding and fusion. However, a more detailed analysis of rhesus ICSI zygotes using transmission electron microscopy (TEM) revealed that the pattern of sperm decondensation is quite altered after the injection of sperm (Figure 2A) although male pronuclear formation is not prevented (Figure 2B). It appears that the persistence of the acrosome over the anterior part of the injected sperm may cause an asynchrony in sperm decondensation (Hewitson et al., 1996; Sutovsky et al., 1996) and may prevent the import of maternal nuclear proteins during male pronucleus formation. For example, NuMA (standing for nuclear mitotic apparatus), a nuclear protein important during interphase and in maintaining spindle architecture (Compton and Cleveland, 1994), is not found in sperm but maternal NuMA enters the sperm nucleus as it decondenses during fertilization. After ICSI, however, NuMA is initially excluded from the regions of paternal chromatin which remain condensed (Hewitson et al., 1999), perhaps because of the physical constraints imposed by the still-intact acrosome (Hewitson et al., 1996; Sutovsky et al., 1996; Figure 3A). The removal of the acrosome does not overcome this asynchronous chromatin decondensation, suggesting an involvement of other structures. A prime candidate would be the perinuclear theca, a structure found on the inner acrosomal membrane of the sperm and typically removed during fertilization at either the zona pellucida or the oocyte's plasma membrane (Sutovsky et al., 1997). Using immunocytochemistry, rhesus oocytes injected with sperm were fixed at various time-points post ICSI to examine the fate of the perinuclear theca. Interestingly, the theca was found overlying the apex of the injected sperm at around 30 minutes post ICSI and remained closely associated with the apical region of the forming paternal pronucleus until at least 16 hours post ICSI (Figure 3B). The close association of the perinuclear theca with the sperm appeared to induce an asynchrony of chromatin decondensation with the basal region of sperm DNA, not in contact with the theca, decondensing much faster than the apical DNA underlying this structure.

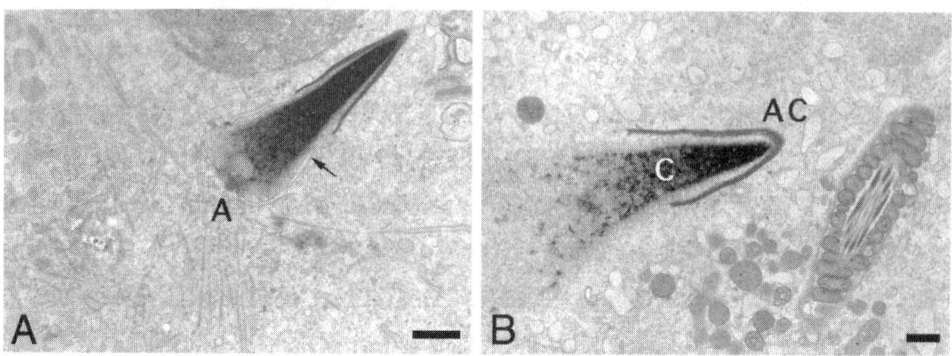

Figure 2. Transmission electron microscopy (TEM) reveals that asynchronous decondensation of sperm chromatin is an unusual feature of ICSI. The apical chromatin is unable to decondense synchronously with the basal region of chromatin, perhaps due to the physical restrictions imposed by the still-intact acrosome. As the sperm continues decondensation, a normal male pronucleus will eventually form. M, male pronucleus; C, chromatin; A, aster; AC, acrosome. Magnification, x5000. Reprinted with permission from Hewitson et al., 1996.

Furthermore, differences between ICSI-fertilized oocytes and those inseminated by IVF have also been reported. For example, vesicle-associated membrane protein (VAMP; Conner et al., 1997), a constituent of the sperm acrosome typically lost at the cell surface during sperm penetration (Ramalho-Santos et al., 2000b), is initially detected as a collar at the equatorial segment of the injected sperm but soon forms a constriction around the decondensing sperm during the first few hours of rhesus ICSI fertilization (Hewitson et al., 1999; Ramalho-Santos et al., 2000a; Figure 2C). The retention of this VAMP collar following ICSI serves to separate the condensed and decondensing chromatin; it persists until paternal pronucleus formation is completed. The significance of this asynchronous decondensation remains to be determined, but it may lead to a diminished ability of the oocyte to express, or be exposed to, important paternal genes or gene products, thus leading to improper embryo formation (Schatten et al., 1998). Furthermore, since the X and Y chromosomes may be preferentially located in the apical region of the human sperm head (Luetjens et al., 1999; Terada et al., 2000), one cannot rule out a possible link between the ICSI technique and chromosome/chromatin integrity since it is this region of sperm DNA that shows altered nuclear remodeling.

We have also recently obtained surprising results using a rhesus ICSI model that demonstrated that exogenous DNA bound to the sperm's surface is retained and transferred into the egg during ICSI but is lost during *in vitro* fertilization (IVF; Chan et al., 2000). Rhodamine-labeled DNA encoding the green fluorescence protein (GFP) gene binds avidly to sperm, and the rhodamine signal, while lost at the egg surface during IVF, remains as a brilliant marker on the microinjected sperm within the egg cytoplasm after ICSI. The transgene is expressed in preimplantation embryos produced by ICSI, but not IVF, as early as the four-cell stage and the percent of expressing embryos and the number of expressing cells increase during embryogenesis to the blastocyst stage. This technique may prove to be a powerful strategy for producing transgenic primates as has been demonstrated for rodents (Perry et al., 1999). However, the ability to transfer foreign DNA into oocytes during ICSI, but not IVF, raises the issue that ICSI could also transmit infectious material since an injected sperm does not undergo the normal nuclear remodeling that occurs during IVF.

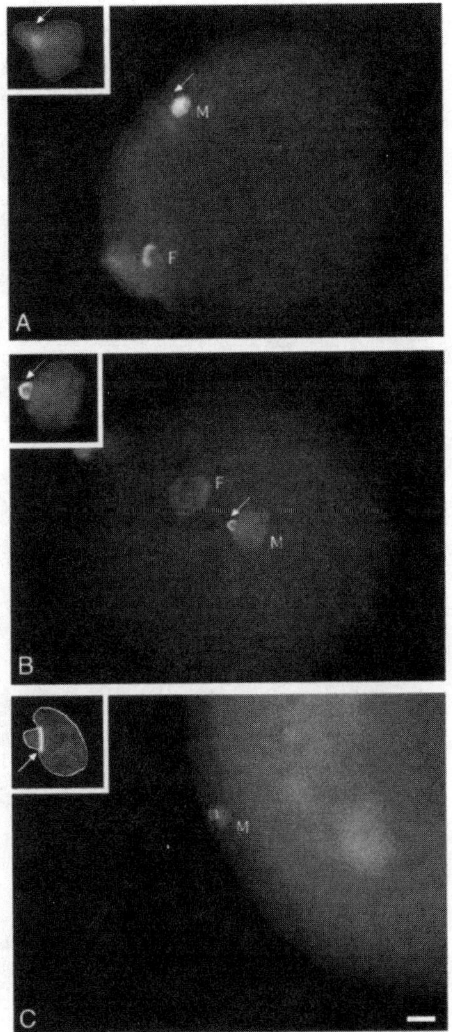

Figure 3. Unusual nuclear remodeling after ICSI. NuMA (A) is excluded from the condensed, apical region of the paternal pronucleus (main arrow), which is showing asynchronous chromatin decondensation (inset, arrow and outline). The perinuclear theca (B and inset) persists for up to 12 h post-ICSI, constricting the apical DNA (arrow). Membrane VAMP (C) is detected as a constricting ring around the paternal pronucleus separating the condensed and decondensing regions (inset, arrow and outline). M, male pronucleus; F, female pronucleus. Magnification A and B x200; C x100. Reprinted with permission from Hewitson et al., 1999.

DNA synthesis, as detected by bromodeoxyuridine (BrdU) incorporation, and pronuclear migration can be delayed by several hours after ICSI in both pronuclei when the paternal pronucleus is still undergoing decondensation in the apical region, identifying a unique G_1/S cell cycle checkpoint (Figure 4A nd B; Hewitson et al., 1999; Ramalho-Santos et al., 2000a). Conversely, after IVF, pronuclear migration has been completed within 12 hours post insemination and DNA synthesis is detected in both pronuclei (Figure 4c). These unique differences in IVF- and ICSI-fertilized oocytes raise concerns about the increasing use of ICSI in fertility clinics and demonstrate the need for further animal experimentation in an attempt to improve the safety of this procedure.

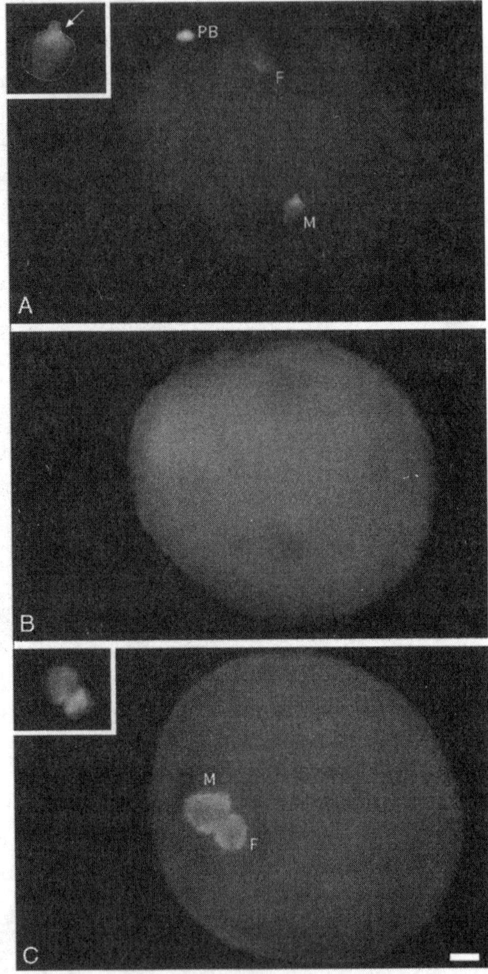

Figure 4. BrdU incorporation in rhesus ICSI and IVF zygotes fixed 16 hours post insemination. In an ICSI zygote, the male pronucleus hasn't completed decondensation in the apical region (inset, arrow and outline) delaying both pronuclear apposition (A) and the onset of DNA synthesis (B). BrdU is not detected in the male or female pronucleus (B). After IVF, BrdU is detected in both pronuclei, which are typically apposed (C). BrdU, bromodeoxyuridine; M, male pronucleus; F, female pronucleus; PB, polar body. Magnification, x100.

CONCLUSIONS

While the introduction of ICSI has revolutionized the treatment of male infertility, there are valid concerns about whether this technique is just perpetuating the same infertility problems in future generations. More importantly, there is still a dearth of information on the long-term health consequences for ICSI offspring although so far, most reports on the incidence of major and minor congenital malformations have been reassuring. The data reviewed here, showing atypical sperm nuclear remodeling after ICSI, highlight other potential concerns, namely, the possible consequences of the ICSI technique itself. Further animal research in an attempt to improve the safety of this procedure, is surely warranted. Non-human primates appear to be an ideal model for these types of studies. Furthermore, the rhesus offspring produced by these techniques can be studied using a mental development index derived for non-human primates with the goal of understanding the outcomes and safety of ICSI and other novel methods of assisted reproduction. Parallel national and international studies to assess ICSI offspring are being undertaken currently to hopefully reassure the parents of ICSI children, and to enable potential parents to consider the evidence of risk of novel methods of assisted reproduction in a more informed manner.

ACKNOWLEDGEMENTS

The authors would like to thank Drs. John Fanton, David Hess, Marc Luetjens, Martha Neuringer, João Ramalho-Santos and Peter Sutovsky; and Kevin Grund, Darla Jacob, Ethan Jacoby, Crista Martinovich, Michelle Miller, Kevin Mueller, Christopher Payne, Tonya Swanson and Diana Takahashi. Hormones for oocyte stimulations were provided by Ares Serono, Inc. and Organon, Inc. Funding for animal studies was provided by the National Institute of Health (NICHD and NCRR), the United States Department of Agriculture and the Mellon Foundation. Animal protocols were approved by the Oregon Health Sciences University's Institutional Animal Care and Use Committee (IACUC).

REFERENCES

Asch, R., Simerly, C., Ord, T., Ord, V.A., and Schatten, G., 1995, The stages at which human fertilization arrests: microtubule and chromosome configurations in inseminated oocytes which fail to complete fertilization and development in humans. *Hum Reprod.* 10:1897-1906.

Bonduelle, M., Joris, H., Hofmans, K., Liebaers, I., and A. Van Steirteghem, 1998a, Mental development of 201 ICSI children at 2 years of age. *Lancet.* 351:1553.

Bonduelle, M., Aytoz, A., and Van Assche, E., 1998b, Incidence of chromosomal aberrations in children born after assisted reproduction through intracytoplasmic sperm injection. *Hum Reprod.* 13:781-782.

Bourne, H., Richings, N., Harari, O., Watkins, W., Speirs, A.L., Johnston, W.I., and Baker, H.W., 1995, The use of intracytoplasmic sperm injection for the treatment of severe and extreme male infertility. *Reprod Fertil Dev.* 7:237-245.

Bowen, J.R., Gibson, F.L., Leslie, G.I. and Saunders, D.M., 1998, Medical and developmental outcome at 1 year for children conceived by intracytoplasmic sperm injection. *Lancet.* 351:1529-1534.

Chan, A.W.S., Luetjens, C.M., Dominko, T., Ramalho-Santos, J., Hewitson, L., Simerly, C., and Schatten, G., 2000, TransgenICSI: Foreign DNA transmission by intracytoplasmic sperm injection: Injection of sperm bound with exogenous DNA results in embryonic GFP expression and live rhesus births. *Mol Hum Reprod.* 6:26-33.

Compton, D. A., and Cleveland, D.W., 1994, NuMA, a nuclear protein involved in mitosis and nuclear reformation. *Curr Opin Cell Biol.* 6:343-346.

Conner, S., Leaf, D., and Wessel, G., 1997, Members of the SNARE hypothesis are associated with cortical granule exocytosis in the sea urchin egg. *Mol Reprod Dev.* 48:106-118.

Cummins, J.M., and Jequier, A.M., 1995, Concerns and recommendations for intracytoplasmic sperm injection (ICSI) treatment. *Hum Reprod.* 10, Suppl 1:138-143.

Devroey, P., Silber, S., Nagy, Z., Liu, J., Tournaye, H., Joris, H., Verheyen, G., and Van Steirteghem, A., 1995, Ongoing pregnancies and birth after intracytoplasmic sperm injection with frozen-thawed epididymal spermatozoa. *Hum Reprod.* 10:903-906.

Dowsing, A.T., Yong, E.L., Clark, M., McLachlan, R.I., de Kretser, D.M., and Trounson, A.O., 1999, Linkage between male infertility and trinucleotide repeat expansion in the androgen-receptor gene. *Lancet.* 354:640-643.

Dumoulin, J., Coonen, E., Bras, M., Bergers-Janssen, J., Ignoul-Vanvuchelen, R., van Wissen, L., Geraedts, J., and Evers, J., 2001, Embryo development and chromosomal anomalies after ICSI: effect of the injection procedure. *Hum Reprod.* 16:306-312.

Fishel, S., Green, S., and Bishop, M., 1995, Pregnancy after intracytoplasmic injection of a spermatid. *Lancet.* 345:1641-1642.

Hamberger, L., Lundin, K., Sjogren, A., and Soderlund, B., 1998, Indications for intracytoplasmic sperm injection. *Hum Reprod.* 13, Suppl 1:128-133.

Harari, O., Bourne, H., Baker, G., Gronow, M., and Johnston I., 1995, High fertilization rate with intracytoplasmic sperm injection in mosaic Klinefelter's syndrome. *Fertil Steril.* 63:182-184.

Hargreave, T.B., 2000, Genetics and male infertility. *Curr Opin Obst Gyn.* 12:207-219.

Hewitson, L., Simerly, C., Tengowski, M., Sutovsky, P., Navara, C., Haavisto, A.J., and Schatten, G., 1996, Microtubule and chromatin configurations during rhesus intracytoplasmic sperm injection: successes and failures. *Biol Reprod.* 55:271-280.

Hewitson, L., Takahashi, D., Dominko, T., Simerly, C., and Schatten, G., 1998, Fertilization and embryo development to blastocysts after intracytoplasmic sperm injection in the rhesus monkey. *Hum Reprod.*13:3449-3455.

Hewitson, L., Dominko, T., Takahashi, D., Martinovich, C., Ramalho-Santos, J., Sutovsky, P., Fanton, J., Jacob, D., Monteith, D., Neuringer, M., Battaglia, D., Simerly C., and Schatten, G., 1999, Unique checkpoints during the first cell cycle of fertilization after intracytoplasmic sperm injection in rhesus monkeys. *Nat Med.* 5:431-433.

Hewitson, L., Martinovich, C., Simerly, C., Takahashi, D., and Schatten, G., 2001, Intracytoplasmic injection of rhesus testicular sperm (TESE-ICSI) or elongated spermatids (ELSI) gives rise to healthy offspring. *Fertil Steril.* 77:794-801.

Kamischke, A., Gromoll, J., Simoni, M., Behre, H.M., and Nieschlag, E., 1999, Transmission of a Y chromosomal deletion involving the deleted in azoospermia (DAZ) and chromodomain (CDY1) genes from father to son through intracytoplasmic sperm injection: case report. *Hum Reprod.* 14:2320-2322.

Keefer, C.L., Younis, A.I., and Brackett, B.G., 1990, Cleavage development of bovine oocytes fertilized by sperm injection. *Mol Reprod Dev.* 25:281-285.

Kent-First, M. G., Kol, S., Muallem, A., Ofir, R., Manor, D., Blazer, S., First, N., and Itskovitz-Eldor, J., 1996, The incidence and possible relevance of Y-linked microdeletions in babies born after intracytoplasmic sperm injection and their infertile fathers. *Mol Hum Reprod.* 2:943-950.

Kent-First ,M., 2000, The critical and expanding role of genetics in assisted reproduction. *Prenat Diagn.* 20:536-551.

Kim, E.D., Bischoff, F.Z., Lipshultz, L.I., and Lamb, D.J., 1998, Genetic concerns for the subfertile male in the era of ICSI. *Prenat Diagn.* 18:1349-1365.

Kim, N.H., Jun, S.H., Do, .J.T., Uhm, S.J., Lee, H.T., Chung, K.S., 1999, Intracytoplasmic injection of porcine, bovine, mouse, or human spermatozoon into porcine oocytes. *Molec Reprod Dev.* 53:84-91.

Kupker, W., Schwinger, E., Hiort, O., Ludwig, M., Nikolettos, N., Schlegel, P.N., and Diedrich K, 1999, Genetics of male subfertility: consequences for the clinical work-up. *Hum Reprod.* 14, Suppl 1:24-37.

Lamb, D.J., and Lipshultz L.I., 2000, Male infertility: recent advances and a look towards the future. *Curr Opin Urol.* 10:359-362.

Liu, J., Nagy, Z., Joris, H., Tournaye, H., Devroey, P., and Van Steirteghem, A., 1995, Successful fertilization and establishment of pregnancies after intracytoplasmic sperm injection in patients with globozoospermia. *Hum Reprod.* 10:626-629.

Luetjens, C.M., Payne, C., and Schatten, G., 1999, Non-random chromosome positioning in human sperm and sex chromosome anomalies following intracytoplasmic sperm injection. *Lancet.* 353:1240.

Meschede, D., Lemcke, B., Behre, H.M., De Geyter, C., Nieschlag, E., and Horst, J., 2000, Clustering of male infertility in the families of couples treated with intracytoplasmic sperm injection. *Hum Reprod.* 15:1604-1608.

Moutel, G., Leroux, N., and Herve, C., 1999, Keeping an eye on ICSI. *Nature Med.* 5:593.

Ng, S.C., Liow, S.L., Ahmadi, A., Yong, E.L., Bongso, A., and Ratnam, S.S., 1995, Intracytoplasmic sperm injection--is there a need for an animal model, especially in assessing the genetic risks involved?. *Hum Reprod.* 10:2523-2525.

Nudell, D., Castillo, M., Turek, P.J., and Pera, R.R., 2000, Increased frequency of mutations in DNA from infertile men with meiotic arrest. *Hum Reprod.* 15:1289-1294.

Palermo, G., Joris, H., Devroey, P., and Van Steirteghem, A., 1992, Pregnancies after intracytoplasmic sperm injection of single spermatozoon into an oocyte. *Hum Reprod.* 340:17-18.

Pedersen, R.A., 2001, Sperm and mammalian polarity. *Nature.* 409:473-474.

Perry, A.C., Wakayama, T., Kishikawa, H., Kasai, T., Okabe, M., Toyoda, Y., and Yanagimachi, R., 1999, Mammalian transgenesis by intracytoplasmic sperm injection. *Science.* 284:1180-1183.

Ramalho-Santos, J., Sutovsky, P., Simerly, C.R., Oko, R., Wessel, G.M., Hewitson,L., and Schatten, G., 2000a, ICSI choreography: Fate of sperm structures after monospermic ICSI and first cell cycle implications. *Hum Reprod.* 15:2610-2620.

Ramalho-Santos, J., Moreno, R.D., Sutovsky, P., Chan, A.W.-S., Hewitson, L., Wessel, G., Simerly, C., and Schatten, G., 2000b, SNAREs in mammalian sperm mediate membrane fusion events during fertilization. *Dev Biol.* 223:54-69.

Rawe, V.Y, Olmedo, S.B., Nodar, F.N., Doncel, G.D., Acosta, A.A., and Vitullo, A.D., 2000, Cytoskeletal organization defects and abortive activation in human oocytes after IVF and ICSI failure. *Mol Hum Reprod.* 6:510-516.

Rybouchkin, A.V., Van der Straeten, F., Quatacker, J., De Sutter, P., and Dhont, M., 1997, Fertilization and pregnancy after assisted oocyte activation and intracytoplasmic sperm injection in a case of round-headed sperm associated with deficient oocyte activation capacity. *Fertil Steril.* 68:1144-1147.

Schatten, G., Simerly, C., and Schatten, H, 1985, Microtubule configurations during fertilization, mitosis, and early development in the mouse and the requirement for egg microtubule-mediated motility during mammalian fertilization. *Proc Natl Acad Sci.* 82:4152-4156.

Schatten, G., 1994, The centrosome and its mode of inheritance: the reduction of the centrosome during gametogenesis and its restoration during fertilization. *Dev Biol.* 165:299-335.

Schatten, G., Hewitson, L., Simerly, C., Sutovsky, P., and Huszar, G., 1998, Cell and molecular biological challenges of ICSI: A.R.T. before science?. *J Law Med Ethics.* 26:29-37.

Schlegel, P.N., Palermo, G.D., Alikani, M., Adler, A., Reing, A.M., Cohen, J., and Rosenwaks Z., 1995, Micropuncture retrieval of epididymal sperm with in vitro fertilization: importance of in vitro micromanipulation techniques. *Urology.* 46:238-241.

Silber, S.J., Van Steirteghem, A.C., Liu, J., Nagy, Z., Tournaye, H., and Devroey, P., 1995, High fertilization and pregnancy rate after intracytoplasmic sperm injection with spermatozoa obtained from testicle biopsy. *Hum Reprod.* 10:148-152.

Silber, S.J., Johnson, L., Verheyen, G., and Van Steriteghem, A., 2000, Round spermatid injection. *Fert Steril.* 73:897-900.

Simerly, C., Wu, G., Zoran, S., Ord, T., Rawlins, R., Jones, J., Navara, C., Gerrity, M., Rinehart, J., Binor, Z., and Schatten, G., 1995, The paternal inheritance of the centrosome, the cell's microtubule-organizing center, in humans and the implications for infertility. *Nature Med.* 1:47-52.

Sofikitis, N., Mantzavinos, T., Loutradis, D., Yamamoto, Y., Tarlatzis, V., and Miyagawa, I., 1998, Ooplasmic injections of secondary spermatocytes for non-obstructive azoospermia. *Lancet.* 351:1177-1178.

St. John, J.C., Cooke, I.D., and Barratt, C.L., 1997, Mitochondrial mutations and male infertility. *Nature Med.* 3:124-125.

Sutovsky, P., Hewitson, L., Simerly, C., Tengowski, M.W., Navara, C.S., Haavisto, A.J., and Schatten, G., 1996, Intracytoplasmic sperm injection for rhesus monkey fertilization results in unusual chromatin, cytoskeletal and membrane events, but eventually leads to pronuclear development and sperm aster assembly. *Hum Reprod.* 11:1703-1712.

Sutovsky, P., Oko, R., Hewitson, L., and Schatten, G., 1997, Binding of oocyte microvilli to the perinuclear theca of fertilizing sperm and subsequent theca removal constitute a previously unrecognized step in mammalian fertilization. *Dev Biol.* 188:75-84.

Tesarik, J., Mendoza, C., and Testart, J., 1995, Viable embryos from injection of round spermatids into oocytes. *N Engl J Med.* 333:525.

Terada, Y., Luetjens, C.M., Sutovsky, P., and Schatten, G., 2000, Atypical decondensation of the sperm nucleus, delayed replication of the male genome, and sex chromosome positioning following intracytoplasmic human sperm injection (ICSI) into golden hamster eggs: does ICSI itself introduce chromosomal anomalies?. *Fertil Steril.* 74:454-460.

te Velde, E.R., van Baar, A.L., and van Kooij, R.J., 1998, Concerns about assisted reproduction. *Lancet.* 351:1524-1525.

Tucker, M., Wright, G., Morton, C., Mayer, M.P., Ingargiola, P.E., and Jones, A.E., 1995, Practical evolution and application of direct intracytoplasmic sperm injection for male factor and idiopathic fertilization failure infertilities. *Fertil Steril.* 63:820-827.

Van Steirteghem, A.C., Nagy, Z., Joris, H., Liu, J., Staessen, C., Smitz, J., Wisanto, A., and Devroey, P., 1993, High fertilization and implantation rates after intracytoplasmic sperm injection. *Hum Reprod.* 8:1061-1066.

Vogt, P.H., 1998, Human chromosome deletions in Yq11, AZF candidate genes and male infertility: history and update. *Mol Hum Reprod.* 4:739-744.

Vogt, P.H., 1999, Risk of neurodegenerative diseases in children conceived by intracytoplasmic sperm injection?. *Lancet.* 354:611-612.

Wennerholm, U.B, Bergh, C., Hamberger, L., Lundin, K., Nilsson, L., Wikland, M., and Kallen, B., 2000, Incidence of congenital malformations in children born after ICSI. *Hum Reprod.* 15:944-948.

Wu, G., Simerly, C., Zoran, S., Funte, L.R., and Schatten, G., 1996, Microtubule and chromatin configurations during fertilization and early development in rhesus monkeys, and regulation by intracellular calcium ions. *Biol Reprod.* 55:260-270.

Yanagimachi, R., 1995, Is an animal model needed for intracytoplasmic sperm injection (ICSI) and other assisted reproductive technologies?. *Hum Reprod.* 10:2525-2526.

INCREASED INCIDENCE OF MALFORMATIONS IN THE OFFSPRING OF MALE MICE PRENATALLY EXPOSED TO SYNTHETIC ESTROGENS

Tetsuji Nagao[1], Nao Kagawa[2], Madoka Nakagomi[1] and Kazuo Fujikawa[2]

[1]Laboratory of Developmental Biology, Department of Life Science, Kinki University, Kowakae 3-4-1, Higashiosaka, Osaka 577-8502, Japan
[2]Laboratory of Animal Genetics, Department of Life Science, Kinki University, Kowakae 3-4-1, Higashiosaka, Osaka 577-8502, Japan

INTRODUCTION

In parallel with increasing concern about the reproductive effects of endocrine disrupters in the environment, evidence has been accumulating that developmental exposure to estrogenic hormones can cause a variety of abnormalities on the male reproductive tract (reviewed by Nagao, 1999). However, little is still known about transgenerational toxicity of estrogens, except for the potential of a transgenerational carcinogenic effect of diethylstilbestrol (DES) reported in mice by Walker (1984) and others. We carried out the present study as an attempt to fill this gap in our knowledge.

In the present study, male ICR mice were exposed as embryos to estradiol benzoate (EB), ethinyl estradiol (EE) or the anti-estrogen, tamoxifen (TAM), and their fetal offspring were examined for evidence of a transgenerational teratogenic effect of the drug. The effect was examined on external characteristics because external malformations are expressed early in life and can be monitored easily either phenotypically or quantitatively. Outbred ICR mice were used because teratogenesis mediated by males has been studied extensively using this strain. Histopathological changes in the reproductive tract of the treated males were also examined. We report here evidence that developmental exposure of males to estrogenic drugs is hazardous not only for development of the reproductive organs but also for embryonic development in the subsequent generation.

MATERIALS AND METHODS

Animals

ICR mice, 6-8 weeks old, were purchased from Charles River (Atsugi, Japan) and housed individually in polycarbonate cages with hard chip bedding in a room in which the temperature and relative humidity were controlled at $24\pm1°$ and $50\pm5\%$, respectively, with

Advances in Male Mediated Developmental Toxicity, edited by Bernard Robaire and Barbara F. Hales.
Kluwer Academic/Plenum Publishers, 2003.

lights on daily from 07:00 to 19:00. Animals were given free access to CE-2 food (CLEA Japan, Inc.) and tap water. To obtain pregnant females, 10 week old virgin females were individually cohabited overnight with single males. The next morning, females with vaginal plugs were regarded as pregnant, and this was designated as day 0 of gestation.

Treatment of Dams with Synthetic Estrogens and Sampling of Treated Males

Synthetic hormones, estradiol benzoate (EB), ethinyl estradiol (EE) and tamoxifen (TAM), were purchased from Sigma Chemical Co. (St. Louis, USA). Immediately before use, they were dissolved in corn oil (Nakalai Tesque, Inc., Kyoto, Japan).

At a defined time (12:00) on gestational days 9 through 16, test solution for EB or TAM was administered subcutaneously to pregnant mice at 2 ml/kg, and EE solution was given orally at 5 ml/kg. The number of dams in each of treated group was 7 to 9. Daily doses used for EB were 0.15 or 0.3 mg/kg, and for EE and TAM were 0.2 mg/kg and 0.5 mg/kg, respectively. In a preliminary study, we found that none of these drugs caused external malformations in fetuses, decreased fetal viability or reduced fetal weight gain when they were administered at these daily doses to pregnant mice for 8 days (data not shown). An adverse effect noticed in the preliminary study was a delay in delivery after treatment with 0.3 mg/kg/day EB. Therefore, all the treated dams and untreated controls in the present study were subjected to cesarean section on day 18 of gestation to obtain treated and control fetuses as live neonates. The day of cesarean section was considered as postnatal day 0.

The neonates obtained by cesarean section were fostered to untreated ICR females, one litter to each female. On postnatal day 4, all female pups were discarded, male pups were weighed, and the number of males per litter was adjusted to 4 or 5 for each foster female. The males were weaned on postnatal day 21 and then individually housed in polycarbonate cages, where they were allowed to grow to sexual maturity.

Examination of Reproductive Toxicity of Synthetic Estrogens in Males of the Treated Generation

At 12 weeks of age, the males were individually mated with single untreated ICR females, and thereafter they were allowed to copulate repeatedly for 2 or 3 weeks with different partners. Copulation was confirmed by the presence of a vaginal plug. Fertility of the males that copulated was confirmed by the presence of implants in the uterus of the partners on day 18 of gestation. Males that failed to copulate or impregnate females were classified as impotent.

At an age of 14 to 15 weeks, all males in the control and the treatment groups were weighed; their testes, epididymides and seminal vesicles were then weighed and fixed. Fixatives used for the testes and epididymides, and for the seminal vesicles were Bouin's solution and phosphate buffered 10% formalin, respectively. Fixed organs were embedded in paraffin, sectioned at a thickness of 5μm and stained with hematoxylin and eosin according to standard procedures. Stained sections were observed microscopically at a magnification of 100x for evidence of pathological changes.

Evaluation of Transgenerational Toxicity of Synthetic Estrogens

Upon examination of the uterine contents of females impregnated by control or treated males on gestational day 18, the numbers of implants, early and late resorptions and live fetuses were recorded. Each live fetus was weighed, sexed and inspected for external malformations including abnormalities of the oral cavity under a dissecting microscope. Fetuses weighing less than 70% of the average of the rest of the litter were classified as showing dwarfism and were included in the category of externally malformed offspring. Frequency of post-implantation loss was calculated as embryonic mortality for each litter by dividing the number summed for early and late resorptions by that of implantation sites. Frequency of malformed fetuses was calculated in litter units as the ratio of malformed to live fetuses. Data not shown in this report are available on request.

RESULTS

Reproductive Function of Males Prenatally Exposed to Synthetic Hormones and Histopathological Changes in the Reproductive Organs

As shown in Table 1, all males in the group prenatally treated as embryos for 8 days with 0.15 mg/kg/day of EB or 0.5 mg/kg/day of TAM succeeded in fertilizing untreated females when mated at 12-14 weeks old. On the other hand, one of 22 males in the group treated with 0.30 mg/kg/day of EB did not copulate. When mated at 12-15 weeks old, two of 22 males in the EE-treated group did not copulate, and one of the 20 males that did copulate in this group failed to impregnate females.

Table 1. Reproductive Functions of Male Mice Prenatally Exposed to Synthetic Hormone EB, EE or TAM and the Number of Males showing Histopathological Changes (HPC) in their Reproductive Organs

Treatment[a]		No. of males			
(mg/kg/day)		mated	copulated	fertile	with HPC
Control		23	23	23	0
EB	0.15	20	20	20	0
	0.30	22	21	21	12[b]
EE	0.2	22	20	19	3[c]
TAM	0.5	20	20	20	0

[a]Subcutaneous (EB and TAM) or oral (EE) administration in dams on gestational days 9 through 16 at the indicated daily dose.
[b]All were fertile males including 7 with dilated seminiferous tubules and epididymal ducts, plus epididymal cysts, 2 with epididymal cysts, and 3 with dilated seminiferous tubules and epididymal ducts.
[c]All were impotent males and showed epididymal cysts and reflux of semen in the seminal vesicle; each had atrophic seminiferous tubules, oviduct-like structures in the epididymis or a uterus-like duct along the vas deferens in addition to these two abnormalities.

As determined histopathologically after the completion of the mating experiments, all males from the groups treated with 0.15 mg/kg/day of EB and 0.5 mg/kg/day of TAM showed no appreciable signs of adverse effects of the drugs in the reproductive organs within the limits of the resolving power of the histological methods used (Table 1). On the other hand, 12 fertile males among 22 males treated with 0.30 mg/kg/day of EB had epididymal cysts, dilated seminiferous tubules and epididymal duct, or both (Table 1); one impotent male from this group showed no abnormalities in the reproductive tract. The three impotent males from the EE-treated group all showed semen reflux in the seminal vesicle and epididymal cysts. All three of these males had atrophic seminiferous tubules, oviduct-like cysts in the epididymis or uterus-like ducts along the vas deferens in addition to these abnormalities. None of the fertile males in this group had any conspicuous changes in the reproductive organs (Table 1). The mean weight of the testes, epididymides, or seminal vesicles measured immediately before fixation of the organs for histopathological examination did not show significant deviation in any of the treated groups from the control weights of 0.25±0.01, 0.11±0.01, and 0.48±0.03 g, respectively.

Reproductive Outcomes of Males Prenatally Exposed to Synthetic Hormones

In any of the treated series, there was no significant deviation from the control in frequency of post-implantation loss or in the number of implants per litter (Table 2). On the other hand, the frequency of externally malformed fetuses increased significantly over the control level among the offspring whose sires had been prenatally treated with 0.30 mg/kg/day of EB or 0.2 mg/kg/day of EE. This was not the case among fetuses from males treated with a lower daily dose of EB or 0.5 mg/kg/day of TAM. The control frequency of malformed fetuses recorded in the present study (0.23 ±0.16%, N=58) does not differ significantly from the historical control value of 0.31±0.08% (N=287) based on information gathered for the ICR strain over the last 10 years in our laboratory at the Hatano Research Institute. Therefore, we concluded that prenatal exposure of ICR males to EB at a high dose or EE caused malformations in the subsequent generation.

Table 2. Frequencies of Postimplantation Loss and Malformed Fetuses in the Offspring of Male Mice Prenatally Exposed to Synthetic Hormone EE, EB or TAM

Treatment[a] (mg/kg/day)		No. of implants per litter	Frequency of postimplantation loss (%)	No. of live fetuses per litter	Frequency of malformed fetuses (%)
Control		14.2±0.3[b](58)[c]	7.5±1.1[b]	13.2±0.3[b]	0.23±0.16[b]
EB	0.15	14.5±0.3 (95)	6.5±0.9	13.7±0.4	0.17±0.11
	0.30	14.6±0.4 (112)	7.7±1.2	13.5±0.4	1.31±0.25**
EE	0.2	14.1±0.3 (212)	9.1±1.0	12.8±0.3	0.85±0.28*
TAM	0.5	14.0±0.3 (87)	7.5±1.5	13.5±0.6	0.24±0.20

[a]Subcutaneous (EB and TAM) or oral (EE) administration in dams for parental males on gestational days 9 through 16 at the indicated dose.
[b]Mean±SE.
[c]Total number of pregnant females examined.
*Significantly different from the control by the Mann-Whitney U test at P = 0.05.
**Significantly different from the control by the Mann-Whitney U test at P = 0.01.

As shown in Table 3, dwarfism, cleft palate and exencephaly were detected as malformations in the fetal offspring of males pre-natally exposed to 0.30 mg/kg/day of EB. In addition to these two types of malformations, open eyelid and polydactyly were observed in the offspring of males exposed in utero to 0.2 mg/kg/day of EE. None of these five types of malformation are atypical, and all occur spontaneously in ICR mice as shown in the last two columns of Table 3. The numerical ratios of cleft palate : dwarfism : exencephaly in the EB-treated, EE-treated and control series were 15:4:1, 15:6:1 and 29:7:2, respectively. These values did not differ significantly from each other ($\chi^2 = 0.67$, d.f.=2, P > 0.05). The frequencies of fetuses with cleft palate in the EB- and EE-treated series were 0.98 (=15/1529) and 0.55% (=15/2688), respectively, both of which were significantly higher than the spontaneous frequency of 0.20% (=28/13329) as tested by Fisher's exact test at $P = 0.05$. When the two sets of data from the treated series were combined, the resulting frequency of dwarfism (10/4217=0.48%) was significantly higher than the spontaneous frequency of 0.06% (=7/13329). These observations reconfirm the transgenerational teratogenic effects of EB and EE.

Table 3. Distribution of Malformations in the Fetal Offspring of Male Mice Exposed Prenatally to Synthetic Hormone EB or EE

Type of Malformations	Abnormal offspring of males exposed to				Untreated control[a]	
	0.30 mg/kg/day of EB		0.2 mg/kg/day of EE			
	No.	F (%)	No.	F (%)	No.	F (%)
Dwarfism	4	0.26	4	0.15	6	0.05
Cleft palate	15	0.98	13	0.48	27	0.20
Cleft palate + dwarfism	_[b]		2	0.07	1	0.01
Cleft palate + brain hernia	-		_[b]		1	0.01
Open eyelid	-		1	0.04	2	0.02
Exencephaly	1	0.07	1	0.04	_[b]	
Exencephaly + polydactyly	-		-		1	0.01
Exencephaly + microphthalmus	-		-		1	0.01
Polydactyly	-		1	0.04	3	0.02
Microphthalmus	-		-		3	0.02
Total	20	1.31	22	0.82	45	0.34
Total no. of fetuses examined	1529		2688		13329	

F = Frequency of malformed fetuses per 100 fetuses.

[a]Data pooled for concurrent and historical controls.

[b]Not detected.

DISCUSSION

The results from the present study have shown that prenatal exposure of ICR males to estrogenic compounds is hazardous not only for the development of the reproductive tract in the treated mice, but also for embryonal development in the subsequent generation. Toxic effects on the reproductive tract were detected as histopathological changes in males treated with EB at 0.3 mg/kg/day or EE at 0.2 mg/kg/day (Table 1). An increased incidence of malformations was recorded in the fetal offspring of these males (Table 2).

With one exception, all reproductive tract abnormalities detected after EB or EE treatment were of types that had not been reported to occur in ICR mice either spontaneously or experimentally. The exception was atrophy of seminiferous tubules in one impotent male from the EE-treated group. This type of abnormality had been reported by Yasuda et al. (1985) after prenatal exposure of ICR males to EE. These authors further reported that prenatal EE exposure can produce ovotestis and intra-abdominal testis with Mullerian and Wolfian ducts. In the present study, a feminization effect of EE was seen as formation of oviduct- and uterus-like structures in the reproductive tract of treated males. Epididymal cysts, which were observed in the EB- and the EE-treated males, have been reported to occur in CD-1 mice after prenatal exposure to DES.

In contrast to the reproductive tract abnormalities, all of the external malformations detected in the fetal descendants of males exposed pre-natally to EB at 0.30 mg/kg/day or EE at 0.2 mg/kg/day were of types known to occur in ICR mice (Table 3). Namely, with regard to the types of malformations, the transgenerational effect observed in the present study is neither atypical for ICR mice nor specific to the drugs tested or the stage of germ-cell development at the time of treatment, as shown by the following observations. The relative ratios of cleft palate: dwarfism: exencephaly in the EB-treated, EE-treated and control series were 1:0.3:0.07, 1:0.4:0.07 and 1:0.2:0.07, respectively. Similarly, cleft palate was the most common, followed by dwarfism and exencephaly in this order among external malformations produced in the offspring of ICR males treated with a mutagen at the spermatogonial stem cell or the post-meiotic stage (Nagao, 1987, 1988; Nagao and Fujikawa, 1990; Nagao and Fujikawa, 1998). Taken together, it seems that heritable damage was induced in embryonic germ cells (i.e., primordial germ cells) after exposure to EB or EE, and the offspring inherited the damage and were sensitized to spontaneous teratogenesis. Primordial germ cells are at risk for chemical induction of transmissible change, as shown by the induction of specific locus mutations and dominant skeletal mutations with ethylnitrosourea (Shibuya et al., 1993; Wada et al., 1994). However, it is not clear whether the germ-cell damage responsible for the effects observed in the present study was genetic or epigenetic in nature.

Despite such uncertainty, the results of the present study on male-mediated teratogenesis agree with those of previous reports on transgenerational carcinogenicity of DES (Turusov et al., 1992; Walker and Kurth, 1995) in suggesting vulnerability of developing germ cells to estrogenic drugs. Considering the differences in the kind of drugs tested, the strain of mice used, and the effects examined between the present and the previous studies, transgenerational effects of estrogenic drugs do not seem to be a rare phenomena in mice. Studies of the molecular mechanisms of the effects are required to evaluate the implications of these observations in human situations.

ACKNOWLEDGEMENTS

These studies were supported by a grant form The Science and Technology Agency of the Japanese Government: Special Coordination Funds for Promoting Science and Technology

REFERENCES

Nagao, T., 1987, Frequency of congenital defects and dominant lethals in the offspring of male mice treated with methylnitrosourea. *Mutation Res.* 177:171-178.

Nagao, T., 1988, Congenital defects in the offspring of male mice treated with ethylnitrosourea. *Mutation Res.* 202:25-33.

Nagao, T., 1999, Multigeneration effects of endocrine disrupting chemicals with special reference to teratogenesis by paternal exposure to synthetic hormones. *Environ Mutagen Res.* 21:267-272.

Nagao, T., and Fujikawa, K., 1990, Genotoxic potency in mouse spermatogonial stem cells of triethylenemelamine, mitomycin C, ethylnitrosourea, procarbazine, and propyl methanesulfonate as measured by F_1 congenital defects. *Mutation Res.* 229:123-128.

Nagao, T., and Fujikawa, K., 1998, Modified susceptibility to teratogenesis in the offspring of male mice exposed to mutagens. *Cong Anom.* 38:1-8.

Shibuya, T., Murota, T., Horiya, N., Matsuda, H., and Hara, T., 1993, The induction of recessive mutations in mouse primordial germ cells with *N*-ethyl-*N*-nitrosourea. *Mutation Res.* 290:273-280.

Shibuya, T., Horiya, N., Matsuda, H., Sakamoto, K., and Hara, T., 1996, Dose-dependent induction of recessive mutations with *N*-ethyl-*N*-nitrosourea in primordial germ cells of male mice. *Mutation Res.* 357:219-224.

Turusov, V.S., Trukhanova, L.S., Parfenov, Y.D., and Tomatis, L., 1992, Occurrence of tumors in descendants of CBA male mice prenatally treated with diethylstilbestrol. *Int J Cancer.* 50:131-135.

Wada, A., Sato, M., Takashima, H., and Nagao, T., 1994, Congenital malformations in the offspring of male mice treated with ethylnitrosourea at the embryonic stage. *Teratogen Carcinogen Mutagen.* 14:271-279.

Waker, B.E., 1984, Tumors in female offspring of mice exposed prenatally to diethylstilbestrol. *J Natl Cancer Inst.* 73:133-140.

Waker, B.E., and Kurth, L.A., 1995, Multigenerational carcinogenesis from diethylstilbestrol investigated by blastocyst transfer in mice. *Int J Cancer.* 61:249-252.

Yasuda, Y., Kihara, T., Tanimura, T., and Nishimura, H., 1985, Gonadal dysgenesis induced by prenatal exposure to ethinyl estradiol in mice. *Teratology.* 32:219-227.

IMPLICATIONS OF RESEARCH IN MALE-MEDIATED DEVELOPMENTAL TOXICITY TO CLINICAL COUNSELLORS, REGULATORS, AND OCCUPATIONAL SAFETY OFFICERS

J. M. Friedman

Department of Medical Genetics
University of British Columbia
Vancouver, BC
Canada V6T 1Z3

DIFFERENT PERSPECTIVES ON MALE-MEDIATED DEVELOPMENTAL TOXICITY

The information discussed at this conference makes it clear that exposure of male mammals to toxic chemicals or radiation can, under some circumstances, adversely affect subsequent reproductive outcomes. This is a very interesting observation, and we need to determine how widespread the agents and circumstances are that produce such effects and to understand the molecular mechanisms that are responsible. However, the question remains: Are these effects of any practical importance to humans?

The answer depends on who is asking the question, and why. For example, consider the effect of exposure to a "doubling dose" of ionizing radiation, i.e., the amount necessary to double the naturally occurring mutation rate in the population. Although there is some controversy over the actual value of the doubling dose of ionizing radiation in humans and the "background" frequencies of birth defects and genetic diseases that result from mutations are only known approximately, the crude comparison shown in Figures 1 and 2 is informative. If this exposure (about 1 Sv or 100 REM) were to affect an entire population, the frequency of children born with genetic disease resulting from new mutations would double (Figure 1). In the United States, this means that more than 15,000 additional children would be born with such diseases each year. This would have a very serious impact on the families involved and important public health implications. On the other hand, the increase in the risk of birth defects in the children of a man who was exposed to this amount of radiation is about the same as the increase that would occur if his partner were 39 instead of 36 years old at the time of delivery (Figure 2); the increased risk associated with maternal age is for Down syndrome and other chromosomal abnormalities, not for autosomal dominant mutations, but the magnitude of the increase in risk is similar

Advances in Male Mediated Developmental Toxicity, edited by Bernard Robaire and Barbara F. Hales.
Kluwer Academic/Plenum Publishers, 2003.

in these two cases, assuming that the man who was exposed to radiation has retained his fertility.

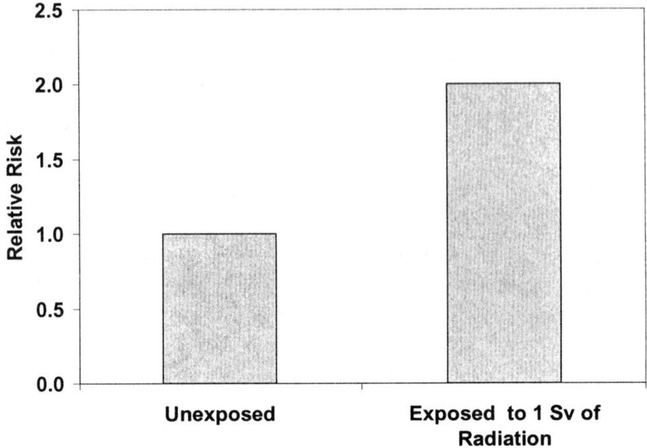

Figure 1. The population perspective: relative risk of genetic disease resulting from new autosomal dominant mutations among the children of men who were exposed to 1 Sv of radiation, based on frequencies given in Vogel and Motulsky (1997) and the assumption that the doubling dose is 1 Sv.

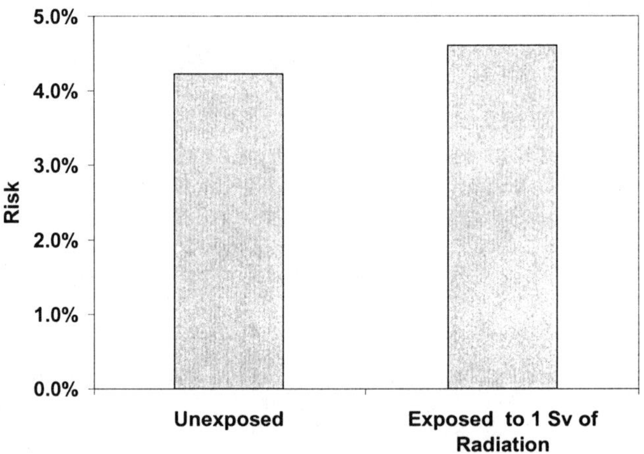

Figure 2. The individual perspective: change in the overall risk of birth defects among the children of a man who was exposed to 1 Sv of radiation, based on frequencies given in Vogel and Motulsky (1997) and the assumption that the doubling dose is 1 Sv.

This paper will consider the implications of what we are learning about male-mediated developmental toxicity from two different perspectives: that of the population as a whole and that of the individual patient. These different perspectives greatly influence the health professionals who deal with issues related to male-mediated developmental toxicity – regulators and public health professionals on one hand, and physicians and counsellors who care for individual patients on the other. I shall consider these two perspectives in some detail, but I shall not discuss occupational health professionals separately because their perspective is a hybrid of the other two. Occupational health professionals strive to protect workers as a group and as individuals. As a consequence, occupational health professionals must adopt both an individual and a population perspective within the context of an efficient workplace.

WHAT REGULATORS NEED TO KNOW

Government regulators wish to make important chemicals (including drugs) available to consumers and to industry, while protecting people from unreasonable harm. Every chemical carries some risk if it is used in an unsafe manner – people can die of water intoxication or table salt poisoning – so regulators of industrial and agricultural chemicals must determine what risks are associated with particular uses of the chemical and the circumstances under which these risks can be considered to be acceptably small. Depending on the nature of the chemical and its distribution and uses, relative safety may need to be considered only for exposed workers, for a subset of the population, for the population as a whole, or for some combination of these groups.

Medicines represent a special case. Most medicines are intentionally taken systemically, often in doses of hundreds of milligrams per day, so the exposures are generally greater than those likely to occur with environmental or occupational exposures. Medicines are also required to demonstrate efficacy in treating the disease or diseases for which they are indicated; this obviously is not a relevant consideration for other chemicals. Some level of toxicity may be considered acceptable for a medication if the anticipated benefit is sufficiently great. Food additives and products such as herbal remedies may also be taken systemically in relatively large doses.

It is important to note that both the benefits and toxicity associated with the use of medications and herbal remedies are almost exclusively limited to the people who take these substances. Third parties, i.e., most members of the population as a whole, who do not intentionally consume a product are not usually subject to its toxicity. This is quite different from exposures that may occur with chemicals such as environmental pollutants, where the benefits may be limited to a particular group, but the risks may be general.

Male-mediated developmental toxicity has had a very limited impact on public health policy. One important reason for this is that few chemicals have been studied thoroughly in humans for male-mediated developmental toxicity (Adler, 1996; Dellarco and Kimmel, 1997). Moreover, the human data that do exist for male-mediated developmental toxicity are negative, inconsistent, and/or tainted by political controversy. Many of the male-mediated developmental toxicity studies that have been done in humans were stimulated by public fears engendered by well-publicized anecdotal reports of birth defects in children of men suspected of having a particular exposure. Almost all other human studies are of known mutagens that might be expected to affect the genetic integrity of germ cells. The studies of children of male soldiers who were exposed to Agent Orange during the Vietnam War are an example of the former, and the studies of children of men who received cancer chemotherapy in childhood, an example of the latter.

Because of the limited availability of human data, regulators usually make decisions regarding the level and conditions of risk that are acceptable for a particular chemical on the basis of studies in rats or other laboratory animals. Extrapolation of the results of these studies to humans must be done with great caution and always involves some degree of uncertainty because of differences that exist between species. A "safety factor", usually a factor of 10, is included in calculations of acceptable exposure levels to account for the uncertainty introduced by extrapolation from laboratory animals to humans. Even if this is done very carefully, many clinicians are not convinced of the relevance of these laboratory data to the human situation because of the lack of a single instance in which the evidence for human male-mediated developmental toxicity is compelling. (See chapter by Savitz, this book). The classic example is shown in Figure 3: Large doses of ionizing radiation are unequivocally mutagenic to the germ cells of rodents and other laboratory animals that have been tested. Nevertheless, no significant increase in the frequency of untoward pregnancy outcomes (congenital anomalies, stillbirth or neonatal death) was found in the extensive data collected on children of men who had previously been exposed to atomic radiation in Hiroshima and Nagasaki. One can argue that this study was not large enough to detect an effect, or that the dose estimates were imprecise, or that the study design was inadequate in some other way. Nevertheless, this study, and others that have been done in humans, generally do not suggest that male exposure to mutagens or other toxic chemicals substantially increases the risk of birth defects in children that are subsequently conceived.

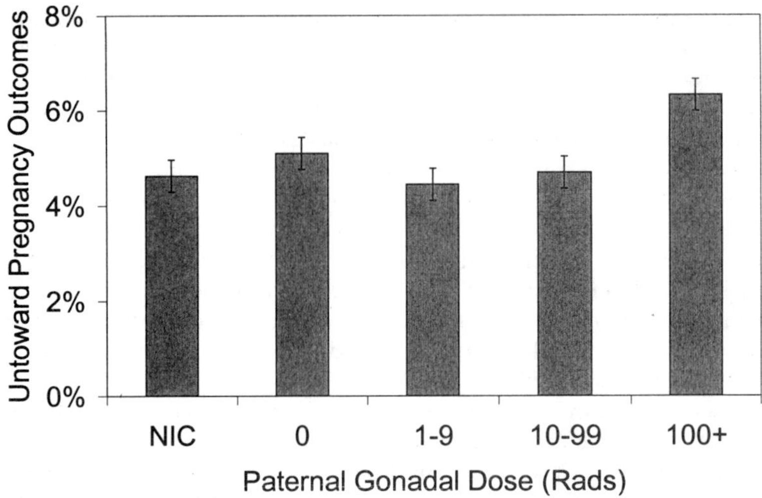

Figure 3. Frequency of untoward pregnancy outcome (a child with a major congenital anomaly, who was stillborn, or who died in the neonatal period) among the subsequent children of men who were previously exposed or unexposed to atomic radiation (NIC = not in city at the time of the atomic bombing). The mothers of all of these children were not in the city at the time of the explosion. Data are from Schull et al., 1981. The differences are not statistically significant.

In making decisions regarding the acceptability of particular exposures on the basis of laboratory animal data, regulatory agencies usually treat male developmental or reproductive toxicity in the same way that they treat neurotoxicity, hepatotoxicity, nephrotoxicity, or toxicity manifested in any other organ system. In the past, this often meant that male-mediated developmental toxicity was not considered at all because the kinds of studies necessary to assess such toxicity were not required and had not been done. There seems to be a general belief that male-mediated developmental effects are rarely the most sensitive indicator of toxicity and therefore rarely the determining factor in regulatory decisions (ICH, 1996). However, this view appears be based more on a paucity of data regarding male-mediated developmental toxicity than on clear evidence regarding the relative susceptibility to male-mediated developmental effects and other kinds of toxicity for a large number of chemicals.

More recently, explicit requirements for male-mediated reproductive toxicity testing in at least one mammalian species have been put forward by regulatory agencies (ICH, 1994, 1996; US EPA, 1996; US FDA, 2000). The U.S. Food and Drug Administration guidelines for food ingredients and the U.S. Environmental Protection Agency guidelines both include a two-generation study that would permit identification of at least some kinds of male-mediated developmental toxicity (Claudio et al., 1999a). The guidelines for pharmaceuticals are more limited and would only permit detection of abnormalities that affect fertility or testicular histology in exposed males (ICH, 1994, 1996). Controversy exists regarding the adequacy and appropriateness of these guidelines for the identification of reproductive and developmental toxicity mediated through the male (Adler, 1996; Sullivan, 1998; Claudio *et al.,* 1999a, b; Lamb and Brown, 2000), but their implementation clearly reflects an increasing recognition of the problem.

In their assessment of toxicity, regulatory agencies may focus not only on the most susceptible organ system but also on the most susceptible individuals (Dellarco and Kimmel, 1997; Lamb and Brown, 2000). This is usually taken to mean children or others who may be at unusually high risk for particular kinds of toxicity because of pre-existing illnesses (e.g., asthmatics may be especially susceptible to certain air pollutants). Occasionally, the embryo is seen as the most susceptible individual. This has been a determining factor in regulation of thalidomide, isotretinoin, and methylmercury (Pless and Risher, 2000; Rice et al., 2000), but there are few other examples (Claudio et al., 1999a, b). As discussed below, new genomic technology may revolutionize regulations based on the principle of protecting the most sensitive individuals within a population.

WHAT EXPOSED MEN WANT TO KNOW

Individual men and their partners look at risks of male-mediated developmental toxicity differently from regulators, whose responsibility is for populations. Regulators focus on prevention; they try to prevent adverse exposures as much as possible within the constraints of practical and economic reality. Regulators must balance the risk of permitting exposures of a certain type against the benefits of doing so and the costs of not doing so.

In contrast, the predominant concern of men who have already suffered a toxic exposure is on possible adverse effects. In the case of developmental toxicity, the concern is about possible effects of the exposure on future or existing children. The relevant comparison here is not with the benefits of the exposure but rather with the "background" risk of similar adverse outcomes.

Although several mutagenic exposures have been clearly demonstrated to produce male-mediated developmental toxicity in laboratory animals, the absolute rate of abnormalities among surviving offspring after such exposures is often low. For example, exposing male mice to a doubling dose of ionizing radiation increases the overall rate of clinically apparent mutations in the offspring from about 0.01% to 0.02% (Neel and Lewis, 1990).

Mutation is an infrequent cause of fetal anomalies in humans, and if mutation is the major mechanism by which male-mediated developmental toxicity is produced, the effect on the overall frequency of birth defects is likely to be small. For example, Sankaranarayanan (2001) has estimated that the total risk of genetic disease and congenital anomalies produced by 1 Gy (100 rads) of chronic low-dose radiation in humans is approximately 3000-4700 cases per million live births. This is an increase of less than 1% in the baseline frequency of these disorders.

A number of other mechanisms of male-mediated developmental toxicity have been proposed (Friedler, 1996; Trasler and Doerksen, 1999) and were discussed at this meeting. None of these mechanisms has yet been shown to produce male-mediated developmental toxicity in humans. Experiments in laboratory animals suggest that even if similar effects occur in men, no more than a few per cent of their offspring are likely to have malformations, even with very large paternal exposures (Trasler and Doerksen, 1999).

The conclusion that male-mediated developmental toxicity is unlikely to produce a high absolute risk of congenital anomalies among the offspring of exposed fathers is supported by empiric studies in humans. No detectable increase in the frequency of congenital anomalies has been demonstrated reproducibly among the children of men exposed to relatively high doses of ionizing radiation, cancer chemotherapeutic agents or any other chemical (Shelby, 1994; Olshan, 1995; Pryor et al., 2000). Male-mediated reproductive toxicity manifested as infertility and/or oligozoospermia does occur in humans after exposure to large doses of ionizing radiation, cyclophosphamide, other cancer chemotherapeutic agents, carbon disulfide, dibromochloropropane, or lead (Pryor et al., 2000), but this is a different matter.

Therefore, the appropriate advice for a man exposed to any potential reproductive toxin in the remote past is clear: if fertility is maintained, the risk of birth defects among the man's children is unlikely to be measurably increased. Such paternal exposures are not an indication for fetal cytogenetic analysis by amniocentesis or chorionic villus sampling. Women who become pregnant with partners who have had such exposures should be offered detailed obstetrical ultrasound examinations at about 18 weeks gestation and serum triple screening (routine screening of some other type for fetal cytogenetic abnormalities), but this is true for every pregnant woman.

MALE-MEDIATED DEVELOPMENTAL TOXICITY IN A WORLD OF SEQUENCED GENOMES

The knowledge and technology that are developing in relationship to sequencing the genome of humans, mice, rats, and other organisms will permit much more precise recognition of populations and individuals who are unusually sensitive to certain toxic effects. Identification of highly-sensitive populations will be important to regulators but will also raise important legal, ethical, and social questions. For example, if a small group of people is found to be 10, 100 or even 1000 times more sensitive to a particular exposure than everyone else in the population, should people from the highly-sensitive group be excluded from occupations in which that exposure occurs, or should safety standards for everyone be adjusted to protect this susceptible subgroup? Should an individual with a known genetic predisposition to an especially severe kind of toxicity be permitted to choose to be exposed, e.g., by taking a particular medication? If such individuals do choose to be exposed, should they be eligible for publicly-funded remedies or for insurance payments if toxicity does, in fact, occur?

For individuals, knowledge of unusual susceptibility to toxicity offers the prospect of avoiding such toxicity, but this will not come without costs. To begin with, there is the cost of testing for individual susceptibility to toxicity. Should this cost be borne by the individual, by his or her health care system, by private insurance, by the manufacturer of a pharmaceutical that is being considered, or by an employer? Even if relatively inexpensive multiplex testing for toxic susceptibility becomes available, for which polymorphisms should testing be provided? Those that occur most commonly in the population? Those that affect the most commonly-encountered exposures? Those for exposures for which avoidance is most practical? Those that can be used to anticipate the most severe kinds of toxicity? Or should different kinds of testing be offered to different people, depending on personal circumstances, age, gender, or financial means?

The routine recognition of subgroups with different susceptibilities to toxic effects will also revolutionize research in male-mediated developmental toxicity. Determining why some people or strains of laboratory animals are especially sensitive to toxic effects may provide important clues to the mechanisms involved. In addition, experiments can be designed that will test possible mechanisms by using laboratory animals modified by specific transgenic or knockout alterations involving a putative pathway. We shall have very powerful new tools to apply to our research, but things are going to get much more complicated!

SUMMARY

In closing, let me return to the somewhat irreverent question that I asked at the beginning: Apart from what we learn about fundamentally important biological processes, does male-mediated developmental toxicity make any practical difference? The answer is no, yes, and we don't know.

- ◆ **No:** The risk of serious birth defects in the future children of a man who is concerned about exposure to radiation or chemicals in the remote past does not appear to be measurably increased in comparison to the background.

- ◆ **Yes:** Male-mediated developmental toxicity may be of sufficient concern to restrict certain exposures in the population as a whole, but in order for male-mediated developmental toxicity to be determinative, these conditions will have to be otherwise non-toxic. To date, no exposures have been identified that produce substantial male-mediated developmental toxicity in the complete absence of other kinds of toxicity.

- ◆ We don't know how big a problem male-mediated developmental toxicity really is because we have not studied a large enough number and variety of exposures. We don't know if our perspective will change as we begin to understand the importance of individual susceptibility better. We don't know all of the mechanisms that may be involved in male-mediated developmental toxicity or their general biological importance. We don't know more than we do know about male-mediated developmental toxicity, and we are anxious to learn more.

REFERENCES

Adler, I.-D., 1996, Future research directions to study genetic damage in germ cells and estimate genetic risk. *Environ Health Perspect.* 104(Suppl. 3):619-624.

Claudio, L., Bearer, C.F., and Wallinga, D., 1999a, Assessment of the U.S. Environmental Protection Agency methods for identification of hazards to developing organisms, Part I: The reproduction and fertility testing guidelines. *Am J Indust Med.* 35:543-553.

Claudio, L., Bearer, C.F., and Wallinga, D., 1999b, Assessment of the U.S. Environmental Protection Agency methods for identification of hazards to developing organisms, Part II: The developmental toxicity testing guideline. *Am J Indust Med.* 35:554-563.

Dellarco, V.L., and Kimmel, C.A., 1997, Risk assessment of environmental agents for developmental toxicity: current and emerging approaches. *Mutat Res.* 396:205-218.

Friedler, G., 1996, Paternal exposures: Impact on reproductive and developmental outcome. An overview. *Pharmacol Biochem Behav.* 55:691-700.

International Conference on Harmonisation of Technical Requirements for Registration of Pharmaceuticals for Human Use (ICH), 1994. Guideline for Industry – Detection of Toxicity to Reproduction for Medicinal Products. (ICH-S5A).

International Conference on Harmonisation of Technical Requirements for Registration of Pharmaceuticals for Human Use (ICH), 1996. Guideline for Industry – Detection of Toxicity to Reproduction for Medicinal Products: Addendum on Toxicity to Male Fertility. (ICH-S5B).

Lamb, J.C., IV, and Brown, S.M., 2000, Chemical testing strategies for predicting health hazards to children. *Reprod Toxicol.* 14:83-94.

Neel, J.V., and Lewis, S.E., 1990, The comparative radiation genetics of humans and mice. *Ann Rev Genetics* 24:327-362.

Olshan, A.F., 1995, Lessons learned from epidemiologic studies of environmental exposure and genetic disease. *Environ Mol Mutagen.* 23(Suppl. 26):74-80.

Pless, R., and Risher, J.F., 2000, Mercury, infant development, and vaccination. *J Pediatr.* 136:571-573.

Pryor, J.L., Hughes, C., Foster, W., Hales, B.F., and Robaire, B., 2000, Critical windows of exposure for children's health: The reproductive system in animals and humans. *Environ Health Perspect.* 108(Suppl 3):491-503.

Rice, G., Swartout, J., Mahaffey, K., and Schoeny, R., 2000, Derivation of U.S. EPA's oral Reference Dose (RfD) for methylmercury. *Drug Chem Toxicol.* 23:41-54.

Sankaranarayanan, K., 2001, Estimation of the hereditary risks of exposure to ionizing radiation: history, current status, and emerging perspectives. *Health Phys.* 80:363-369.

Schull, W.J., Otake, M., and Neel, J.V., 1981, Genetic effects of the atomic bombs: A reappraisal. *Science.* 213:1220-1227.

Shelby, M.D., 1994. Human germ cell mutagens. *Envrion Mol Mutagen.* 23 (Suppl. 24):30-34.

Trasler, J.M., and Doerksen, T.. Teratogen update: paternal exposures-reproductive risks. *Teratology.* 60:161-172.

Sullivan, F.M., 1998. Reproductive toxicity tests: retrospect and prospect. *Hum Toxicol.* 7:423-427.

U.S. Environmental Protection Agency (US EPA), Office of Research and Development, 1996. Guidelines for reproductive toxicity risk assessment. (EPA/630/R-96/009a).

U. S. Food and Drug Administration (US FDA), Center for Food Safety & Applied Nutrition, Office of Premarket Approval, 2000, Redbook 2000: Toxicological Principles for the Safety of Food Ingredients, Section IV.C.9.a. Guidelines for Reproduction Studies.

Vogel, F., and Motulsky, A.G., 1997, Chapter 11, Mutation: Induction by ionizing radiation and chemicals, in *Human Genetics: Problems and Approaches*, 3rd edition, Springer-Verlag, Berlin, pp. 457-493.

RESTORATION OF SPERMATOGENESIS AFTER EXPOSURE TO TOXICANTS: GENETIC IMPLICATIONS

Marvin L. Meistrich, Gene Wilson, Gunapala Shetty, and Gladis A. Shuttlesworth

Department of Experimental Radiation Oncology
The University of Texas M. D. Anderson Cancer Center
Houston, TX 77030, U.S.A.

INTRODUCTION

Many men have been sterilized by exposure to radiation and chemotherapy agents as treatment for cancer; radiation and almost all of the chemotherapeutics are also mutagens. In the United States alone, 17,000 men 15 to 45 years old are diagnosed each year with Hodgkin's disease, lymphoma, bone and soft tissue sarcomas, testicular cancer, or leukemia (Landis et al., 1999). Of these, over 2,400 are long-term survivors that were treated with doses of alkylating agents, platinum drugs, or radiation sufficient to induce prolonged azoospermia. In addition, 6,000 boys under 15 are diagnosed each year with cancer, including leukemia, nervous system tumors, lymphomas, and other solid tumors. About 80% of them receive chemotherapy or gonadal irradiation, and about 550 of the long-term survivors are azoospermic when they reach adulthood.

Environmental and occupational toxicants can also produce prolonged azoospermia. This was most dramatically shown with dibromochloropropane (DBCP), which is also a potential mutagen, as highly exposed manufacturing (Whorton et al., 1979) and agricultural workers (Slutsky et al., 1999) had an increased incidence of azoospermia.

In response to this problem, our primary goal has been to develop methods to prevent or reverse the induction of azoospermia in those men, thereby maintaining or restoring fertility. In rats, methods involving hormonal pretreatment with gonadotropin releasing hormone (GnRH) analogs or steroids have been shown to enhance the recovery of spermatogenesis after exposure to radiation or chemotherapeutic agents (Delic et al., 1986; Kurdoglu et al., 1994). However, attempts to directly translate these observations to reduce the damaging effects of cytotoxic therapies on human spermatogenesis have had only limited success (Masala et al., 1997; Morris and Shalet, 1990).

A contributing factor to the failure of the clinical studies is the lack of knowledge of the mechanism by which the hormonal treatment works in the rat. It had previously been believed that the hormone treatment protected stem spermatogonia from the damaging effects of radiation or chemotherapy. However, all tests of possible mechanisms for stem cell protection failed to support any of the proposed mechanisms (Meistrich et al., 1994; Meistrich et al., 1997; Wilson et al., 1999). Furthermore, we observed that some stem cells survived cytotoxic treatment, yet failed to differentiate (Kangasniemi et al., 1996). Moreover, we recently showed that treating the rats with the same hormones after irradiation or chemotherapy could stimulate spermatogenic recovery (Meistrich and Kangasniemi, 1997; Meistrich et al., 1999). Although we do not know the exact molecular steps involved in hormonal stimulation of spermatogenic recovery given after cytotoxic therapy, we do know that it is achieved by a transient reduction of testosterone and probably FSH, which otherwise inhibit the differentiation process (Shetty et al., 2000; Shuttlesworth et al., 2000).

If we were able to apply these procedures effectively to restore spermatogenesis and fertility in men, the genetic consequences to the offspring must be considered. As shown in other presentations in this volume, the known heritable genetic risks to the offspring from the currently used anticancer treatments that allow recovery of spermatogenesis are minimal. However, if we were to induce recovery of fertility after higher doses of mutagenic toxicants than have been previously compatible with recovery of fertility, it might pose an increased genetic risk.

In this chapter, we will first discuss the characteristics of spermatogenic failure after cytotoxic damage. Secondly, we will review the hormonal treatments that restore spermatogenesis and what we know about their mechanisms. Then we will present some limited data on genetic damage after restoration of spermatogenesis with hormonal treatment. Finally we will discuss the chances for clinical application of these principles.

FAILURE OF SPERMATOGENESIS AFTER CYTOTOXIC EXPOSURE

Previous studies in the mouse showed that, after cytotoxic insults, surviving stem cells retained the ability to largely repopulate the seminiferous tubules and to produce cells that differentiated into spermatozoa (Meistrich, 1982). Therefore, the killing of stem spermatogonia was necessary to induce prolonged failure of spermatogenesis. However in rats treated with radiation (Kangasniemi et al., 1996) or toxic chemicals, such as the chemotherapeutic drug procarbazine (Meistrich et al., 1999) or hexanedione (Boekelheide and Hall, 1991), which is a metabolite of the common solvent hexane, there is a failure of recovery of spermatogenesis despite the presence of type A stem spermatogonia. Furthermore, this inhibition appears to be a general phenomenon, as it is observed after a variety of toxicants given to the rat acutely or subchronically, during aging in the rat, and in cryptorchid mice (Table 1).

We have studied this process in most detail after irradiation. At 4 and 6 weeks after irradiation, there is some recovery of spermatogenesis, as evidenced by the differentiation of type A into type B spermatogonia and the appearance of some spermatocytes (Maiti et al., 2001). But after 6 weeks, the ability of the A spermatogonia to differentiate is progressively lost. Whereas at 6 weeks after a dose of 3.5 Gy, 75% of the tubules contained differentiating cells derived from surviving stem cells, that number progressively decreased until, by 60 weeks after irradiation, differentiation had completely ceased.

The only germ cells in the atrophic tubules were undifferentiated type A spermatogonia. The loss of the ability to support differentiation was not due to loss of A spermatogonia, as their number remained relatively constant at between 1 and 2 spermatogonia per 100 Sertoli cells (Kangasniemi et al., 1996). It was surprising, given their constant number, to observe that these spermatogonia were actively proliferating as demonstrated by high bromodeoxyuridine-labeling indices and mitotic indices (Allard et al., 1995). Their number was limited by apoptosis, which counterbalanced the increase caused by mitotic divisions (Allard and Boekelheide, 1996; Shuttlesworth et al., 2000).

It was not expected that the failure of spermatogenic recovery would be due to hormone deficiency, as spermatogonial differentiation in the rat is independent of testosterone and only quantitatively dependent on FSH, with at most a factor of 2 reduction in differentiating cells observed upon FSH withdrawal (Meachem et al., 1999). When we did measure hormone levels in the toxicant-treated rats, we found that, whereas serum testosterone levels were unchanged, there were 3-, 1.6-, and 2.5-fold increases, respectively, in serum LH, serum FSH, and intratesticular testosterone levels (Meistrich et al., 1999).

Table 1. Toxicants and other phenomena that result in the failure of spermatogenesis despite the presence of type A spermatogonia and their prevention or reversal by GnRH-analog or steroid treatment.

Species: Agent	Category	Treatments[1] causing prevention / reversal	Reference
Rat:			
Hexanedione	Metabolite of hexane	GnRH-Ag	(Blanchard et al., 1998; Boekelheide and Hall, 1991)
Boric Acid	Consumer product	Not tested	(Ku et al., 1993)
Radiation	Anticancer agent	PreRx – T+E; PostRx – GnRH-Ag or -Ant	(Kurdoglu et al., 1994; Meistrich and Kangasniemi, 1997)
Indenopyridine	Potential contraceptive	Pre+PostRx – GnRH-Ag	(Hild et al., 2001)
Procarbazine	Anticancer drug	PreRx – T+E, GnRH-Ag or Ant; PostRx – GnRH-Ag	(Kangasniemi et al., 1995a; Kangasniemi et al., 1995b; Meistrich et al., 1999; Parchuri et al., 1993)
Dibromo-chloropropane	Pesticide	PostRx – GnRH-Ag	(Meistrich, Wilson, and Shuttlesworth, unpublished)
Aging	Natural phenomenon	At 27 months – GnRH-Ag	(Schoenfeld et al., 2001)
Mouse:			
Cryptorchidism	Birth defect	Not tested	(de Rooij et al., 1999)

[1] T+E : testosterone plus estradiol; GnRH-Ag: GnRH agonist ; GnRH-Ant: GnRH antagonist; PreRx: prior to drug treatment; PostRx: after drug treatment.

PREVENTION OF SPERMATOGENIC FAILURE AND RESTORATION OF SPERMATOGENESIS AFTER CYTOTOXIC EXPOSURE

It was originally proposed that it might be possible to protect spermatogenesis with hormonal treatment by interruption of the pituitary-gonadal axis, reducing the rate of spermatogenesis and rendering the resting testis more resistant to the effects of chemotherapy (Glode et al., 1981). Indeed it has been demonstrated that pretreatment of rats with steroids or GnRH analogs to suppress gonadotropins and testosterone, and hence the completion of spermatogenesis, results in a greater recovery of spermatogenesis after procarbazine administration or irradiation (Delic et al., 1986; Kangasniemi et al., 1995a; 1995b; Kurdoglu et al., 1994; Parchuri et al., 1993). However, the mechanism originally proposed cannot be correct since hormonal suppression does not alter the kinetics of spermatogenesis. Furthermore, the enhancement of spermatogenic recovery is achieved under conditions that do not modify the kinetics of spermatogonial proliferation prior to cytotoxic therapy (Meistrich et al., 1997). We have considered other possible mechanisms for this apparent protective effect and have ruled them out (Wilson et al., 1999) except for the possibility that the hormonal pretreatment could act by enhancing the recovery of spermatogenesis from surviving stem spermatogonia.

To demonstrate that the hormonal treatment was not protecting the survival of the stem spermatogonia, we counted the numbers of A spermatogonia remaining in the tubules of irradiated-only rats and rats receiving irradiation after a 'protective' hormone treatment (Meistrich et al., 2000). The similar spermatogonial counts at 6 weeks after irradiation in both groups of rats indicated that the hormone treatment was not acting to protect stem cells from killing by the cytotoxic agent. To examine whether the effect of the hormones was to enhance the recovery of spermatogenesis from surviving spermatogonia, we counted the percentages of tubules containing differentiating cells in both groups at various times after irradiation. Although at 6 weeks after irradiation, the percentages were similar in both groups, dramatic differences were observed at later times. In the rats given irradiation alone, spermatogenesis steadily declined after 6 weeks due to loss of the ability of stem cells to differentiate, as described above, but when rats were pretreated with hormones prior to irradiation, this decline was prevented, and there was a progressive recovery of spermatogenesis. Thus the enhancement of spermatogenic recovery by hormone treatment results from modification of an injury-induced change in spermatogonia or their environment that would have otherwise resulted in failure of spermatogonial differentiation.

We considered the possibility that, if hormonal pretreatment could affect later spermatogenic recovery, hormonal posttreatment might also work. When rats were given 3.5 Gy of irradiation, but no hormone treatment, the percentage of tubules showing recovery of spermatogenesis at 10 weeks after irradiation was 37%. When GnRH-agonist treatment was started immediately after irradiation and maintained for 10 weeks, differentiation was observed in 91% of tubules (Meistrich and Kangasniemi, 1997). Since our first report that recovery of spermatogenesis can be stimulated by GnRH analogs after irradiation, we have been amazed at the range of exposure situations to which this treatment has been successfully applied (Table 1).

However, during GnRH-analog treatment, differentiation could not proceed past the round spermatid stage since testosterone is required for maturation of spermatids. In order to complete spermiogenesis, it is necessary to stop the hormone treatment so that testosterone levels may be restored. Initially we did not know whether spermatogonial differentiation would continue after stopping the treatment or whether it would again be blocked. In the above experiment, when an additional 10 weeks without further treatment was allowed after the 10-week GnRH treatment, 100% of the tubules showed differentiating cells and the sperm counts recovered to 50% of the levels in unirradiated controls (Meistrich et al., 1999). Furthermore, all of these rats were fertile.

To demonstrate that recovery of spermatogenesis could be stimulated after regression of spermatogenesis had occurred, rats were treated with GnRH agonist beginning at 20 weeks after 3.5 Gy of irradiation, when only 30% of tubules contained germ cells past the A spermatogonial stage (Meistrich et al., 1999). Treatment with GnRH agonist for 10 weeks increased the percentage of tubules with differentiating cells to 54%, and even 10 weeks after cessation of GnRH-agonist treatment, this percentage further increased to 78%.

More recent studies have demonstrated that both GnRH agonists and antagonists were effective in restoring spermatogenesis and fertility in rats given sterilizing doses of irradiation (Meistrich et al., 2001). For example, when a 10-week GnRH-agonist treatment was initiated 10 weeks after 5-Gy irradiation, at which time spermatogonial differentiation had already ceased, fertility was restored in 83% of the male rats within 10 weeks after the cessation of the hormone treatment.

The exact mechanisms involved in the inhibition of spermatogonial differentiation after these toxicant treatments, and the means by which steroid or GnRH-analog treatments prevent or reverse this inhibition, are not known. However, we now have convincing evidence that GnRH analogs stimulate the recovery of spermatogonial differentiation after cytotoxic insult by reducing the levels of testosterone, which appears to be the major inhibitory factor (Shetty et al., 2000). In addition to testosterone, FSH also appears to be an inhibitory factor and its suppression also contributes to the stimulation of recovery. Since spermatogonia lack androgen and FSH receptors, these hormones must be acting on a somatic cell to enhance its presentation of proapoptotic factors or reduce the levels of growth and differentiation factors that affect the survival of the spermatogonia. Direct damage to the spermatogonia themselves from the toxic insult may contribute to their sensitivity to or requirement for certain factors. Although it is still puzzling that hormonal treatment given for 6 weeks before the toxic insult exerts its effect on spermatogonial differentiation more than 6 weeks after the insult, it appears that there may be a common mechanism that could explain the beneficial effects of hormonal treatment either before or after cytotoxic damage.

GENETIC CONSEQUENCES OF RESTORATION OF SPERMATOGENESIS

Since many of these cytotoxic agents are also genotoxicants and known germ cell mutagens in rodents, it is important to consider the question of whether restoration of fertility in males, which would have otherwise been sterilized, would lead to an increased risk of genetic damage in the offspring. Although our primary interest has been to gain an understanding of the mechanisms regulating the inhibition of spermatogonial differentiation by testosterone, we can use some of the data obtained to make inferences regarding fertility. We also will review and interpret the available literature in this area.

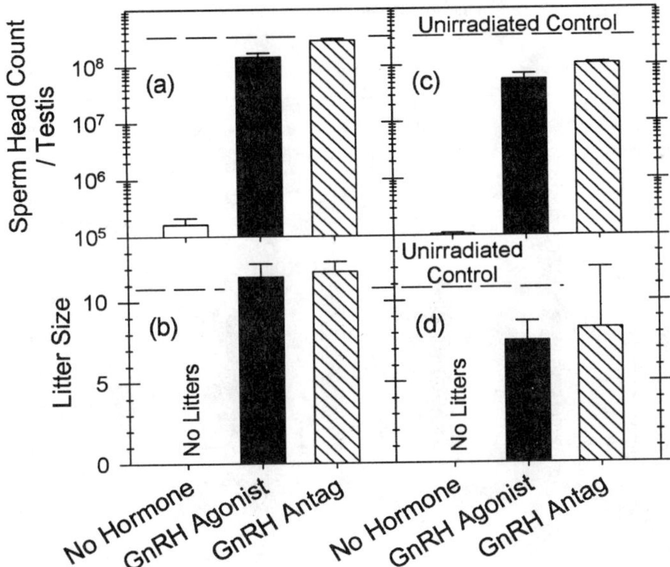

Figure 1. Effects of GnRH-agonist or -antagonist treatment on maintenance (a,b) or recovery (c,d) of spermatogenesis in irradiated rats. (a,b) Rats irradiated with 3.7 Gy were treated immediately with depot doses of the indicated GnRH analog, which lasted for about 5-10 weeks. They were mated between weeks 18-22 and were killed at week 22. (c,d) Rats were irradiated with 5 Gy and 10 weeks later they were treated with the GnRH analog for about 10 weeks. They were mated on weeks 30 and 31, and killed at week 32. (a,c) Sperm head count per testis. (b,d) Average size of litters produced by fertile males. The dashed lines indicate levels found in unirradiated control rats.

Irradiated rats were treated with GnRH agonist or antagonist and mated with females as described in the caption of Figure 1 (Meistrich et al., 2001). Although this experiment was designed to assess fertility and not embryonic loss, some information can be obtained from the sizes of the litters that were born and in one group (GnRH antagonist, Fig. 1d) from analysis of the uterine contents of the females killed at 14 days after the end of mating.

Whereas none of the rats irradiated with 3.7 Gy (Fig 1a,b) and not given any hormone treatment were fertile, all those treated with either the GnRH agonist or GnRH antagonist were fertile. Litter sizes were unaffected (Fig. 1b), indicating that there could only be very minimal pre- or post-implantation loss. All of the live-born pups appeared normal. There was only one dead pup in 422 produced by the GnRH-antagonist-treated group and none in 321 pups produced by the GnRH-agonist-treated group, which was similar to the frequency of dead pups (3 in 292) produced by the control rats.

Similarly, after 5 Gy of irradiation and delayed treatment with GnRH analogs (Fig 1c,d), the irradiated-only rats were all sterile. Hormonal treatments produced a significant improvement in fertility, as 5 of 6 rats treated with GnRH agonist and 3 of 6 of those treated with GnRH antagonist were fertile. The litter sizes appeared to be slightly reduced but not significantly different from control levels. Since the numbers of litters was small, no strong conclusions could be reached regarding litter size. Furthermore, since sperm counts were reduced to 18% and 32% of control levels in the GnRH-agonist and antagonist treated rats, respectively, some reduction in litter size could have been a result of failure of fertilization rather than pre- or post-implantation loss. All of the live-born pups appeared normal. In only one case, in the GnRH-agonist-treated group, 2 pups (of a total of 86 in the group) were killed by the mother. In the GnRH-antagonist-treated group in which the uterine contents of the females were examined, there were no dead embryos among the 25 observed, indicating minimal post-implantation loss.

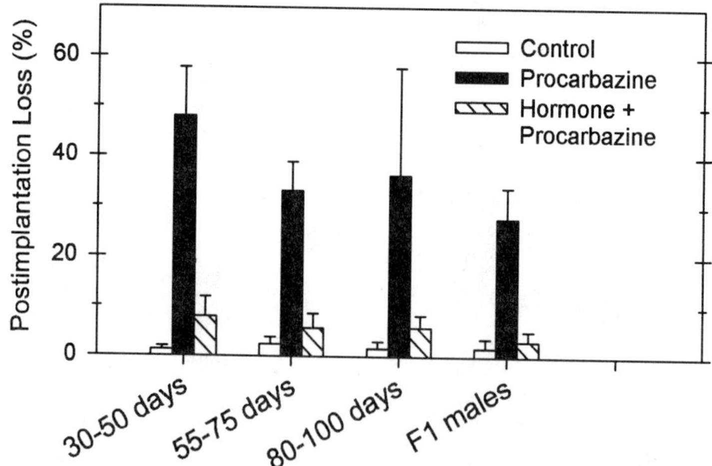

Figure 2. Effects of hormonal pretreatment on procarbazine-induced post-implantation loss in rats (Velez de la Calle and Jégou, 1990). Control rats received sham injections. Procarbazine was given as 3 injections of 150 mg/kg at weekly intervals. Daily injections of testosterone and medroxyprogesterone acetate were given for 55 days, starting 40 days before the first injection of procarbazine and until the final procarbazine injection. The time intervals indicate when, after the last procarbazine injection, the males were mated, with 30-50 days, 55-75 days, and 80-100 days representing sperm derived from cells that were spermatogonia and spermatocytes, undifferentiated and differentiated spermatogonia, and stem spermatogonia, respectively, during at least one of the procarbazine injections. F_1 males were obtained from offspring from the 80-100-day matings of the different groups of F_0-treated males; they themselves received no treatment.

Studies of hormonal modification of testicular effects of cytotoxic agents, which were more directed towards genetic endpoints, were performed by Jegou and co-workers (Jégou et al., 1991; Velez de la Calle and Jégou, 1990). They treated rats with testosterone plus a progestin to suppress gonadotropins and intratesticular testosterone levels before exposing them to radiation or before and during 3 injections of procarbazine. This hormonal treatment enhanced the eventual levels of sperm production observed 14 to 19 weeks after the cytotoxic treatment. In addition they mated the treated males at various times after exposure and monitored post-implantation loss.

There was a large increase (to 30-50%) in post-implantation loss in embryos derived matings with the procarbazine-treated males, which appeared independent of the stage of spermatogenic cell that was exposed. In contrast, the embryos derived from matings with rats that had been hormonally treated before and during the procarbazine treatments showed minimal increases in post-implantation loss. The level of post-implantation loss in these hormone-treated rats was significantly lower than in those treated with procarbazine alone.

In addition, there was a transgenerational effect. When F_1 males derived from procarbazine-treated fathers were subsequently mated, the frequency of post-implantation loss in the embryos that they produced was just about as high as those produced by the treated F_0 rats themselves.

Very similar results were obtained with radiation-treated rats (Fig. 3). In fact, the frequency of post-implantation loss was extremely high, at between 70% to 100%, during the first few cycles of sperm produced from stem cells (days 66-106). Subsequently, the frequency of post-implantation loss fell to about 20%. The levels of post-implantation loss in the litters sired by rats pretreated with hormones before irradiation were slightly but not significantly increased above control but were markedly lower than the levels observed in the litters produced by the irradiated-only rats.

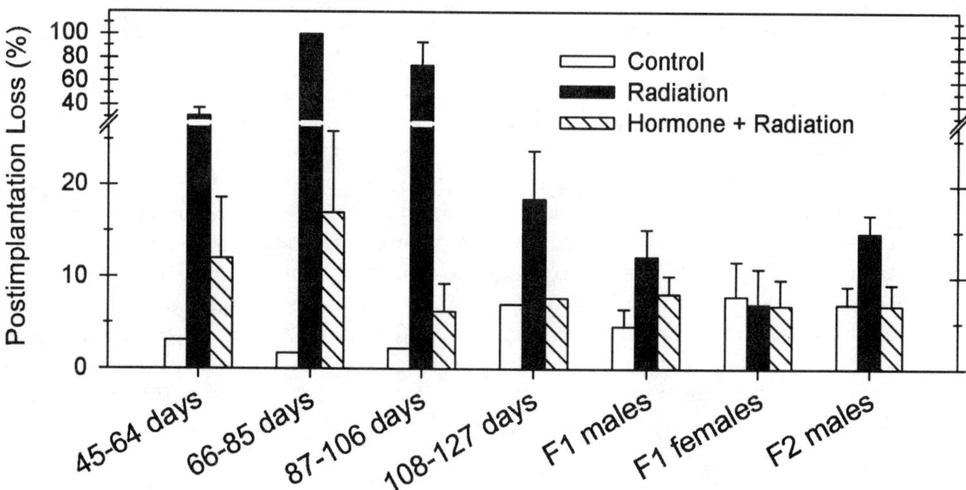

Figure 3. Effects of hormonal pretreatment on radiation-induced post-implantation loss in rats (Jégou et al., 1991). Control rats received sham treatments. Radiation was given as a single dose of 3 Gy of X-rays. Daily injections of testosterone and medroxyprogesterone acetate were given for 55 days prior to irradiation. The time intervals indicate the time after irradiation at which the males were mated, with 45-64 days representing sperm derived from cells that were differentiating spermatogonia and early primary spermatocytes, and greater than 66 days cells that were stem spermatogonia during irradiation. F_1 offspring were obtained from the 80-100 day matings of the different groups of F_0-treated males; they themselves received no treatment, and F_2 males were derived from F_1 males that were descendents of the different F_0-male treatment groups.

There also appeared to be transgenerational effects, but they were only expressed in the males. The F_1 males from irradiated fathers produced a marginally increased frequency of post-implantation embryo loss. However, the F_2 males from irradiated F_0 rats did produce a significantly increased frequency of post-implantation embryo loss and this frequency was significantly reduced if the F_0 males had been given hormone treatment before irradiation.

These data can be compared to results obtained in the mouse, a more widely studied genetic model. The incidence of post-implantation loss after irradiation of stem spermatogonia in the rat appeared to be only slightly higher than the value of about 8% observed in the mouse after similar radiation doses (Ehling, 1971). However, in the mouse procarbazine produces dominant lethal mutations only when late primary spermatocytes or spermatids are exposed (Ehling, 1974). In contrast to the rat, no dominant lethal mutations were observed in any of the spermatogonial stages from mice given procarbazine or other alkylating agents, nor were heritable translocations, which are also indicative of the type of damage that results in embryonic loss (Witt and Bishop, 1996). Since post-implantation loss contributes to the dominant lethal frequency, the absence of dominant lethals in these mouse studies indicates a level of post-implantation loss that was below detection in those studies. There must be an interspecies difference since values for post-implantation loss of about 30%, as were observed in rat spermatogonia, would have been detectable in the mouse studies.

Based on the observations that the hormone treatment does not protect the survival of stem spermatogonia but rather acts to enable their recovery after irradiation, it is surprising that the hormone treatment produces this major reduction in genetic damage manifest in the embryos produced. One way to reconcile this apparent discrepancy is that the chromosomal damage resulting in this embryo loss is not produced directly in the germ cells themselves by the radiation or procarbazine.   Rather, the resulting alterations in their environment, e.g. Sertoli and other somatic cell factors necessary for normal germ cell development, affect the genetic development of germ cells derived from genetically normal stem cells, resulting in the continuous generation of sperm with chromosomal damage. Because the hormone treatment enhances the ability of the somatic cells to support spermatogonial differentiation, it may also enhance the ability of these cells to support the process without generating additional genetic damage.

The persistence of the effects in the F_1 and F_2 generations of male rats treated with radiation or procarbazine is harder to explain and falls under the still vague concept of genomic instability. One possible explanation could be that a large percentage of the post-implantation loss is attributable to unbalanced products of reciprocal translocations, so that the survivors are carrying balanced products and give rise to offspring carrying reciprocal translocations. These offspring will produce a high frequency of gametes that carry an unbalanced chromosome complement, resulting in further embryonic loss. Other purely speculative explanations could involve failure of imprinting or the transmission of a system that continues to generate reactive oxygen species in the testis or germ cells of the offspring.

POSSIBLE APPLICATION TO HUMAN AZOOSPERMIA

At present, it is not known if GnRH-agonist treatment will stimulate recovery of spermatogenesis in men with toxicant-induced azoospermia because we do not know whether or not the mechanism that stops spermatogenesis is the same in humans and rats. One important criterion is whether or not men have surviving stem spermatogonia, as do rats.

There is histological evidence that at least some of the men who presented with prolonged azoospermia after exposure to cytotoxic agents have surviving stem spermatogonia (Kreuser et al., 1989), but others have no germ cells, at least in the biopsy sample analyzed (Van Thiel et al., 1972). In some cases the block to spermatogenesis in the human testis appears, as in the rat, to be at the level of spermatogonial differentiation (Kreuser et al., 1989); in other cases, however, it is at the spermatocyte stage (Meistrich and van Beek, 1990).

Further support for the existence of surviving stem cells is the observation that some men, without any medical treatment, recover spermatogenesis after two or more years of azoospermia. This has been observed after exposure to radiation (Rowley et al., 1974), chemotherapy (Marmor et al., 1992; Meistrich et al., 1992; Pryzant et al., 1993), or dibromochloropropane (Potashnik and Porath, 1995). In order for sperm production to recover, these men must have had surviving stem spermatogonia in their testes that had been unable to differentiate into spermatozoa. The trigger for this apparently spontaneous recovery of spermatogenesis in men is not known.

Therefore, testosterone-suppressive treatment, such as with a GnRH agonist, might be beneficial to men who have azoospermia resulting from cytotoxic therapy, provided there are not other associated risks. GnRH-agonist treatment is widely used in men with prostate cancer, and the only side effects are those associated with reduction of testosterone. Such suppression would be transient in a suggested 6-month treatment protocol based on the studies in rats, and the side effects would be reversible when the treatment is stopped.

One must also consider the possible risks of transmitting increased levels of genetic damage to the offspring. Since there is as yet no evidence for increased genetic risk in the offspring of men who were treated in the past with chemo- or radiotherapy (Byrne et al., 1998; Senturia et al., 1985), it is not a major concern unless there is evidence that the hormonal treatment would enhance the risk in rodent models. Our data on litter size in rats, whose fertility was restored by hormone treatment after irradiation, fail to indicate any increased risk, but these studies were not designed to examine genetic risk (Meistrich et al., 2001). The data using hormonal pretreatment, which we have argued above might be acting by the same mechanism as the post-treatment (Meistrich et al., 2000), show a dramatic reduction in genetic risk (Jégou et al., 1991; Velez de la Calle and Jegou, 1990). This genetic risk was measured by post-implantation loss of the embryos produced by the male treated with chemo- or radiotherapy and hereditary or transgenerational effects involving the post-implantation loss of embryos produced by the F_1 and F_2 offspring of the treated males. Thus, there is no evidence for a genetic risk as a mitigating factor for instituting clinical trials of hormone treatment to attempt to maintain or restore human male fertility after cytotoxic therapy for cancer or environmental or occupational exposure to a reproductive toxicant.

ACKNOWLEDGEMENTS

We thank Walter Pagel for editorial assistance and the National Institute of Environmental Health Sciences of the NIH (Grant ES-08075) for support of most of the research presented from the authors' laboratory.

REFERENCES

Allard, E.K., and Boekelheide, K., 1996, Fate of germ cells in 2,5-hexanedione-induced testicular injury. II. Atrophy persists due to a reduced stem cell mass and ongoing apoptosis. *Toxicol. Appl. Pharmacol.*, 137: 149-156.

Allard, E.K., Hall, S., and Boekelheide, K., 1995, Stem cell kinetics in rat testis after irreversible injury induced by 2,5-hexanedione. *Biol. Reprod.*. 53:186-192.

Blanchard, K.T., Lee, J., and Boekelheide, K., 1998, Leuprolide, a gonadotropin-releasing hormone agonist, reestablishes spermatogenesis after 2,5-hexanedione-induced irreversible testicular injury in the rat, resulting in normalized stem cell factor expression. *Endocrinology.* 139:236-244.

Boekelheide, K., and Hall, S.J., 1991, 2,5-Hexanedione exposure in the rat results in long-term testicular atrophy despite the presence of residual spermatogonia. *J Androl.* 12:18-26.

Byrne, J., Rasmussen, S.A., Steinhorn, S.C., Connelly, R.R., Myers, M.H., Lynch, C.F., Flannery, J., Austin, D.F., Holmes, F.F., Holmes, G.E., Strong, L.C., and Mulvihill, J.J., 1998, Genetic disease in offspring of long-term survivors of childhood and adolescent cancer. *Am J Hum Genet.* 62:45-52.

de Rooij, D.G., Okabe, M., and Nishimune, Y., 1999, Arrest of spermatogonial differentiation in jsd/jsd, Sl17H/Sl17H, and cryptorchid mice. *Biol Reprod.* 61:842-847.

Delic, J.I., Bush, C., and Peckham, M.J., 1986, Protection from procarbazine-induced damage of spermatogenesis in the rat by androgen. *Cancer Res.* 46:1909-1914.

Ehling, U.H., 1971, Comparison of radiation- and chemically-induced dominant lethal mutations in male mice. *Mutat Res.* 11:35-44.

Ehling, U.H., 1974, Differential spermatogenic response of mice to the induction of mutations by antineoplastic drugs. *Mutat Res.* 26:285-295.

Glode, L.M., Robinson, J., and Gould, S.F., 1981, Protection from cyclophosphamide-induced testicular damage with an analogue of gonadotropin-releasing hormone. *Lancet.* 1:1132-1134.

Hild, S.A., Meistrich, M.L., Blye, R.P., and Reel, J.R., 2001, Lupron depot prevention of antispermatogenic/antifertility activity of the indenopyridine, CDB-4022, in the rat. *Biol Reprod.* 65:165-172.

Jégou, B., Velez de la Calle, J.F., and Bauche, F., 1991, Protective effect of medroxyprogesterone acetate plus testosterone against radiation-induced damage to the reproductive function of male rats and their offspring. *Proc Natl Acad Sci USA.* 88:8710-8714.

Kangasniemi, M., Huhtaniemi, I., and Meistrich, M.L., 1996, Failure of spermatogenesis to recover despite the presence of A spermatogonia in the irradiated LBNF$_1$ rat. *Biol Reprod.* 54:1200-1208.

Kangasniemi, M., Wilson, G., Huhtaniemi, I., and Meistrich, M.L., 1995a, Protection against procarbazine-induced testicular damage by GnRH-agonist and antiandrogen treatment in the rat. *Endocrinology.* 136:3677-3680.

Kangasniemi, M., Wilson, G., Parchuri, N., Huhtaniemi, I., and Meistrich, M.L., 1995b, Rapid protection of rat spermatogenic stem cells against procarbazine by treatment with a gonadotropin-releasing hormone antagonist (Nal-Glu) and an antiandrogen (flutamide). *Endocrinology.* 136:2881-2888.

Kreuser, E.D., Kurrle, E., Hetzel, W.D., Heymer, B., Porzsolt, R., Hautmann, R., Gaus, W., Schlipf, U., Pfeiffer, E.F., and Heimpel, H., 1989, Reversible germ cell toxicity after aggresive chemotherapy in patients with testiclular cancer: Results of a prospective study. *Klinische Wochenschrift.* 67:367-378.

Ku, W.W., Chapin, R.E., Wine, R.N., and Gladen, B.C., 1993, Testicular toxicity of boric acid (BA): Relationship of dose to lesion development and recovery in the F344 rat. *Reprod Toxicol.* 7:305-319.

Kurdoglu, B., Wilson, G., Parchuri, N., Ye, W.-S., and Meistrich, M.L., 1994, Protection from radiation-induced damage to spermatogenesis by hormone treatment. *Radiat Res.* 139:97-102.

Landis, S.H., Murray, T., Bolden, S., and Wingo, P.A., 1999, Cancer statistics, 1999. *CA Cancer J Clin.* 49:8-31, 1.

Maiti, S., Meistrich, M.L., Wilson, G., Shetty, G., Marcelli, M., McPhaul, M.J., Morris, P.L., and Wilkinson, M.F., 2001, Irradiation selectively inhibits expression from the androgen-dependent *Pem* homeobox gene promoter in Sertoli cells. *Endocrinology.* 142:1567-1577.

Marmor, D., Grob-Menendez, F., Duyck, F., and Delafontaine, D., 1992, Very late return of spermatogenesis after chlorambucil therapy: Case reports. *Fertil Steril.* 58:845-846.

Masala, A., Faedda, R., Alagna, S., Satta, A., Chiarelli, G., Rovasio, P.P., Ivaldi, R., Taras, M.S., Lai, E., and Bartoli, E., 1997, Use of testosterone to prevent cyclophosphamide-induced azoospermia. *Ann Intern Med.* 126:292-295.

Meachem, S.J., McLachlan, R.I., Stanton, P.G., Robertson, D.M., and Wreford, N.G., 1999, FSH immunoneutralization acutely impairs spermatogonial development in normal adult rats. *J Androl.* 20:756-762.

Meistrich, M.L., 1982, Quantitative correlation between testicular stem cell survival, sperm production, and fertility in the mouse after treatment with different cytotoxic agents. *J Androl.* 3:58-68.

Meistrich, M.L., and Kangasniemi, M., 1997, Hormone treatment after irradiation stimulates recovery of rat spermatogenesis from surviving spermatogonia. *J Androl.* 18:80-87.

Meistrich, M.L., and van Beek, M.E.A.B., 1990, Radiation sensitivity of the human testis. In J.R. Lett and K.I. Altman (Eds.), Adv. Rad. Biol. (pp. 227-268). New York: Academic Press.

Meistrich, M.L., Wilson, G., Brown, B.W., da Cunha, M.F., and Lipshultz, L.I., 1992, Impact of cyclophosphamide on long-term reduction in sperm count in men treated with combination chemotherapy for Ewing's and soft tissue sarcomas. *Cancer.* 70:2703-2712.

Meistrich, M.L., Wilson, G., and Huhtaniemi, I., 1999, Hormonal treatment after cytotoxic therapy stimulates recovery of spermatogenesis. *Cancer Res.*, 59: 3557-3560.

Meistrich, M.L., Wilson, G., Kangasniemi, M., and Huhtaniemi, I., 2000, Mechanism of protection of rat spermatogenesis by hormonal pretreatment: stimulation of spermatogonial differentiation after irradiation. *J. Androl.*, 21: 464-469.

Meistrich, M.L., Wilson, G., Shuttlesworth, G., Huhtaniemi, I., and Reissmann, T., 2001, GnRH agonists and antagonists stimulate recovery of fertility in irradiated LBNF$_1$ rats. *J. Androl.*, 22: (in press).

Meistrich, M.L., Wilson, G., Ye, W.-S., Kurdoglu, B., Parchuri, N., and Terry, N.H.A., 1994, Hormonal protection from procarbazine-induced testicular damage is selective for survival and recovery of stem spermatogonia. *Cancer Res.*, 54: 1027-1034.

Meistrich, M.L., Wilson, G., Zhang, Y., Kurdoglu, B., and Terry, N.H.A., 1997, Protection from procarbazine-induced testicular damage by hormonal pretreatment does not involve arrest of spermatogonial proliferation. *Cancer Res.*, 57: 1091-1097.

Morris, I.D., and Shalet, S.M., 1990, Protection of gonadal function from cytotoxic chemotherapy and irradiation. *Ballieres Clin. Endocrinol. Metab.*, 4: 97-118.

Parchuri, N., Wilson, G., and Meistrich, M.L., 1993, Protection by gonadal steroid hormones against procarbazine-induced damage to spermatogenic function in LBNF$_1$ hybrid rats. *J. Androl.*, 14: 257-266.

Potashnik, G., and Porath, A., 1995, Dibromochloropropane (DBCP): A 17-year reassessment of testicular function and reproductive performance. *J. Occup. Environ. Med.*, 37: 1287-1292.

Pryzant, R.M., Meistrich, M.L., Wilson, E., Brown, B., and McLaughlin, P., 1993, Long-term reduction in sperm count after chemotherapy with and without radiation therapy for non-Hodgkin's lymphomas. *J. Clin. Oncol.*, 11: 239-247.

Rowley, M.J., Leach, D.R., Warner, G.A., and Heller, C.G., 1974, Effect of graded doses of ionizing radiation on the human testis. *Radiat. Res.*, 59: 665-678.

Schoenfeld, H.A., Hall, S.J., and Boekelheide, K., 2001, Continuously proliferative stem germ cells partially repopulate the aged, atrophic rat testis after gonadotropin-releasing hormone agonist therapy. *Biol. Reprod.*, 64: 1273-1282.

Senturia, Y.D., Peckham, C.S., and Peckham, M.J., 1985, Children fathered by men treated for testicular cancer. *Lancet*, 2: 766-769.

Shetty, G., Wilson, G., Huhtaniemi, I., Shuttlesworth, G.A., Reissmann, T., and Meistrich, M.L., 2000, Gonadotropin-releasing hormone analogs stimulate and testosterone inhibits the recovery of spermatogenesis in irradiated rats. *Endocrinology*, 141: 1735-1745.

Shuttlesworth, G.A., de Rooij, D.G., Huhtaniemi, I., Reissmann, T., Russell, L.D., Shetty, G., Wilson, G., and Meistrich, M.L., 2000, Enhancement of A spermatogonial proliferation and differentiation in irradiated rats by GnRH antagonist administration. *Endocrinology*, 141: 37-49.

Slutsky, M., Levin, J.L., and Levy, B.S., 1999, Azoospermia and oligospermia among a large cohort of DBCP applicators in 12 countries. *Int. J. Occup. Environ. Health*, 5: 116-122.

Van Thiel, D.H., Sherins, R.J., Myers, G.H., and De Vita, V.T., 1972, Evidence for a specific seminiferous tubular factor affecting follicle-stimulating hormone secretion in man. *J. Clin. Invest.*, 51: 1009-1019.

Velez de la Calle, J.F., and Jegou, B., 1990, Protection by steroid contraceptives against procarbazine-induced sterility and genotoxicity in male rats. *Cancer Res.*, 50: 1308-1315.

Whorton, D., Milby, T.H., Krauss, R.M., and Stubbs, H.A., 1979, Testicular function in DBCP exposed pesticide workers. *J. Occup. Med.*, 21: 161-166.

Wilson, G., Kangasniemi, M., and Meistrich, M.L., 1999, Hormone pretreatment enhances recovery of spermatogenesis in rats after neutron irradiation. *Radiat. Res.*, 152: 51-56.

Witt, K.L., and Bishop, J.B., 1996, Mutagenicity of anticancer drugs in mammalian germ cells. *Mutat. Res.*, 355: 209-234.

EPIGENETICS: ROLE OF GERM CELL IMPRINTING

Marisa S. Bartolomei

Howard Hughes Medical Institute
Department of Cell & Developmental Biology
University of Pennsylvania School of Medicine
Philadelphia, PA 19104

INTRODUCTION

Epigenetic modification of the genome includes those processes such as DNA methylation and chromatin remodeling that effect changes in gene expression. These processes are dynamic with dramatic changes occurring during the growth and development of eucaryotic organisms. In mammals, these modifications are also essential since perturbations can lead to altered growth or lethality. For example, DNA methylation is essential to proper mouse development since deletion of any one of 3 DNA methyltransferases (DNMT) results in extensive disruption in DNA methylation patterns and embryonic lethality (Li et al., 1992; Okano et al., 1999). Similarly, in humans, defects in the gene encoding DNMT3B are associated with a syndrome which leads to immunodeficiency, centromere instability and facial anomalies (ICF syndrome) (Xu et al., 1999). Finally, defects in chromatin proteins also cause lethality in mice (O'Carroll et al., 2001).

It has recently been demonstrated that changes in DNA methylation are associated with an increasing number of cancers (Jones and Laird, 1999). The CpG-rich promoters of a number of tumor suppressor genes are hypermethylated in tumors, suggesting that DNA methylation can serve as an alternative mechanism of tumor suppressor inactivation in cancer. Interestingly, mice predisposed towards the development of intestinal neoplasia exhibit a suppression in tumor formation when bred to mice harboring a *DNA methyltransferase 1* gene mutation (Laird et al., 1995), again supporting a role for DNA methylation in the development of cancer.

While significant experimental evidence exists to support the claim that DNA methylation and chromatin protein perturbations can lead to cancer, disease and developmental defects, the effect of environmental influences on epigenetic control of gene regulation remains less well understood. Nevertheless, the environment is certain to play a key role in such processes. Furthermore, it is possible that environmental perturbations could affect imprinted gene expression. It is the purpose of this chapter to define genomic imprinting, with particular emphasis on the epigenetic regulation of the *H19* and *Igf2* gene locus, and discuss how external influences could disrupt imprinting of this locus, thereby leading to developmental defects.

IMPRINTING

A small number of autosomal genes in the mammalian genome are inherited in a silent state from one of the two parents and in an active state from the other. These genes are said to be subject to parental or genomic imprinting (Bartolomei and Tilghman, 1997). With the existence of imprinted genes comes the requirement for both a maternal and a paternal pronuclear contribution to the developing embryo. Elegant nuclear transplantation experiments pioneered by Solter and Surani in the 1980's showed that diploid androgenetic embryos derived from two sperm pronuclei or diploid gynogenetic embryos derived from two female pronuclei fail to thrive (McGrath and Solter, 1983; Surani and Barton, 1983). The failure of these uniparental embryos is caused by the loss of function of genes expressed from a single parental genome. In addition to demonstrating the presence of imprinted genes, these experiments also showed that imprinted genes encode products that are essential for appropriate mammalian development.

Currently, approximately 50 genes are imprinted in mammals (for a complete listing of imprinted genes http://www.mgu.har.mrc.ac.uk/research/imprinted/imprin.html) and it is highly likely that an additional hundred genes will be identified in the coming years. These genes do not fall into any one specific class, but the presence of genes encoding growth factors and growth regulators is noteworthy. Furthermore, a number of genes that do not encode protein products (i.e. the active product is the RNA) are imprinted. One particularly striking property of imprinted genes is that many (if not all) are located in clusters that are distributed throughout the genome. It is certain that the clustering of these genes plays a role in their regulation.

Two of the first three imprinted genes that were identified are the mouse *H19* and *Insulin-like growth factor II* (*Igf2*) genes (Bartolomei et al., 1991; DeChiara et al., 1991; Zemel et al., 1992). The *H19* gene encodes an abundant nontranslated RNA that is expressed exclusively from the maternally-inherited chromosome and the *Igf2* gene encodes a fetal mitogen that is expressed from the paternally-inherited chromosome. These genes, which constitute the first identified cluster, are just 75 kb apart on the distal end of mouse chromosome 7 and their imprinting and linkage is conserved in other mammalian species (Rainier et al., 1993). It has been also shown that the genes share 3' enhancers; these enhancers reside in proximity to *H19* and were first identified as *H19*-specific enhancers (Yoo-Warren et al., 1988). Gene targeting experiments demonstrated that *H19* has access to the enhancers on the maternal allele, and *Igf2* utilizes the enhancers on the paternal allele (Leighton et al., 1995).

EPIGENETIC REGULATION OF THE *H19/IGF2* LOCUS

The mechanism by which imprinted genes are marked or designated with their parental identity remains poorly understood and is a matter of intense investigation. The parental alleles must be differentially modified such that transcriptional machinery can distinguish between the alleles. Perhaps the best time to achieve such marking is in germ cells or during the first few embryonic cleavages as these are the only times that the parental genomes are in distinct compartments and can be differentially modified. The most widely studied mechanism for achieving the parental-specific marks is DNA methylation. In mammals, the cytosine in CpG dinucleotides is methylated by one of a series of DNA methyltransferases (Bestor, 2000). The use of DNA methylation as an allele-specific mark is especially attractive since DNA hypermethylation can differentially modify parental alleles as well as stably repress the inactive alleles (Jones et al., 1998; Nan et al., 1998).

Early experiments in somatic cells demonstrated that the *H19* gene is hypermethylated on the inactive allele over a 7 kb region that includes the promoter and a 4 kb region upstream from the start of transcription (Bartolomei et al., 1993; Ferguson-Smith et al., 1993). A small part of the somatic hypermethylation is found on the inactive allele in sperm and throughout development, including during preimplantation development when the mammalian genome goes through a period of generalized demethylation. The region includes over 50 CpG dinucleotides and spans from –2 kb to –4 kb relative to the start of transcription (Tremblay et al., 1997; Tremblay et al., 1995). Thus, the differential methylation in this region, designated at the differentially methylated domain (DMD) or the imprinting control region (ICR), represents an excellent candidate for conferring the allele-specific mark. Interestingly, this same region is hypersensitive to nucleases on the active, maternal allele in somatic cells (Hark and Tilghman, 1998; Khosla et al., 1999). While this chromatin epigenetic modification could also represent an allele-specific mark, it remains to be determined whether the DMD/ICR is hypersensitive to nucleases in oocytes and preimplantation embryos.

To determine whether the DMD is important for imprinted expression of *H19*, we deleted it at the endogenous locus using gene targeting in embryonic stem (ES) cells and then generated mice from the homologously targeted ES cells by blastocyst injection. The first deletion that we tested, designated ΔDMD, removed 1.6 kb of the 2 kb DMD element and resulted in an extensive perturbation of imprinting at the *H19/Igf2* locus (Thorvaldsen et al., 1998). It should be noted that other deletions that removed all of the DMD resulted in no further disruption of imprinting at the locus. We first tested the consequence of transmission of the deleted allele through the father and determined that the normally silent paternal *H19* allele was activated. Furthermore, the linked and oppositely expressed *Igf2* gene experienced a decrease in the level of expression on the mutant allele. These experiments are consistent with a role for the DMD in the repression of the inactive *H19* allele. Surprisingly, when the DMD deletion was transmitted to the progeny by the mother, the expression of the normally active *H19* allele was diminished and the *Igf2* gene was activated. These results suggested that the DMD has an additional role that is required for exclusive expression of *H19* on the maternal allele. Thus, not only does the DMD act to differentiate the maternal and paternal alleles of *H19*, it is also required to repress paternal-specific expression of *H19* and to mediate exclusive maternal-specific *H19* expression.

Initially, it was unclear how the DMD exerted its effect on the maternal allele, but soon thereafter it was found that the DMD could act as an insulator or boundary in vitro. Insulators are DNA elements that can block promoter and enhancer interactions when placed between them and can protect a gene and its regulatory elements from position effects. A number of groups found that the DMD could block promoter and enhancer interactions by generating the appropriate plasmids and testing them in vitro by cell transfection (Bell and Felsenfeld, 2000; Hark et al., 2000; Kaffer et al., 2000; Kanduri et al., 2000; Szabo et al., 2000). Interestingly, the DMD acted similarly to the recently defined chicken globin insulator (Bell et al., 1999), and it was also shown that the DMD could bind the same insulator protein as the globin element, the ubiquitous transcription regulatory protein CTCF. Moreover, CTCF binding was sensitive to DNA methylation, suggesting that the insulator might only work on the hypomethylated maternal allele in vivo. The story came full circle when it was shown that the CTCF binding sites mapped to the previously reported, maternal-specific nuclease hypersensitive sites. These sites correspond to the only conserved sequences within the DMD of the human and mouse *H19* genes (Frevel et al., 1999; Stadnick et al., 1999). Taken together these results lead to the following model for how imprinted expression operates at the *H19/Igf2* locus. On the maternal allele, CTCF (and presumably other unidentified boundary accessory proteins) bind to sites on the DMD. The resulting formation of the boundary allows the promoter of *H19* exclusive access to its enhancers. In contrast, the DMD on the paternal allele is hypermethylated, thereby excluding CTCF. Since the *H19* promoter is also hypermethylated, it cannot compete for shared enhancers and *Igf2* has exclusive access to these enhancers. Two final comments should be made regarding this model. The first is that the model has not yet been tested and verified in vivo. For this, we await the appropriate gene targeting experiments. Second, it is likely that other cis-acting elements are required for the appropriate imprinting and expression at this locus (Constancia et al., 2000).

ANALYSIS OF *H19* METHYLATION PATTERNS IN GERM CELLS

To determine how imprinting is set up in the developing embryo, it is first necessary to determine when the imprint is imposed. Mouse primordial germ cells are specified between 6.5 and 7.5 days post coitum (dpc) in the posterior end of the primitive streak and migrate extensively between 8-11 dpc, eventually populating the embryonic gonads (Anderson et al., 2000). In female embryos, the germ cells enter meiosis by 13.5 dpc and arrest as diplotene oocytes in meiosis I. In contrast, germ cells in the developing testis enter mitotic arrest between 13.5-15.5 dpc. The pro-spermatogonia remain non-dividing until after birth, when they resume mitosis and differentiate into spermatogenic stem cells (McCarrey, 1993). We isolated germ cells from the embryonic gonad of 13.5, 14.5, 15.5 and 18.5 dpc embryos and characterized the methylation of the DMD (Davis et al., 2000). We found a complete erasure of methylation on both parental *H19* alleles in embryonic germ cells, followed by temporally distinct patterns of methylation acquisition on the maternal and paternal alleles. Our results indicate that the unmethylated maternal and paternal alleles of *H19* are not identical and that parental identity is retained in the absence of methylation.

To account for these results, we preposed the following model to explain differential acquisition of methylation in male germ cells (Davis et al., 2000). The maternal allele of *H19* is epigenetically marked by CTCF or other proteins in the female germline. This maternal allelic imprint is maintained in the early embryo and in embryonic germ cells. Although DNMT is able to methylate the paternal allele of *H19* as well as other genomic DNA in pro-spermatogonia, proteins that bind to the boundary element initially prevent access of the DNA methyltransferase to the maternal *H19* allele. Later in gestation, maternal allelic identity is lost, possibly by a reduction in affinity or degradation of the proteins that bind to the maternal allele, thereby allowing access of DNMT. Further experiments examining chromatin structure and CTCF in germ cells will be required to test this model.

EFFECTS OF CELL CULTURE ON IMPRINTING

A number of studies in mice have suggested that in vitro culture and manipulation of eggs and preimplantation embryos can lead to reduced viability and growth, as well as developmental abnormalities (Bowman and McLaren, 1970; Dean et al., 1998; Khosla et al., 2001; Reik et al., 1993). Where analyzed, these defects are linked to epigenetic changes in the embryo (Dean et al., 1998; Reik et al., 1993). For example, Dean and colleagues found that developmentally compromised fetuses derived exclusively from the culture of embryonic stem cells exhibited perturbations in the expression and methylation of four imprinted genes, including *H19* (Dean et al., 1998). Furthermore, in one case, the phenotypic abnormalities were transmitted to the progeny of the manipulated parental mice, suggesting that some epigenetic changes could be stably propagated through the germline (Roemer et al., 1997).

We studied the lability of the *H19* imprint in culture using various conditions (Doherty et al., 2000). After culture of 2-cell embryos to the blastocyst in Whitten's medium, the normally silent paternal *H19* allele was aberrantly expressed, whereas little paternal expression was observed following culture in an optimized medium, KSOM containing amino acids. Analysis of the methylation status of a CpG dinucleotide in the DMD revealed a loss in methylation in embryos cultured in Whitten's but not in embryos cultured in KSOM medium. Thus, *H19* expression and methylation were adversely affected by culture in Whitten's medium, while the response of *H19* to culture in KSOM plus amino acids approximated more closely the in vivo situation. Not all imprinted genes are similarly affected since the imprinting of the *Snprn* gene was unperturbed. These studies suggest that in vitro culture and other environmental insults can dramatically, but selectively, affect imprinted gene expression.

The observed effects of mouse culture may be directly related to the unusual phenomenon known as large calf syndrome. Culture of bovine and ovine embryos to the blastocyst stage prior to embryo transfer results in a greater fraction of offspring with a significant increase in birth weight, as well as in higher incidences of fetal and perinatal loss (Walker et al., 1996). In cattle, these abnormalities are not observed if the embryos are first transferred to the oviducts of sheep and then transferred at the bastocyst stage to the reproductive tracts of recipient heifers (Behboodi et al., 1995). Thus, these abnormalities are likely attributable to embryo culture. Since the culture conditions are probably suboptimal for these species, loss-of-imprinting of specific genes may occur during the culture period and could contribute to the observed differences in birth weight, as well as fetal and perinatal loss.

Recently, an intriguing study was reported by Fleming and colleagues (Kwong et al., 2000). In rats, maternal low protein diet fed exclusively during the preimplantation period of development induced programming of altered birthweight, postnatal growth rate and hypertension. Thus, long-term programming of postnatal growth and physiology can be irreversibly induced during preimplantation development by maternal protein under-nutrition. While these results support Barker's fetal origins of adult disease hypothesis, in which cardiovascular and related disorders that occur later in life can be attributed to maternal undernutrition in utero (Barker, 1994), it was suprising that long lasting effects could be induced in rats at the preimplantation stage.

CONCLUSIONS

The experiments described in this chapter suggest that the epigenetic programming that occurs in germ cells late in development and in the preimplantation embryo could be highly susceptible to environmental stresses. In fact, the environment stress of suboptimal culture does, in fact, cause disruptions in imprinted gene expression and DNA methylation patterns. Furthermore, such stresses during preimplantation development may not be immediately obvious; it may be years before the consequences are manifest. Thus, women intending to become pregnant and clinicians involved in artificial reproductive technology should treat embryos and fetuses carefully at all times of development.

REFERENCES

Anderson, R., Copeland, T.K., Scholer, H., Heasman, J., and Wylie, C., 2000, The onset of germ cell migration in the mouse embryo. *Mech Dev.* 91:61-68.

Barker, D. J. P., 1994, Mothers, Babies and Disease in Later Life (London, BMJ Publishing).

Bartolomei, M.S., and Tilghman, S.M., 1997, Genomic imprinting in mammals. *Annu Rev Genet.* 31:493-525.

Bartolomei, M.S., Webber, A.L., Brunkow, M.E., and Tilghman, S.M., 1993, Epigenetic mechanisms underlying the imprinting of the mouse *H19* gene. *Genes Dev.* 7:1663-1673.

Bartolomei, M.S., Zemel, S., and Tilghman, S.M., 1991, Parental imprinting of the mouse *H19* gene. *Nature.* 351:153-155.

Behboodi, E., Anderson, G. B., BonDurant, R. H., Cargill, S. L., Kreuscher, B. R., Medrano, J. F., and Murray, J. D., 1995, Birth of large calves that developed from in vitro-derived bovine embryos. *Theriogenology.* 44:227-232.

Bell, A.C., and Felsenfeld, G., 2000, Methylation of a CTCF-dependent boundary controls imprinted expression of the *Igf2* gene. *Nature.* 405:482-485.

Bell, A.C., West, A.G., and Felsenfeld, G., 1999, The protein CTCF is required for the enhancer blocking activity of vertebrate insulators. *Cell.* 98:387-396.

Bestor, T.H., 2000, The DNA methyltransferases of mammals. *Hum Mol Genet.* 9:2395-402.

Bowman, P., and McLaren, A., 1970, Viability and growth of mouse embryos after *in vitro* culture and fusion. *J Embryol Exp Morph* 23:693-704.

Constancia, M., Dean, W., Lopes, S., Moore, T., Kelsey, G., and Reik, W., 2000, Deletion of a silencer element in Igf2 results in loss of imprinting independent of H19. *Nature Genet.* 26:203-236.

Davis, T.L., Yang, G.J., McCarrey, J.R., and Bartolomei, M.S., 2000, The *H19* methylation imprint is erased and reestablished differentially on the parental alleles during male germ cell development. *Hum Mol Genet.* 9:2885-2894.

Dean, W., Bowden, L., Aitchison, A., Klose, J., Moore, T., Meneses, J.J., Reik, W., and Feil, R., 1998, Altered imprinted gene methylation and expression in completely ES cell-derived mouse fetuses: associaton with aberrant phenotypes. *Development.* 125:2273-2282.

DeChiara, T.M., Robertson, E.J., and Efstratiadis, A., 1991, Parental imprinting of the mouse insulin-like growth factor II gene. *Cell.* 64:849-859.

Doherty, A.S., Mann, M.R. W., Tremblay, K.D., Bartolomei, M.S., and Schultz, R.M., 2000, Differential effects of culture on imprinted *H19* expression in the preimplantation mouse embryo. *Biol Reprod.* 62:1526-1535.

Ferguson-Smith, A.S., Sasaki, H., Cattanach, B.M., and Surani, M.A., 1993, Parental-origin-specific epigenetic modification of the mouse *H19* gene. *Nature.* 362:751-754.

Frevel, M.A.E., Hornberg, J.J., and Reeve, A.E., 1999, A potential imprint control element: identification of a conserved 42 bp sequence upstream of *H19*. *Trends Genet.* 15:216-218.

Hark, A.T., Schoenherr, C.J., Katz, D.J., Ingram, R.S., Levorse, J.M., and Tilghman, S.M., 2000, CTCF mediates methylation-sensitive enhancer-blocking activity at the *H19/Igf2* locus. *Nature.* 405:486-489.

Hark, A.T., and Tilghman, S.M., 1998, Chromatin conformation of the *H19* epigenetic mark. *Hum Mol Genet.* 7:1979-1985.

Jones, P.A., and Laird, P.W., 1999, Cancer epigenetics comes of age. *Nature Genet.* 21:163-167.

Jones, P.L., Veenstra, G.J.C., Wade, P.A., Vermaak, D., Kass, S.U., Landsberger, N., Strouboulis, J., and Wolffe, A.P., 1998, Methylated DNA and MeCP2 recruit histone deacetylase to repress transcription. *Nature Genet.* 19:187-191.

Kaffer, C.R., Srivastava, M., Park, K.-Y., Ives, E., Hsieh, S., Batle, J., Grinberg, A., Huang, S.-P., and Pfeifer, K., 2000, A transcriptional insulator at the imprinted *H19/Igf2* locus. *Genes Dev.* 14:1908-1919.

Kanduri, C., Holmgren, C., Pilartz, M., Franklin, G., Kanduri, M., Liu, L., Ginjala, V., Ulleras, E., Mattsson, R., and Ohlsson, R., 2000, The 5' flank of mouse *H19* in an unusual chromatin conformation unidirectionally blocks enhancer-promoter communication. *Curr Biol.* 10:449-457.

Khosla, S., Aitchison, A., Gregory, R., Allen, N.D., and Feil, R., 1999, Parental allele-specific chromatin configuration in a boundary/imprinting-control element upstream of the mouse *H19* gene. *Mol Cell Biol.* 19:2556-2566.

Khosla, S., Dean, W., Brown, D., Reik, W., and Feil, R., 2001, Culture of preimplantation mouse embryos affects fetal development and the expression of imprinted genes. *Biol Reprod.* 64:918-926.

Kwong, W.Y., Wild, A.E., Roberts, P., Willis, A.C., and Fleming, T.P., 2000, Maternal undernutrition during the preimplantation period of rat development causes blastocyst abnormalities and programming of postnatal hypertension. *Development.* 127:4195-4202.

Laird, P.W., Jackson-Grusby, L., Fazeli, A., Dickinson, S.L., Jung, W.E., Li, E., Weinberg, R.A., and Jaenisch, R., 1995, Suppression of intestinal neoplasia by DNA hypomethylation. *Cell.* 81:197-205.

Leighton, P.A., Saam, J.R., Ingram, R.S., Stewart, C.L., and Tilghman, S.M., 1995, An enhancer deletion affects both *H19* and *Igf2* expression. *Genes Dev.* 9:2079-2089.

Li, E., Bestor, T.H., and Jaenisch, R., 1992, Targeted mutation of the DNA methyltransferase gene results in embryonic lethality. *Cell.* 69:915-926.

McCarrey, J.R., 1993, Development of the germ cell. In Cell and Molecular Biology of the Testis, C. Desjardins, and L. L. Ewing, eds. (Oxford, Oxford Univ. Press), pp. 58-89.

McGrath, J., and Solter, D., 1983, Nuclear transplantation in mouse embryos. *J Exp Zool.* 228:355-362.

Nan, X., Ng, H.-H., Johnson, C.A., Laherty, C.D., Turner, B.M., Eisenman, R.N., and Bird, A., 1998, Transcriptional repression by the methyl-CpG-binding protein MeCP2 involves a histone deacetylase complex. *Nature.* 393:386-389.

O'Carroll, D., Erhardt, S., Pagani, M., Barton, S.C., Surani, M.A., and Jenuwein, T., 2001, The polycomb-group gene ezh2 is required for early mouse development. *Mol Cell Biol.* 21:4330-4336.

Okano, M., Bell, D.W., Haber, D.A., and Li, E., 1999, DNA methyltransferases Dnmt3a and Dnmt3b are essential for de novo methylation and mammalian development. *Cell.* 99:247-257.

Rainier, S., Johnson, L.A., Dobry, C.J., Ping, A.J., Grundy, P.E., and Feinberg, A.P., 1993, Relaxation of imprinted genes in human cancer. *Nature.* 362:747-749.

Reik, W., Rîmer, I., Barton, S.C., Surani, M.A., Howlett, S.K., and Klose, J., 1993, Adult phenotype in the mouse can be affected by epigenetic events in the early embryo. *Development.* 119:933-942.

Roemer, I., Reik, W., Dean, W., and Klose, J., 1997, Epigenetic inheritance in the mouse. *Curr Biol.* 7:277-280.

Stadnick, M.P., Pieracci, F.M., Cranston, M.J., Taksel, E., Thorvaldsen, J.L., and Bartolomei, M.S., 1999, Role of a 461 bp G-rich repetitive element in *H19* transgene imprinting. *Dev Genes Evo.* 209:239-248.

Surani, M.A.H., and Barton, S.C., 1983, Development of gynogenetic eggs in the mouse: implications for parthenogenetic embryos. *Science.* 222:1034-1036.

Szabo, P.E., Tang, S.-H., Rentsendorj, A., Pfeifer, G.P., and Mann, J.R., 2000, Maternal-specific footprints at putative CTCF sites in the *H19* imprinting control region give evidence for insulator function. *Curr Biol.* 10:607-610.

Thorvaldsen, J.L., Duran, K.L., and Bartolomei, M.S., 1998, Deletion of the *H19* differentially methylated domain results in loss of imprinted expression of *H19* and *Igf2*. *Genes Dev.* 12:3693-3702.

Tremblay, K.D., Duran, K.L., and Bartolomei, M.S., 1997, A 5' 2-kilobase-pair region of the imprinted mouse *H19* gene exhibits exclusive paternal methylation throughout development. *Mol Cell Biol.* 17:4322-4329.

Tremblay, K.D., Saam, J.R., Ingram, R.S., Tilghman, S.M., and Bartolomei, M.S., 1995, A paternal-specific methylation imprint marks the alleles of the mouse *H19* gene. *Nature Genet.* 9:407-413.

Walker, S.K., Hartwich, K.M., and Seamark, R.F., 1996, The production of unusually large offspring following embryo manipulation: concepts and challenges. *Theriogenology.* 45:111-120.

Xu, G.L., Bestor, T.H., Bourc'his, D., Hsieh, C.L., Tommerup, N., Bugge, M., Hulten, M., Qu, X., Russo, J. J., and Viegas-Pequignot, E., 1999, Chromosome instability and immunodeficiency syndrome caused by mutations in a DNA methyltransferase gene. *Nature.* 402:187-191.

Yoo-Warren, H., Pachnis, V., Ingram, R.S., and Tilghman, S.M., 1988, Two regulatory domains flank the mouse H19 gene. *Mol Cell Biol.* 8:4707-4715.

Zemel, S., Bartolomei, M.S., and Tilghman, S.M., 1992, Physical linkage of two mammalian imprinted genes. *Nature Genet.* 2:61-65.

EFFICIENCY AND SAFETY OF ANIMAL CLONING

Ryuzo Yanagimachi

The Institute for Biogenesis Research
University of Hawaii School of Medicine
Honolulu, Hawaii 96822

INTRODUCTION

Developmental totipotency of the cell nuclei in the early embryonic stage was first clearly demonstrated by Spemann (1938). He ligated a recently fertilized newt egg so that cell division occurred in only one of two halves. When the nucleated portion cleaved to 8 to 16 cells, one of these nuclei was allowed to enter the uncleaved portion of the egg before it was completely separated from the rest. This resulted in the development of two normal larvae, indicating that the nuclei of an 8-16 cell embryo retain full developmental potential. Theoretically, 8-16 genetically identical offspring can be obtained from a single 8-16 cell embryo. Since most frontier experiments of animal cloning were carried out with the frog (Gurdon, 19622; Di Berardinom 1997), it was somewhat unexpected that the first animal cloned with an adult somatic cell was a mammal (sheep), not a frog. Almost all investigators thought that nuclei of adult somatic cells of the frog (and all other higher animals) were differentiated irreversibly and had lost their developmental totipotency. Wilmut et al. (1997) surprised the world by reporting the birth of a sheep (Dolly) that was cloned with an adult mammary gland cell. We (Wakayama et al., 1998) and Kato et al. (1998) soon confirmed that cloning with adult somatic cells was indeed possible, at least in the mouse and cattle. Subsequently, goats, pigs rabbits and a cat were cloned from adult somatic cells (Polejaeva et al., 2000; Behboodi et al., 2001; Chesne et al., 2002; Shin et al., 2002). The cells used for cloning include: cumulus cells, oviductal cells and skin cells (Table 1).

METHODS AND THE EFFICIENCY OF MAMMALIAN CLONING

The principle of animal cloning is simple and straightforward: remove the nucleus from an oocyte and replace it with the nucleus of a somatic cell from the embryo, fetus, juvenile or adult animal. Since somatic cells (nuclei) lack the ability to "activate" oocytes, reconstructed oocytes must be activated artificially to allow them to develop. The introduction of donor nuclei into enucleated oocytes is most commonly achieved by

Advances in Male Mediated Developmental Toxicity, edited by Bernard Robaire and Barbara F. Hales.
Kluwer Academic/Plenum Publishers, 2003.

electrofusion of an oocyte with a donor cell, but it can be done by fusing the two cells with the fusogenic Sendai virus or by mechanically injecting a donor cell nucleus into an enucleated oocyte. Electric pulse (single or multiple) has been most commonly used for this purpose, but chemical stimuli or a combination of electric and chemical stimuli can be used as well (Wakayama et al., 1998; Loi et al., 1998; Cibelli et al., 1998; Kishikawa et al., 1999; Wells et al., 1999; Wakayama and Yanagimachi, 2001).

Table 1. Comparison of cloning success rates[1] in various animals

Type of Donor Cell	Species	Total number of reconstructed (nuclear-transferred) oocytes	Number (%) of live offspring	Notes	Ref[2]
FETAL					
Fibroblast	Mouse	3057	5 (0.2)		1
	Bovine	276	4 (1.4)	1 died	2
		1896	6 (0.3)		3
	Goat	285	3 (1.1)		4
	Pig	210	1 (0.5)		5
	Sheep	417	14 (3.4)	11 died within 6 months	6
JUVENILE					
Skin cell	Bovine	175	1 (0.6)	Died 51 days later	7
Sertoli cell	Mouse	1190	7 (0.6)		8
ADULT					
Mammary gland cell	Sheep	227	1 (0.4)		9
Cumulus cell	Mouse	2468	31 (1.3)		10
Fibroblast	Bovine	440	6 (1.4)	2 died	11
		664	8 (1.2)		12
Granulosa cell	Bovine	557	10 (1.8)		13
	Pig	977	5 (0.5)		14

[1] As many investigators did not describe the original numbers of the oocytes used for nuclear transfer, success rates of cloning shown here are based on the number of reconstructed oocytes.
[2] Ref (1) Ono et al., 2001; (2) Cibelli et al., 1998; (3) Lanza et al., 2000; (4) Baguisi et al., 1999; (5) Onishi et al., 2000; (6) McCreath et al., 2000; (7) Renard et al., 1999; (8) Ogura et al., 2000; (9) Wimmut et al., 1997; (10) Wakayama et al., 1998; (11) Kubota et al., 2000; (12) Lacham-Kaplan et al., 2000; (13) Wells et al., 1999; (14) Polejaeva et al., 2000.

Some investigators calculated the success rate of cloning with the proportion of midterm fetuses that developed from the blastocysts transferred into surrogate mothers. Others calculate it from the proportion of live offspring developed from all nuclear transferred oocytes. As some investigators never reported the numbers of nuclear transferred oocytes, it was impossible for this reviewer to calculate the overall success rates in their experiments. To ease comparisons, I calculated the success rate based on the proportion of live offspring developed from all manipulated (nuclear transferred) oocytes. Table 1 shows success rates of cloning when somatic cells of fetuses, immature, and mature animals were used. Note that cloning efficiency is no more than 3%, regardless of the developmental age of the cell (from embryonic to senile adult) and the types of the cell used (see also Solter, 2000). Inner cell mass (ICM) cells, embryonic stem (ES) cells, as well as embryonic germ (EG) cells are considered developmentally "totipotent", yet cloning success rates with these cells are again less than 3% (Wells et al., 1999; Rideout et al., 2000; Brink et al., 2000). What does this 3% mean? Are less than 3% of cells of any embryonic or adult tissues developmentally totipotent? If so, will it be possible to increase the cloning success rate substantially (10 times or more) by pre-selecting cells? Are there some tissues in which a relatively higher proportion of cells retain developmental totipotency? These questions remain to be answered.

It is important to emphasize that the majority of cloned embryos die before and after implantation (Fig. 1). We noticed that cloned mouse fetuses at term had unusually large placentas (Wakayama and Yanagimachi, 1999; Ogura et al., 2000; Ono et al., 2001), the meaning of which is unclear at the present time. Abnormal placentae were also noted in cattle (Cibelli et al., 1998; De Lille et al., 2001). Large fetus syndrome, which was first reported in cloned cattle (Cibelli et al., 1998), was also seen in some cloned mice (Eggan et al., 2001). Large offspring syndrome could be due, in part, to in vitro handling of the oocytes and preimplantation embryos (Young et al., 1998) rather than to the cloning procedure itself. Some cloned mice died of respiratory problems at birth (Wakayama et al., 1999; Eggan et al., 2001). It is also known that many cloned cattle offspring died of respiratory as well as immunological and multiple systemic dysfunctions (Cibelli et al. 1998; Renard et al., 1999; Wells et al. 1999; Pace et., 2001; Taneja et al., 2001). Thus far, cloned offspring that survived birth and reached adulthood were the exception rather than the rule. We found that >90% of mouse embryos cloned with cumulus cells had normal chromosome constitutions (Yanagimachi et al., unpublished data). Thus, it is very likely that most cloned embryos/fetuses died of faulty epigenetic reprogramming rather than genomic problems.

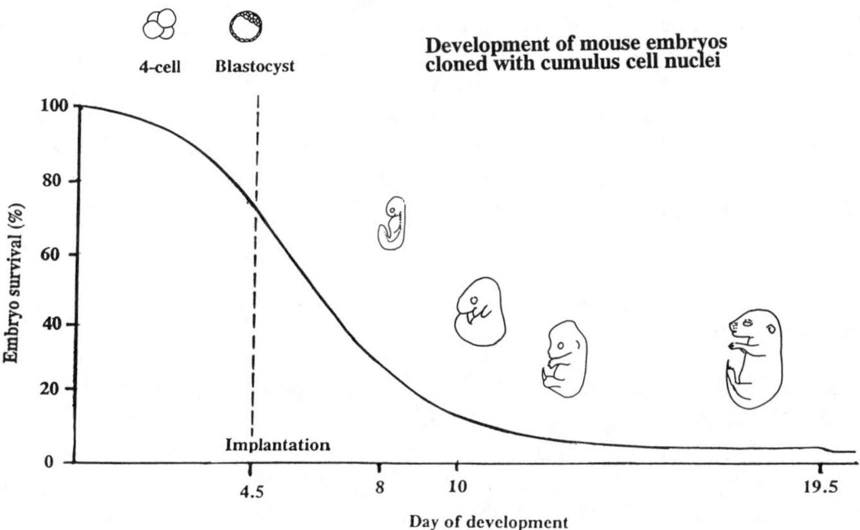

Figure 1. Development and survival of mouse embryos cloned with cumulus cells. Note that many embryos die soon after implantation.

FACTORS CONTRIBUTING TO CLONING SUCCESS AND FAILURE

Enucleated oocytes receiving adult somatic cell nuclei are able to develop into live offspring, suggesting that molecular functions of the adult somatic cell nuclei are somehow "reprogrammed" back to the zygotic state. Although the nature and mechanism of nuclear reprogramming are unknown, we obtained convincing evidence that at least some imprinted genes (e.g., genes responsible for X chromosome inactivation) are reprogrammable by nuclear transfer (Eggan et al., 2000).

The vast majority of cloned embryos die during embryonic development despite their normal chromosome constitution. This seems to suggest that epigenetic reprogramming of the reconstructed oocytes is incomplete in most cases. We compared the genome-wide methylation status of CpG islands in placenta, skin and kidney cells of cloned and normal (control) mice and found that in all cloned offspring, several CpG islands were aberrantly methylated or unmethylated. Interestingly, the extent of aberrant DNA methylation varies among phenotypically normal offspring (Ohgane et al., 2001). Apparently, slight aberrations in DNA methylation in somatic cells are not life-threatening. Extensive aberrations, on the other hand, must be fatal to cloned embryos. Although cloned offspring that survived birth may be genomically identical with the original animal, they may not be phenotypically identical with the original animal or their siblings because of random errors in epigenetic gene expression.

We learned, somewhat unexpectedly, that the so-called "totipotent" embryonic stems (ES) cells were not as easy to use for cloning as we had originally anticipated. In our experiments, an average of less than 3% of nucleus-transferred mouse oocytes developed into live offspring (Wakayama et al. 1999; Rideout et al., 2000; Humphrey et al., 2001) (Fig.1). We found that the epigenetic state of the ES cells cultured in vitro was unstable and varied even between clonally related cells. Imprinted gene expression of ES cell populations seems to be inherently unstable. It is rather amazing that some cloned embryos survived the birth despite a wide range of imprinting errors (Humphreys et al., 2001). We found that the fetuses cloned with ES cells of hybrid mice survived birth, whereas those cloned with inbred mouse ES cells seldom did (Rideout et al., 2000; Eggan et al., 2001). Newborns had the best chance of survival when both donor cells and recipient oocytes were from hybrid mice (Wakayama and Yanagimachi, 2001). These facts seem to suggest that epigenetic errors in cloned offspring are overcome, to some extent, by genomic heterogeneity. We found that the offspring of cloned female mice that had mated with cloned male mice were all normal and fertile (Tamashiro et al., 2002). Apparently, all epigenetic problems in cloned parents are resolved when cell nuclei go through the germ line.

SAFETY OF CLONING

Cloning has great potential in the area of the production of medically and pharmaceutically valuable farm animals (e.g., Baguisi et al., 1999; Polejaeva et al., 2000). Production of cloned animals with transgenic fetal fibroblasts, for example, has already been successful (Baguisi et al., 1999). As cloned animals are likely to have "dormant" epigenetic problems, it would be advisable to allow them to reproduce themselves by natural mating before they advance in age. Passing genomes through the germ plasm would erase all epigenetic errors. Cloning pet animals is appealing to some people, even though cloned animals may not be 100% identical phenotypically with the original animals. Cloning of highly inbred animals could be difficult as has already been shown in inbred mice. Cloning endangered species is a noble idea, but may not be as easy as we think. The production of a herd of cloned animals with very limited or no genomic diversity could quickly lead to the second extinction crisis.

It seems to be possible to allow embryonic and adult stem cells to differentiate, in vitro or in vivo, into various cell types (Robertson, 1987; Thompson et al., 1998; McDonald et al., 1999; Fuchs and Segre, 2000). Someday, it will become possible to "reprogram" nuclei of fully differentiated cells without using oocyte cytoplasm and allow them to re-differentiate into entirely different types of cell. However, we must be fully aware that incomplete epigenetic reprogramming of the donor cell nucleus may lead to the production of cells and tissues with dormant pathological problems. It is important to eliminate epigenetic problems entirely, or as much as possible, before we apply cloning technology to reproduction and therapeutic purposes on a large scale.

REFERENCES

Baguisi, A., Behboodi, E., Melican, D.T., Pollock, J.S., Destrempes, M.M., Cammuso, C., Williams, J.L., Nims, S.D., Porter, C.A., Midura, P., Palacios, M.J., Ayres, S.L., Denniston, R.S., Hayes, M.L., Ziomek, C.A., Meade, H.M., Godke, R.A., Gavin, W.G., Overstrom, E.W., Echelard, Y., 1999, Production of goats by somatic cell nuclear transfer. *Nature Biotech.* 17:456-461.

Behboodi, E., Melican, D.T., Liem, H., Chen, J.H. et al., 2001, Production of transgenic goats by adult somatic cell nuclear transfer. *Theriogenology.* 55:254(abst.)

Brink, M.F., Bishop, M.D., Pieper, F.R., 2000, Developing efficiency strategies for the generation of transgenic cattle which produce biopharmaceuticals in milk. *Theriogenology.* 53:139-148.

Chesne, P., Adenot, P.G., Viglietta, C., Baratte, M., Boulanger, L., Renaud, J.P. 2002, Cloned rabbits produced by nuclear transfer from adult somatic cells. Nat. Biotechnol. 20: 366-369.

Cibelli, J.B., Stice, S.L., Goleuke, P.J., Kane, J.K., Jerry, J., Balackwell, C., de Leon, A., Robl, J.M., 1998, Cloned transgenic calves produced from nonquiescent fetal fibroblasts. *Science.* 280:1256-1258.

De Lille, A.J., Anthony, R.V., Seidel, G.E., 2001, Characteristics of placental and fetal tissues from Day 75 nuclear cloned bovine pregnancies. *Theriogenology.* 55:263 (abst.).

Di Bernardino, M.A., 1997, Genomic Potential of Differentiated Cells. Columbia University Press, New York.

Eggan, K., Akutsu, H., Lorings, J., Jackson-Grusby, L., Klemm, M., Rideout, W.M., Yanagimachi, R., Jaenisch, R., 2001, Hybrid-Vigor, fetal overgrowth and viability of mice derived by nuclear cloning and tetraploid embryo complementaries. *Proc Natl Acad Sci USA.* in press.

Eggan, K., Akutsu, H., Hochelinger, K., Rideout, W., Yanagimachi, R., Jaenisch, R., 2000, X-chromosome inactivation in cloned mouse embryos. *Science.* 290:1578-1581.

Fuchs, E., Segre, J.A., 2000, Stem cells: A new leave on life. *Science.* 100:143-155.

Gurdon, J.B., 1962, Adult frogs derived from the nuclei of single somatic cells. *Dev Biol.* 4:256-273.

Humphreys, D., Eggan, K., Akutsu, H., Hochedlinger, K., Rideout, W., Biniszkiewicz, D., Yanagimachi, R., Jaenisch, R., 2001, Epigenetic instability in ES cells and cloned mice. *Science.* in press.

Kato, Y., Tani, T., Sotomaru, Kurokawa, K., et al., 1998, Eight calves cloned from somatic cells of a single adult. *Science.* 282:2095-2098.

Kishikawa, H., Wakayama, T., Yanagimachi, R., 1999, Comparison of oocyte-activating agents for mouse cloning. *Clone.* 1:153-159.

Lacham-Kaplan, 0., Diamente, M., Pushett, D., Lewis, I., Trounson, A., 2000, Developmental competence of nuclear transfer cow oocytes after direct injection of fetal fibroblast nuclei. *Cloning.* 2:55-62.

Lanza, R. P., Cibelli, J.B., Blackwell, C., Cristofalo, V.J. et al., 2000, Extension of celllife-span and telomere length in animals cloned from senescent somatic cells. *Science.* 288:665-669.

Loi, P., Ledda, S., Fulka, J., Cappari, P., Moore, R.M., 1998, Development of parthenogenetic and cloned ovine embryos: Effect of activation protocols. *Biol Reprod.* 58:1177-1187.

McCreath, K.J., Howcroft, J., Campbell, H.S., Colman, S., Schneike, A.E., Kind, A.J., 2000, Production of gene-targetted sheep by nuclear transfer from cultured somatic cells. *Nature.* 405:1066-1069.

McDonald, J.W., Liu, X., Qu, Y., Liu, S., Mickey, S.K., Turetsky, D., Gottlieb, D.I., Choi, D.W., 1999, Transplanted embryonic stem cells survive, differentiate and promote recovery in injured rat spinal cord. *Nature Genet.* 5:1410-1412.

Ogura, A., Inoue, K., Ogonuki, N., Suzuki, 0., Lee, J., Ishino, J., Matsuda, J., 2000, Production of male cloned mice from fresh, cultured, and cryopreserved immature Sertoli cells. *Biol Reprod.* 62:1579-1584.

Ohgane, J., Aikawa, J., Ogura, A., Hat tori, N., Ogawa, T., Shiota, K., 1998, Analysis of CpG islands of trophoblast giant cells by restriction landmark genome scanning. *Dev Genet.* 22:132-140.

Ohgane, J., Wakayama, T., Koga, Y., Sendai, S., Hat tori, N., Tanaka, S., Yanagimachi, R., Shiota, K., 2001, DNA methylation variations in cloned mice. *Genesis.* 30:45-50.

Onishi, A., Iwamoto, M., Akita, T., Mikawa, S., Takeda, K., Awata, T., Hanada, H., Perry, A.C., 2000, Pig cloning by microinjection of fetal fibroblast nuclei. *Science.* 289:1188-1190.

Ono, Y., Shimozawa, N., Ito, M., Kano, T., 2001, Cloned mice from fetal fibroblast cells arrested at metaphase by a serial transfer. *Biol Reprod.* 64:44-50.

Pace, M.M., Mell, G., Forsberg, E.J., Bathauser, J., et al., 2001, Cloning using somatic cells: Analysis of 75 calves. *Theriogenology.* 55:281 (abst.).

Polejaeva, I.A., Chen, S., Vaught, T.D., Page, R.L., et al., 2000, Cloned pigs produced by nuclear transfer from adult somatic cells. *Nature.* 407:505-508.

Renard, J., Chastant, S., Chesne, P., Richard, C., Marchal, J., 1999, Lymphoid hypoplasia and somatic cloning. *Lancet.* 359:1489-1491.

Rideout, W.M., Wakayama, T., Wutz, A., Eggan, *K.,* Dausman, J., Yanagimachi, R., Jaenisch, R., 2000, Generation of mice from wild-type and targeted ES cells by nuclear cloning. *Nature Genet.* 24:109-110.

Robertson, E.J. (ed.), 1987, Teratocarcinomas and Embryonic Stem Cells: A practical Approach. IRL Press, Oxford.

Shin, T., Kraemer, D., Pryor, J., Liu, L., Ruglia, J. Howe, L., Buck, S., Murphy, K., Lyons, L., Westhusin, M. 2002, A cat cloned by nuclear transplantation. Nature. 415: 859

Solter, D., 2000, Mammalian cloning; Advances and limitation. *Nature Rev Genet.* 1:199-207.

Spemann, H., 1938, Embryo Development and Induction. Yale Univ. Press, New Haven, Conn.

Tamashiro, K., Wakayam, T., Akutsu, H., Yamazaki, Y., Lachey, J.L., Wortman, M.D., Seeley, R.J., D'Alessio, D.A., Woods, S.W., Yanagimachi, R., Sakai, R. 2002, Cloned mice have an obese phenotype not transmitted to their offspring. Nature Med. 8: 262-267.

Teneja, M., French, R., Levine, H., Tauro-Miller, D., Yang, X., 2001, Clinical and pathological status of cloned calves born pre-term. *Theriogenology.* 55:293 (abst.).

Thompson, J.A., Itskovitz-Eldor, J., Shapiro, S.S., Waknitz, M.A., Swiergiel, J .J., Marshall, V.S., Jones, J .M., 1998, Embryonic stem cell lines derived from human blastocysts. *Science.* 282:1145-1147.

Wakayama, T., Perry, A.C., Zuccotti, M., Johnson, K.R., Yanagimachi, R., 1998, Full-term development of mice from enucleated oocytes injected with cumulus cell nuclei. *Nature.* 394:369-374.

Wakayama, T., Rodriguez, I., Perry, A.C., Yanagimachi, R., Mombaerts, P., 1999, Mice cloned from embryonic stem cells. *Proc Natl Acad Sci USA.* 96:14984-14989.

Wakayama, T., Yanagimachi, R., 1999, Cloning of male mice from adult tail-tip cells. *Nature Genet.* 22:127-128.

Wakayama, T., Yanagimachi, R., 2001, Mouse cloning with nucleus of donor cells of different age and type. *Mol Reprod Dev.* 58:376-383.

Wlls, D.N., Misica, P., Tervit,H.R., 1999, Production of cloned calves following nuclear transfer with culture adult mural granulosa cells. *Biol. Reprod.* 60: 996-1005.

Wilmut, I., Schnieke, A.E., Whir, J., Kind, A.J., Campbell, K.H.S., 1997, Viable offspring derived from fetal and adult mammalian cells. *Nature.* 385:810-813.

Young, L.E., Sinclair, K.D., Wilmut, I., 1998, Large offspring syndrome in cattle and sheep. *Rev Reprod.* 3:155-163.

INTEGRATING NEW TESTS OF SPERM GENETIC INTEGRITY INTO SEMEN ANALYSIS: BREAKOUT GROUP DISCUSSION

Sally D. Perreault[1], R. John Aitken[2], H.W.Gordon Baker[3],
Donald P. Evenson[4], Gabor Huszar[5], D. Stewart Irvine[6],
Ian D. Morris[7], Rebecca A. Morris[1], Wendie A. Robbins[8],
Denny Sakkas[5], Marcello Spano[9], and Andrew J. Wyrobek[10]

[1] U.S. EPA, ORD, NHEERL, Reproductive Toxicology Division, Research Triangle Park, NC 27711
[2] School of Biological & Chemical Sciences, University of Newcastle, Newcastle, NSW, Australia, 2308
[3] Department of Obstetrics and Gynaecology, Royal Women's Hospital, University of Melbourne, Australia
[4] Olson Biochemistry Laboratories, ASC 136, South Dakota State University, Brookings, SD 57007
[5] Department of Obstetrics and Gynaecology, Yale University School of Medicine, New Haven, CT 06520
[6] Centre for Reproductive Biology, MRC Human Reproductive Sciences Unit, 37 Chalmers Street, Edinburgh, Scotland, UK, EH3 9ET
[7] Biological Sciences, Stopford Building, University of Manchester,Oxford Road, Manchester, M13 9PT. UK.
[8] UCLA Center for Occupational and Environmental Health, 10822 LeConte Ave., Los Angeles, CA 90095
[9] Section of Toxicology & Biomedical Sciences, ENEA CR Casaccia, Via Anguillarese 301, Rome, Italy 00060
[10] Biology and Biotechnology Research, Lawrence Livermore National Laboratory, 7000 East Avenue, L-448, Livermore, CA 94550

INTRODUCTION

The First International Conference on Male-Mediated Developmental Toxicity, held in September 1992, reported that the spermatozoon can bring genetic damage into the oocyte at fertilization and thereby contribute to subsequent abnormal pregnancy outcomes (Olshan and Mattison, 1994). At that time, laboratory tests for genetic defects in sperm were at an early stage of development and were relatively untested in the clinic and the field. A breakout group at that meeting discussed the need for improved sperm biomarkers of adverse reproductive effects and concluded that sensitive, reliable, and practical methods

Advances in Male Mediated Developmental Toxicity, edited by Bernard Robaire and Barbara F. Hales.
Kluwer Academic/Plenum Publishers, 2003.

for detecting DNA damage in animal and human spermatozoa were needed (Wyrobek et al., 1994).

Since the first conference, our understanding of the structure, function and vulnerability of the sperm nucleus has increased significantly, and tests to measure DNA breakage, abnormal chromatin packaging, and sperm chromosomal aberrations have been developed and improved (reviewed by Hassold, 1998; Evenson, 1999; Sakkas et al., 1999; Shi and Martin, 2000; Wyrobek et al., 2000). Therefore, a major objective of the Second International Conference on Male-Mediated Developmental Toxicity was to elaborate on the utility of these new and improved markers of genetic damage in sperm. Accordingly, the conference organizers charged one of the breakout groups with debating the relative merits of these new tests and making recommendations regarding their integration into routine semen analysis for the purpose of diagnosing infertility and/or evaluating risks of exposures to pharmaceuticals or environmental contaminants.

Specific tests were targeted for consideration based on their prevalence in the literature and the presence, at the breakout session, of technical experts on their implementation. These included: assays for sperm aneuploidy and chromosome aberrations based on the use of chromosome-specific molecular probes; tests for chromatin/DNA integrity based on the detection of denatured or broken DNA; and, briefly, tests that detect sperm maturity insofar as they may correlate with DNA damage. The intent of the discussions, and this summary, is not to review each method in great detail. That has been done in many of the chapters in this volume. Rather this chapter summarizes our deliberations as to the practicality of including them in a semen analysis and interpreting the results from both clinical and epidemiological perspectives.

FRAMEWORK FOR DISCUSSIONS

To help focus the workgroup efforts, each test was discussed with reference to a list of practical questions/issues (Table 1). Two general concepts soon emerged. First, since these tests are still evolving and some have been used more extensively than others, it was not possible to answer these questions in a thorough or definitive manner. Therefore, we did not attempt to reach consensus regarding details of assay implementation. Rather, we used this opportunity to point out where further research is needed. Second, the usefulness of a given test is relative, and depends upon the intended application. The clinician, for example, wants to know whether a certain test will predict or explain infertility or estimate genetic risks for a specific couple. Thus, in order to make a diagnosis, the clinician needs a validated test that has a high degree of sensitivity and specificity. On the other hand, the epidemiologist wants to know if one population of men differs from another with respect to semen endpoints. Here, estimation of risk is to the group, and by extrapolation, to populations of men. In this case, a less specific test may actually be better able to detect group differences. Similarly, a more costly test may be justified for clinical applications, e.g., to assure a couple that it is safe for them to attempt conception. However, low cost may be a requirement for an epidemiology study wherein many samples must be assayed.

With these general concepts in mind, we proceeded to discuss each test, attempting to address the questions listed in Table 1 as thoroughly as possible. We limited our discussions to tests that can be performed directly on a semen sample, once routine measures of semen volume, sperm concentration, and sperm motility and morphology have been obtained.

Table 1. Issues Affecting the Relative Utility of Tests of Sperm Genetic Damage/ Integrity.

---

1) What is the purpose of the test?
    - What is being measured?
    - What is the specificity of the test?

2) What is the underlying principle of the test?
    - Is the test a direct or indirect indicator of chromosome/DNA damage?

3) What are the advantages or disadvantages of the test?
    - Objectivity, accuracy, robustness

4) What is the relative "cost" of incorporating the test into an epidemiological study?
    - Sample preparation time?
    - Storage time allowed prior to processing?
    - Experiment running time?
    - Equipment necessary?
    - Expertise required?

5) For what types of applications is the technique most useful?
    - Clinical diagnosis
    - Epidemiology investigations
    - Animal toxicology studies

6) Is the particular test redundant to any other tests of genetic integrity?

TESTS FOR SPERM ANEUPLOIDY AND CHROMOSOME ABERRATIONS

Based on the principle of fluorescence in situ hybridization (FISH), chromosome-specific probes can be used to identify chromosome regions in individual spermatozoa. Briefly, a typical probe is designed to recognize a relatively large section of a particular chromosome (usually 0.2 - 2 Mb), and then labeled with a fluorochrome. After hybridization of the probe with a sample of spermatozoa, the labeled portion of the chromosome appears as a fluorescent domain within the sperm nucleus and can be identified using fluorescence microscopy (reviewed by Shi and Martin, 2000; Wyrobek et al., 2000). A normal haploid sperm nucleus should contain a single domain for each labeled autosome and a domain for either of the sex chromosomes. When two or more probes are used, it is possible to distinguish between disomy (two domains present for one of the probes) and diploidy (two domains present for each probe).

Earlier FISH assays for sperm aneuploidy were restricted to evaluating those chromosomes for which reliable probes were available. Now, multi-color probes are available for all human chromosomes (and many rodent chromosomes). This has allowed selection of specific chromosomes for study. For example, recent attention has been directed towards surveying those chromosomes for which sperm aneuploidy would result in trisomic or monosomic pregnancies that are likely to be compatible with postnatal life (Wyrobek et al 2000). Thus, probes for chromosomes X, Y, 13, 18, and 21 allow us to detect sperm that may lead to various important aneuploidy syndromes: triple X, Klinefelter, Turner, XYY, Patau, Edward, or Down. Multi-probe assays also allow for the enumeration of diploid sperm, and the distinction between meiosis I and II errors. Due to limitations of the eye for detecting color differences, assays are usually limited to three or four probes at a time.

Multi-color sperm FISH methods have also been developed to detect human sperm that carry various types of structural chromosomal aberrations. These are consequences of breakage events in germ cells, and therefore occur through fundamentally different mechanisms than those that lead to sperm aneuploidy or diploidy. The first generation assay for structural damage was developed by VanHummelen et al. (1996) using DNA probes to detect partial chromosomal duplications and deletions of chromosome 1p. This assay has been applied to studies of healthy men (Baumgartner et al 1999) and men who received chemotherapeutic drugs. A second generation assay, termed the sperm ACM assay (for the three satellite probes it uses: alpha, classical, and midi satellite probes), incorporated tandem labeling to detect sperm carrying chromosomal breaks, partial chromosomal duplications and deletions, as well as numerical abnormalities (Sloter et al., 2000). The advantages of the ACM assay are that it can detect recent post-meiotic chromosomal breaks, rearrangements and breakage events that occur during meiosis, plus sperm products of stem cell reciprocal translocations. Stem-cell translocations are known to be induced by a variety of environmental agents including ionizing radiation in human and mouse germ cells. Once such damage is induced it is expected to persist indefinitely for the remainder of the reproductive life of the affected individual. Recently, in a study on HIV anti-retroviral therapy effects, a classic three-color FISH assay was used to evaluate sperm aneuploidy for chromosomes X, Y and 18, and a tandem probes assay (two probes for chromosome 1) was used to evaluate chromosome breakage in the same semen samples (Robbins et al., 2001).

The breakout group was in agreement that a major advantage of FISH is its high degree of specificity. Each probe detects a specific part of a particular chromosome with little or no error. The approach is flexible, allowing combinations of probes to survey different chromosomes of interest, and to detect translocations and chromosome breakage. This specificity makes the test suitable for clinical diagnosis. For example, the test can be used to evaluate semen of couples with multiple miscarriages to detect possible translocations or other risk factors for aneuploid conceptions or spontaneous abortions. Similarly, it can be a valuable assay for monitoring long-term consequences of radiation or chemotherapy on the genetic integrity of sperm in cancer survivors to estimate his risk of conceiving an aneuploid child.

From a logistic standpoint, it is also feasible to incorporate a FISH assay into an epidemiologic or field study (Perreault et al., 2000). Only a small volume of semen is needed (0.1-0.2 ml being sufficient except for oligospermic samples), and the aliquot can be frozen in the field, without cryopreservative, and stored frozen for long periods of time before analysis. Thus, samples can be accumulated over the course of a study and later selected for analysis based on clinical conditions, or exposures of interest.

On the other hand, sperm FISH is labor-intensive. Because aneuploidy events are rare in sperm, a large number of sperm cells must be screened to generate accurate estimates. The number of cells to screen is dependent on several factors including the expected magnitude of an effect, the number of men in a study, and the number of samples evaluated per man. For example, the minimum requirements to detect a doubling in the frequency of aneuploid sperm might include control and test groups of at least 10 donors each, scoring ~10,000 sperm per donor. The most time consuming aspect of sperm FISH is the microscope scoring, which can be estimated to require ~40 working days to analyze ~200,000 spermatozoa for such a hypothetical study.

Test accuracy requires close attention to strict scoring criteria and requires rigorous quality control to insure precision and minimize inter-technician variability. Moreover, the probes can be relatively expensive on a per sample basis, especially when probing multiple chromosomes. The number of chromosomes that can be scored at one time is limited to the number of color-specific dyes that can be viewed at once – usually 3 or 4. The workgroup agreed that methods to automate the assay are needed to make it more practical. Promising results in this regard were reported recently by Baumgartner et al. (2001a) who used Laser-Scanning Cytometry to automate the FISH analysis of the X chromosome and a selected autosome in human and mouse sperm. While such instrumentation has great potential, it is also expensive and requires a highly skilled operator. Clearly, further research is needed to develop practical and reliable automated methods.

To date, this assay has been applied in a limited number of epidemiology studies (see Robbins et al., this volume). Preliminary findings indicate that it is important to control for factors such as age and life style factors (use of tobacco products, consumption of alcohol or caffeine) when analyzing effects of pharmaceuticals or environmental contaminants on this outcome (Shi and Martin, 2000; Perreault et al, 2000; Robbins, this volume). Because of the expense of assaying a large number of samples, it may be advisable to prioritize the samples and test those from individuals with the highest exposures first. Also, if some chromosomes are more susceptible to nondysjunction in sperm, they may be selected as sentinels in population-based studies.

FISH techniques have been adapted for use with animal sperm including mice and rats (common test species for toxicology studies) and domestic species. Therefore, experiments in these test species can be implemented to identify aneugens, characterize the dose response, and identify mechanisms of drug or toxicant action (Adler et al., 1996; Lowe et al., 1996; Lowe et al., 1998; Schmid et al., 1999; Baumgartner et al., 2001b). Such data should lead to valuable models for interspecies risk extrapolation.

Any balanced decision on risk assessment requires study-specific information regarding (a) the size of the exposed population, (b) the nature of the exposure regime (e.g., chronic vs. acute), (c) the fraction of exposed men in their reproductive years, and (d) the types and relevance of the chromosomal changes detected in sperm of affected men. For example, even small increases in disomy 21 sperm or XY sperm should be considered a significant potential risk when large numbers of young men are exposed to significant doses of the agent of concern. However, very few exposures have been surveyed for their potential to induce aneuploid sperm, and more research is needed to identify potential high risk factors.

TESTS FOR SPERM CHROMATIN STRUCTURAL INTEGRITY

The Sperm Chromatin Structure Assay (SCSA^{TM})

The Sperm Chromatin Structure Assay (SCSA^{TM}) measures the susceptibility of sperm nuclear DNA to denaturation *in situ* (reviewed by Evenson, 1999; Evenson et al., 2002). It is an indirect indicator of DNA damage because it measures the amount of single stranded DNA after treatments that normally do not denature sperm DNA (heat or acid pH). The test employs the unique metachromatic and equilibrium staining properties of acridine orange (AO), a dye that fluoresces green when intercalated into double stranded "native" DNA and red when bound to single stranded "denatured" DNA (or RNA). Using flow cytometry, the relative amount of red vs. green fluorescence is measured on a per sperm basis, in large numbers of sperm (typically 5,000 to 10,000 per sample) in only a few minutes. Each sperm is classified as normal or abnormal based on the amount of single stranded DNA it contains, and the percentage of abnormal cells is calculated for each sample. Once the flow cytometer is aligned, and standards are run, an experienced user can run about 6 samples per hour in duplicate. With additional time included for statistical analysis and report output, about three samples per hour can be analyzed. As with many tests, SCSA requires careful attention to quality control to insure that each sample is run under identical conditions. It also requires an understanding of the complexity of the data in order to derive valid and consistent interpretation of the results. The costs of purchasing and running a flow cytometer (cell sorter) are viewed as prohibitive for some laboratories and may limit the use of the assay by individual clinicians. However, the availability of central laboratories to which samples can be shipped for analysis should enhance the feasibility of conducting this assay on a routine basis.

Slide-based AO sperm assays have been used to mimic the flow cytometric SCSA. However, the reliability of the slide technique is considered to be relatively poor for two reasons. First, the AO staining conditions required for the metachromatic shift from green to red fluorescence are poor because the glass surface adsorbs the AO. Second, it is difficult for a reader to score cells that contain both red and green fluorescence as normal or abnormal in an accurate and consistent manner

The SCSA is less specific relative to the DNA damage tests described below in that it may detect alterations in protamine content and disulfide cross-linkage within the protamine, as well as sperm DNA damage. Sperm with altered proteins show up as a subpopulation with high DNA stainability that is distinct from subpopulations with increased DNA denaturation (Evenson et al., 2000). Indeed, SCSA data include a) the percentage of sperm with non-detectable, medium and high levels of DNA fragmentation, b) the percentage of sperm with a high level of DNA stainability (immature chromatin) and c) relative amounts of seminal debris, bacteria and broken cells. As mentioned above, this lower specificity can be an advantage when predicting infertility since sperm that are defective in one or more ways will be detected.

An advantage of the SCSA that it shares with FISH and some of the other tests described below, is the flexibility to run the test on either fresh or frozen semen samples. An aliquot of semen can be frozen at a collection site, stored over the course of a clinical or field study, and assayed at the end of the study. Samples appear to be stable for years if stored in an ultracold (-80 to -100°C) freezer, and even decades in LN_2 (Evenson, unpublished observation). Being able to store samples until the end of a field study minimizes assay-to-assay variation, allows the researcher to prioritize samples for analysis, and makes it feasible to send samples to a central (or core) laboratory for analysis. .

Clinical studies with this assay have demonstrated that a percentage of abnormal cells per sample that reaches about 30% or higher is highly predictive of infertility (Evenson et al., 1999, 2002; Spano et al., 2000). The test also predicts failure to initiate a pregnancy after *in vitro* fertilization (Larson et al., 2000). Current data on >1000 semen samples from various studies show that when the % of abnormal sperm per sample reached our threshold of 30%, the pregnancy rates ranged from zero to about half of normal (but with a significant increase in spontaneous abortions). Of these cases, 28% fell into the 0-15% DNA fragmented category, 34% had 15-25% DNA fragmentation, 11% were between 25 and 30%, and 27% had > 30% sperm with DNA fragmentation (Evenson et al., 2002).

SCSA has been used in a number of epidemiology studies to compare semen quality in exposed and unexposed groups of men (e.g., Selevan et al., 2000; Larsen et al., 1998; Kolstad et al., 1999; Bonde et al., 2001). Some of the confounders that might impact the results are being identified (Spano et al., 1998). With respect to applications in toxicology, SCSA can be readily adapted for use in rodents and agriculturally important species. Thus, like FISH, SCSA can be considered a bridging biomarker such that experimental (hypothesis-driven) data obtained in animal studies can be used to model human exposures and strengthen human risk assessment.

The workgroup participants agreed that the available literature supports the following conclusions: SCSA parameters appear to be generally independent of the routine semen parameters (volume, sperm concentration, sperm motility and sperm morphology); within person SCSA measurements are highly repeatable (i.e., there is less within individual variability than for other semen measures); this property coupled with the remarkable consistency of the methodology makes SCSA an ideal technique to detect sperm cell damage in longitudinal surveys; and, SCSA is a good predictor, relative to other sperm measures, for the clinical diagnosis of male infertility.

SPECIFIC TESTS FOR BROKEN DNA IN SPERM

Terminal Deoxynucleotidyl Transferase (TUNEL) and in situ Nick Translation (NT) Assays

The principle of these tests is the addition of labelled DNA precursors (nucleotides), driven by DNA polymerases, at sites of single and/or double stranded breaks in DNA. By using fluorescently labeled precursors (or fluorescently labeled reporter probes), sperm can be scored as positive or negative under fluorescence microscopy, or can be evaluted perhaps more conveniently and with more sensitivity by flow cytometry. As with SCSA, flow cytometry allows rapid analysis of a large number of sperm in a short period of time and provides a highly sensitive means of defining "positive" in an objective manner.

In its conventional form the flow cytometry-based TUNEL assay is convenient and generates results that correlate with the functionality of the spermatozoa (Sun et al., 1997), and with other tests such as the SCSA (Gorczyca et al, 1993; Aravindan et al., 1997). The NT assay has also been used to demonstrate the existence of endogenous DNA breaks in human spermatozoa (Manicardi et al, 1995).

One potential problem with these assays is that they depend upon enzymes (terminal transferase in the TUNEL assay, and DNA polymerase in the NT assay) gaining access to the nicks in the highly compacted sperm nucleus. In order to overcome this problem, variations on the nick translation theme have been introduced in which the sperm chromatin is decondensed prior to the initiation of the assay (Twigg et al., 1998a,b). Unfortunately, this version of the assay cannot be used in conjunction with flow cytometry.

To date, TUNEL has most frequently been conducted under a fluorescent microscope and its application has been limited to the detection of abnormal sperm cells in the ejaculates of individuals attending fertility clinics, as opposed to being used in epidemiological studies. Several reports have demonstrated that an increased fraction of human spermatozoa showing DNA strand breaks has a negative impact on the success rate of assisted fertilization techniques (Sakkas et al., 1996; Sun et al., 1997; Lopes et al., 1998a, b; Sakkas et al., 1998).

These assays are specific in that they detect DNA strand breaks, but the origin of the breaks is not always clear. In somatic cells, TUNEL has been reported to be more selective in detecting DNA degradation typical of apoptosis, whereas nick translation is thought to be indicative of necrosis (Gorczyca et al., 1993). Although spermatozoa, by their very nature, do not exhibit the morphologic changes characteristic of apoptosis in somatic cells, it has been suggested that DNA breaks detected by TUNEL in sperm may reflect apoptosis in their germ cell precursors, or even in mature sperm. Alternatively, the DNA nicks detected by the TUNEL assay might reflect direct damage to DNA via oxidative stress or deficiencies in physiological processes such as recombination and chromatin packaging. The fact that DNA nicks can be created post ejaculation by the induction of oxidative stress (Twigg et al., 1998 a,b) indicates that not all TUNEL positive cells are the result of abortive apoptosis. Moreover, because these nicks can be detected in 'swim-up' populations of spermatozoa (Twigg et al., 1998b). such damage is not confined solely to dead or moribund spermatozoa. Indeed, the results obtained with the TUNEL and NT assays are highly correlated in spermatozoa (Manicardi et al., 1998). Thus, although evidence exists that TUNEL positive sperm also show abnormal levels of apoptotic proteins, conclusive evidence for an induced apoptotic response in mature spermatozoa is lacking (see chapter by Sakkas et al., this volume; Evenson et al., 2002).

In the context of male-mediated developmental toxicity it is important to emphasize that sperm with DNA breaks/nicks may be fully capable of fertilization. For example, Twigg et al. (1998a) showed that sperm exposed to hydrogen peroxide, NADPH or activated leucocytes, treatments that induce DNA strand breakage, decondensed and formed morphologically normal pronuclei after being microinjected into hamster oocytes. Furthermore, sperm with radiation-induced DNA damage were capable of fertilizing oocytes and thereby transmitting the damage to the embryos, which show developmental abnormalities (Ahmadi and Ng, 1999a,b). Thus, the breakout group reaffirmed the proposition that practical and reliable tests to detect damaged DNA in spermatozoa are needed in order to optimize assisted reproductive technologies so as to minimize adverse developmental defects in the embryos (Aitken, 1999).

Single cell electrophoresis (Comet) Assay

The single cell electrophoresis (Comet) assay was developed to measure single and double strand breaks in somatic cells and is widely used in genotoxic research and more recently in biomonitoring (Tice et al. 2000). In principle, cells are cast in agarose gel upon a microscope slide, lysed and subjected to electrophoresis. DNA freed from the nucleus by virtue of DNA breaks, migrates under the electrophoresis conditions to create the tail of the "comet" while intact DNA remains in the nucleus. The "comet" is visualized with a DNA-specific fluorescent dye. Measurement of DNA damage is done by measuring the length of DNA migration, the percentage of DNA in the tail (or that remaining in the head) or a function of both called the "tail moment." Image analysis software allows this to be measured in a quick and objective manner. While no large piece of equipment, such as a flow cytometer, is needed for this assay, it does require an electrophoresis set up, and a fluorescence microscope equipped with a suitable imaging system (camera, computer and software package) and a trained operator.

Although the assay seems simple in principle, applying it to spermatozoa is not straightforward, mainly because it is first necessary to open up the highly compact sperm nucleus so that the DNA is releasable during electrophoresis. This requires harsh chemical treatment with detergents, enzymes (RNase and/or proteinase K), and/or disulfide reducing agent that may induce DNA breaks. Furthermore, the stainability of sperm DNA is dynamic, increasing as the sperm DNA opens up and its associated proteins (protamine) are degraded/removed. Therefore direct comparisons between the amount of DNA remaining in the sperm (comet) head with that moving into the comet tail are difficult to make.

Early on, it was shown that the "alkaline" method for detecting single strand breaks produced large comets in all cells, presumably due to the presence of alkaline labile sites in spermatozoa (Singh et al. 1989). This makes it difficult to distinguish between endogenous and induced breaks or between single and double stranded breaks. Several labs now conduct the electrophoresis at a more neutral pH in order to detect only double strand breaks (Haines et al., 1998; Singh and Stevens, 1998). Many protocol variations have been reported, with differences not only in pH, but also composition of the cell lysis solution, the timing/conditions of the electrophoresis, and most recently, the application of newer, more sensitive DNA dyes such as YOYO-1. Once comets are generated they have been measured in different ways and with different software programs. Finally, the outcome(s) has been reported differently: some labs calculate the percentage of sperm forming comets, and others report average measures of the extent of DNA migration for a given sperm population (Hughes et al., 1996; Irvine et al., 2000, Evenson et al., 2002). The breakout group agreed that the method is still evolving and the current lack of standardized, detailed protocols for conducting this test and reporting the results limits the ability to compare sperm Comet studies across laboratories, and to compare sperm Comet with other tests for sperm DNA damage.

Even so a number of studies have found an association between increased DNA damage measured by the Comet assay and measures of infertility (Irvine et al. 2000; Donnelly et al. 2001). The origin of such DNA damage is not clear. However it may be significant that reactive oxygen species normally found in the male reproductive tract induce DNA damage detected by Comet (Aitken 1999). Additionally the level of DNA damage in sperm increases after treatment of both men and mice with genotoxins such as chemotherapy and radiation (Chatterjee et al. 2000; Haines et al. 2001). Thus, this test has potential as a means for evaluating DNA strand breaks in sperm as an indicator for male mediated developmental toxicity.

TESTS FOR SPERM MATURITY THAT MAY CORRELATE WITH DNA DAMAGE

Normal sperm chromatin structure depends on the correct replacement of somatic histones with sperm-specific protamines during spermiogenesis. It is becoming increasingly apparent that sperm function, i.e., both the ability to fertilize the oocyte and to support embryonic development, is also related to proper nuclear remodeling in the mature sperm (Braun, 2001). Indeed, deviations from the normal protamine 2/protamine 1 or protamine-to-histone ratio may be responsible for certain conditions of sub/infertility (De Yebra et al., 1993). Furthermore, aberrant protamine 1 and 2 ratios have been documented in sperm from infertile males (Balhorn et al., 1988; Belokopytova et al., 1993), and differences in protamine 1 and 2 mRNA levels have been reported in the spermatids of infertile men (Steger et al., 2001). The latter study showed that the protamine 1 and 2 mRNA ratio correlated to fertilization rates after ICSI. Importantly, recent gene targeting experiments showed that haploinsufficiency of protamine 1 or 2 causes infertility in mice (Cho et al., 2001). Therefore, there is biologic motivation for the development of tests that detect underprotamination (or excess histones) in sperm nuclei. Whether such tests are indirect indicators of damaged DNA or perhaps of DNA that is at increased risk of being damaged (due to improper packaging) remains to be determined.

The tests developed to date are microscopic slide-based, histochemical tests. Since they are scored visually, they are labor intensive, and suffer from inconsistencies arising from subjective appraisal. On the other hand, because they are based on relatively simple staining methods and can be done on smears of sperm prepared at the same time as the morphology smears, they are practical and relatively inexpensive to include in an andrologic workup, or to add to a field study.

Although tests for underprotamination are rather nonspecific with regard to predicting genetic damage in sperm, they provide some indication of clinical infertility, are correlated with poor semen quality, and add to the prediction of *in vitro* fertilization (Liu and Baker, 1992a, b). To date, their application has been confined largely to the human andrology clinics, and no major application in epidemiological studies has been reported. For example, aniline (toluidine) blue, which may pick up sperm with an excess of residual histones, has been demonstrated to be valuable in the assessment of male fertility prior to assisted reproduction (Hammadeh et al., 2001). Results of two variations of a slide-based sperm chromatin structure assay (using acridine orange) correlated highly with aniline blue or toluidine blue staining (Erenpreiss et al., 2001). This observation suggests that the simpler stains (not requiring fluorescence microscopy or flow cytometry) may be a useful adjunct to routine semen measures, despite their lack of specificity, and that AO staining can detect immature sperm that do not necessarily have DNA damage.

Chromomycin A_3 (CMA_3) is another slide-based test that detects under-protaminated sperm (Manicardi et al., 1995; Bizarro et al., 1998). This fluorochrome appears to bind to protamine binding sites on DNA such that underprotaminated sperm become fluorescent. The percentage of CMA_3 positive sperm was positively correlated with the presence of endogenous nicks in sperm DNA (Manicardi et al., 1995), and showed significant associations with low sperm count, high percentages of abnormal sperm morphology and low fertilization rates in vitro (Lolis et al., 1996). According to recent reports, it has been shown to be more sensitive and specific than aniline blue and slide-based acridine orange tests in predicting fertilization rates in vitro (Nasr-Esfahani, et al., 2001).

Another marker for sperm immaturity is cytoplasmic retention. Biochemical markers of cytoplasmic retention such as increased CK activity and decreased HspA2 (a chaperone protein associated with synaptonemal complex) have been shown to be associated with immature sperm, sperm with abnormal chromatin structure (detected with other tests including aniline blue staining), and sperm with diminished fertilizing ability and fertility (Huszar and Vigue, 1993; Huszar et al, 1997, 2000). Importantly, the proportion of immature sperm identified by these assays also correlated with the incidence of sperm chromosome aneuploidy (Kovanci et al., 2001). The authors concluded that the common factor underlying sperm immaturity and aneuploidies may be diminished expression of HspA2, a protein that is essential for the completion of meiosis and is detectable in mature sperm. Relatively simple, slide-based immunocytochemistry assays have been developed for evaluating human semen in this regard (Huszar et al., 2000), and are currently being adapted for field use.

DISCUSSION AND RESEARCH NEEDS

This overview describes an ever-growing list of biomarkers that have been advanced to evaluate the genetic and chromosomal integrity of sperm. However, few have found acceptance and routine integration into clinical settings or field studies because of technical concerns and our generally limited understanding of their predictive value for abnormal reproductive outcomes of clinical importance. Most of the techniques are highly specialized and are performed by only a few focused laboratories. Several are technically difficult and expensive. Although it is generally agreed that the research community would benefit if additional labs evaluated and developed these methods, there are few reliable training programs. The SCSA and sperm FISH assays are examples of assays that have been successfully performed in several laboratories, but experience with these methods has underscored the importance of quality control and harmonization protocols.

In terms of predictive value, past emphasis has been placed on relating the test to clinical fertility/infertility as opposed to predicting abnormal pregnancy outcomes. In general, correlations have been found between infertility, poor semen quality, and increased numbers of sperm with chromosome or DNA damage by any of these tests. For example, abnormal semen quality has been reported to be associated with higher incidences of sperm aneuploidy (e.g. McInnes et al., 1998). Poor semen quality, especially low sperm concentration, has also been associated with high levels of DNA fragmentation detected by Comet and two variations of an NT assay (Irvine et al., 2000). Poor semen quality is an indication for in vitro fertilization by intracytoplasmic sperm injection. As discussed by Aitken (1999), there is, therefore, concern that in such cases, the chance of picking up and microinjecting a sperm with fragmented DNA into the oocyte is increased, with obvious consequences pertinent to the generation of male-mediated developmental deficits. Because some men with good semen quality may nevertheless have high levels of sperm chromosome or DNA damage, it is clearly important to have tests to detect the problem. Likewise, exposure to testicular toxicants (environmental or pharmaceutical) may result in poor semen quality with associated risks not only of infertility but also of transmissible genetic damage in sperm. Again, the workgroup agreed that it is important to be able to screen for and interpret such effects.

Further research is needed to determine the predictive value of each sperm biomarker for the various categories of abnormal reproductive outcomes. The sperm FISH assays for aneuploidy have the conceptual advantage of being highly specific to identify risk factors for paternally transmitted aneuploidy in pregnancy and in offspring. Through careful selection of probes, these assays can be used to determine whether a male has an increased risk of producing sperm that may lead to specific aneuploidy syndromes such as Down, Edward, Patau, Klinefelter, Turner, triple X, or XYY. Other chromosomes can be used to determine whether there are risks for chromosomes that terminate early or late in development.

On the other hand, the sperm ACM FISH assay for structural aberrations utilizes chromosome 1 as a sentinel chromosome to assess the genomic burden of chromosome breakage and structural aberrations and is therefore less specific in its predictive value for abnormal reproductive outcomes. Structural aberrations in sperm can result in developmental defects due to abnormal gene dosage, or by breakage events involving specific genes. Thus, men with increased frequencies of sperm with chromosomal aberrations are expected to be at increased risk for a broad range of abnormal reproductive outcomes ranging from spontaneous abortion to morphological and/or genetic defects in the offspring.

Most of the other assays described in this chapter measure some abnormality or change in sperm chromatin structure, cytoplasm, or DNA. These endpoints have been most commonly linked with reductions in fertility, as was previously done for the conventional semen parameters. More research is needed to examine their predictive value for spontaneous abortions, malformations, developmental defects, or chromosomal damage in offspring.

In recommending which, if any, of these tests to offer an infertile couple, or to include in an epidemiology study, we need to know more about what each test is actually measuring. We recognize that the FISH assays for aneuploidy are providing information distinct from tests for DNA damage. However, some of the tests for DNA damage may be redundant with each other. For example, SCSA was strongly correlated with results of TUNEL and Comet (Aravindan et al., 1997). However, that study scored the presence or absence of a comet tail using ethidium bromide staining and visual scoring under low magnification microscopy. Current data also from Evenson's laboratory (Tritle et al., unpublished), generated by analyzing digital images using the VisCOMET program, indicate that correlations between Comet and SCSA are weaker than previously reported ($r\sim0.4$), and that Comet shows more intra-assay variability than SCSA. In general, comparisons among the tests are few in number and the sample size is often small; therefore, more extensive side by side comparisons across methods are needed.

To a certain extent, SCSA, TUNEL and Comet may complement each other. SCSA can detect abnormalities in chromatin packaging and/or DNA strand breaks; TUNEL can detect single- and double-strand DNA breaks; Comet (under "neutral" conditions) can detect double-strand DNA breaks (Singh and Stevens, 1998). Although it is likely that sperm cells characterized by an increased sensitivity to denaturation also have extensive DNA strand breakage, the correlation level was different when comparing SCSA vs. TUNEL, SCSA vs. Comet, and TUNEL vs. Comet, thus reinforcing current opinions that each test could address a particular facet of the complex processes involved in sperm nuclear packaging.

The heterogeneous nature of a semen sample, particularly in the human, may also argue for use of more than one test. Evidence for the existence of subpopulations of human spermatozoa was obtained by applying both the NT and CMA_3 tests to the same semen samples (Manicardi et al., 1995). Results suggested that most of the healthy sperm contained compact chromatin (highly protaminated CMA_3 negative sperm) and unbroken DNA. However, subpopulations of abnormal sperm contained loosely packaged chromatin (probably under-protaminated, CMA_3 positive sperm) with or without nicked DNA. Knowing the relative proportion of these classes of spermatozoa in a patient's ejaculate might help predict both his fertility and risk of transmitting genetic damage to the embryo.

In general, these sperm-based tests of genetic and chromosomal integrity have inherent advantages as biomarkers of paternal risk factors for transmitted genetic disease. Specifically: (a) they can be used to identify high risk groups and high risk exposures for abnormal reproductive outcomes; (b) in cases where men receive acute or subacute exposures, they can help to identify the sensitive germ cell stages; and, (c) in cases where men receive differing doses of exposure, they can help to determine the shape of the dose response curve and help estimate risks at low doses.

Major research problems that need to be addressed now and in the long term are listed below.

Technical advances

* Standardize the protocols used to detect chromosomal abnormalities and DNA damage in sperm, and develop training protocols to reliably export and compare the results of assays among laboratories, including quality control and harmonization protocols.
* Develop efficient automated systems for the detection of chromosomally defective sperm by multi-color FISH

Mechanistic studies

* Determine how the results generated by the different assays correlate with each other among healthy men, clinical populations, and men exposed to various specific exogenous agents.
* Understand the biochemical, toxicological, and genetic pathways that lead to the various types of chromosomal and DNA damage in human sperm.

Identification of risk factors and definition of risk

* Conduct focused studies using sperm FISH assays to identify and prioritize potential risk factors for the induction of aneuploid and chromosomally abnormal sperm.
* Optimize and implement corollary rodent sperm assays for toxicology and for studies of genetic and biochemical mechanisms.
* Understand the predictive value of individual sperm biomarkers for infertility, spontaneous abortions, malformations, birth with a chromosomal defect, and other transmitted genetic diseases.
* Develop multi-assay batteries and apply them in large scale epidemiological studies to identify risk factors for paternally transmitted genetic damage (i.e., environmental or occupational exposures, inherent factors such as age, and lifestyle factors).
* Use batteries of sperm assays in combination with genomic methods to identify genetically susceptible subpopulations of men at risk for infertility, spontaneous abortions, and birth defects.

This workgroup represented a starting point for developing collaborative validation studies that could be carried out with good quality control on sufficiently large numbers of samples and using common reference materials. A proposed long-term goal is to develop consensus documents regarding test protocols and standards, such as those that have been developed for other semen outcomes (e.g. ESHRE, 1996).

DISCLAIMER: This document has been reviewed in accordance with the U.S. Environmental Protection Agency policy and approved for publication. Mention of trade names or commercial products does not constitute endorsement or recommendation for use.

REFERENCES

Adler, I.D., Bishop, J., Lowe, X., Schmid, T.E., Schriever-Schwemmer, G, Xu, W., and Wyrobek, A.J., 1996, Spontaneous rates of sex chromosomal aneuploidies in sperm and offspring of mice: a validation of the detection of aneuploidy sperm by fluorescence in situ hybridization, *Mutat Res.* 372:259-268.

Ahmadi, A. and Ng, S.C., 1999a, Developmental capacity of damaged spermatozoa, *Hum Reprod.* 14:2279-2285.

Ahmadi, A. and Ng, S.C., 1999b, Fertilizing ability of DNA-damaged spermatozoa. *J Exp Zool.* 284:696-704.

Aravindan, C.R., Bjordahl, J., Jost, L.K., and Evenson, D.P., 1997, Susceptibility of human sperm to *in situ* DNA denaturation is strongly correlated with DNA strand breaks identified by single-cell electrophoresis. *Exp Cell Res.* 236:231-237.

Aitken, R.J., 1999, The Amoroso Lecture: The human spermatozoon – a cell in crisis? *J Reprod Fertil.* 115:1-7.

Balhorn, R., Reed, S., and Tanphaichitr, N., 1988, Aberrant protamine 1/protamine 2 ratios in sperm of infertile human males, *Experientia.* 44:52-55.

Baumgartner, A., Schmid, T.E., Maaerz, H.K., Adler, I.-D., Tarnok, A., and Nuesse, M., 2001a, Automated evaluation of frequencies of aneuploid sperm by laser-scanning cytometry (LSC), *Cytometry.* 44:156-160.

Baumgartner, A., Schmid, T.E., Schutz, C.G., and Adler, I.-D., 2001b, Detection of aneuploidy in rodent and human sperm by multicolor FISH after chronic exposure to diazepam, *Mutat Res.* 490:11-19.

Baumgartner, A., Van Hummelen, P., Lowe, X.R., Adler, I.-D., and Wyrobek, A.J., 1999, Numerical and structural chromosomal abnormalities detected in human sperm with a combination of multicolor FISH assays. *Environ Mol Mutagen.* 33:49-58.

Belokopytova, I.A., Kostyleva, E.I., Tomilin, A.N., and Vorob'ev, V.I., 1993, Human male infertility may be due to a decrease of the protamine P2 content in sperm chromatin, *Mol Reprod Dev.* 34:53-57.

Bizzaro, D., Manicardi, G.C., Bianchi, P.G., Bianchi, U., Mariethoz, E., and Sakkas, D., 1998, In-situ competition between protamine and fluorochromes for sperm DNA, *Mol Hum Reprod.* 4:127-132

Bonde, J.P., Joffe, M., Apostoli, P., Dale, A., Kiss, P., Spano, M., Giwercman, A., Bisanti, L., Porru, S., Vanhoorne, M., Comhaire, F., and Zschiesche, W., 2001, Sperm count and chromatin structure in men exposed to inorganic lead: lowest adverse effect levels, *Occup Environ Med.* 59:234-242.

Braun, R.E., 2001, Packaging paternal genome with protamine, *Nature Genet.* 28:10-12.

Chatterjee, R., Haines, G.A., Perera, D.M., Goldstone, A. and Morris, I.D., 2000, Testicular and sperm DNA damage after treatment with fludarabine for chronic lymphocytic leukaemia. *Hum Reprod.* 15:762-766.

Cho, C., Willis, W.D., Goulding, E.H., Jung-Ha, H., Choi, Y-C., Hecht, N.B., and Eddy, M.M., 2001, Haploinsufficiency of protamine-1 or -2 causes infertility in mice, *Nature Genet.* 28:82-86.

De Yebra, L., Ballesca, J.L., Vanrell, J.A., Bassas, L., and Oliva, R., 1993, Complete selective absence of protamine P2 in humans. *J Biol Chem.* 268:10553-10557.

Donnelly, E.T., Steele, E.K., McClure, N., and Lewis, S.E., 2001, Assessment of DNA integrity and morphology of ejaculated spermatozoa from fertile and infertile men before and after cryopreservation. *Hum Reprod.* 16:1191-1199.

Erenpreiss, J., Bars, J., Lipatnikova, V., Erenpreisa, J., and Zalkalns, J., 2001, Comparative study of cytochemical tests for sperm chromatin integrity, *J Androl.* 22:45-53.

ESHRE (European Society of Human Reproduction and Embryology) Andrology Special Interest Group, 1996, Consensus workshop on advanced diagnostic andrology techniques, *Hum Reprod.* 11:1463-1479.

Evenson, D.P., 1999, Alterations and damage of sperm chromatin structure and early embryonic failure, in: *Towards Reproductive Certainty: Fertility and Genetics Beyond 1999*, R. Jannsen and D. Mortimer, eds., Parthenon Publishing Group Ltdl, New York, p. 313.

Evenson, D.P., Jost, L.K., Marshall, D., Zinaman, M.J., Clegg, E., Purvis, K., Deangelis, P., and Claussen, O.P., 1999, Utility of the sperm chromatin structure assay as a diagnostic and prognostic tool in the human fertility clinic. *Fertil Steril.* 14:1039-1049.

Evenson, D.P., Jost, L.K., Corzett, M., Balhorn, R., 2000, Characteristics of human sperm chromatin structure following an episode of influenza and high fever: a case study. *J Androl.* 21:739-746.

Evenson, D.P., Larson, K., and Jost, L.K., 2002, The sperm chromatin structure assay (SCSA™): clinical use for detecting sperm DNA fragmentation related to male infertility and comparisons with other techniques. Andrology Lab Corner. *J Androl.* 23:25-43.

Gorczyca, W., Traganos, F., Jesionowska, H., and Darzynkiewicz, Z., 1993, Presence of DNA strand breaks and increased sensitivity of DNA in situ to denaturation in abnormal human sperm cells: analogy to apoptosis of somatic cells. *Exp Cell Res.* 207:202-205.

Haines, G.A., Hendry, J.H., Daniel, C.P., and Morris, I.D., 2001, Increased levels of Comet-detected spermatozoal DNA damage following in vivo isotopic- or X-irradiation of spermatogonia. *Mutat Res.* 495:21-32.

Haines, G., Marples, B., Daniel, P. and Morris, I, 1998, DNA damage in human and mouse spermatozoa after in vitro-irradiation assessed by the comet assay, in: *Reproductive Toxicology*, ed, J. del Mazo, Plenum Press, New York.

Hammadeh, M.E., Zeginiadov, T., Rosenbaum, P., Georg, T., Schmidt, W., and Strehler, E., 2001, Predictive value of sperm chromatin condensation (aniline blue staining) in the assessment of male fertility, *Arch Androl.* 46:99-104.

Hassold, T.J., 1998, Nondysjunction in the human male, in: *Meiosis and Gametogenesis*, M.A. Handel, ed., Academic Press, New York, p. 383.

Hughes, C.M., Lewis, S.E.M., McKelvey-Martin, J., and Thompson, .W., 1996, A comparison of baseline and induced DNA damage in human spermatozoa from fertile and infertile men, using a modified comet assay, *Mol Hum Reprod.* 2: 613-619.

Huszar, G. and Vigue, L.., 1993, Incomplete development of human spermatozoa is associated with increased creatine phosphokinase concentrations and abnormal head morphology, *Mol Reprod Dev.* 34:292-298.

Huszar, G., Sbracia, M., Vigue, L., Miller, D., and Shur, B., 1997, Sperm plasma membrane remodeling during spermiogenetic maturation in men: relationship among plasma membrane β-1,4,-galactosyltransferase, cytoplasmic creatine phosphokinase, and creatine phosphokinase isoform ratios, *Biol Reprod.* 56:1020-1024.

Huszar, G., Stone, K., Dix, D., and Vigue, L., 2000, Putative creatine kinase M-isoform in human sperm is identified as the 70-kilodalton heat shock protein HspA2, *Biol Reprod.* 63:925-932.

Irvine, D.S., Twigg, J.P., Gordon, E.L., Fulton, N., Milne, P.A., and Aitken RJ., 2000, DNA integrity in human spermatozoa: relationships with semen quality, *J Androl.* 21:33-44.

Kolstad, H.A., Bonde, J.P., Giwercman, A., Spano, M., Zschiesche, W., and ASCLEPIOS, 1999, Change in semen quality or DNA denaturation patterns during occupational styrene exposure. A longitudinal study, *Int. Arch. Occup Environ Health.* 72:135-141.

Kovanci, E., Kovacs, T., Moretti, E., Vigue, L., Bray-Ward, P., Ward, D., Huszar, G., 2001, FISH assessment of aneuploidy frequencies in mature and immature human spermatozoa classified by the absence or presence of cytoplasmic retention, *Hum Reprod.* 16:1209-1217.

Larsen, S.B., Giwercman, A., Spano, M., and Bonde, J.P., 1998, A longitudinal study of semen quality in pesticide spraying Danish farmers. The ASCLEPIOS Study Group, *Reprod Toxicol.* 12:581-589.

Larson, K.L., DeJonge, C.J., Barnes, A.M., Jost, L.K., and Evenson, D.P., 2000, Sperm chromatin structure assay parameters as predictors of pregnancy following assisted reproductive techniques, *Hum Reprod.* 15:1717-1722.

Liu, D.Y. and Baker, H.W.G., 1992a, Tests of human sperm function and fertilization in vitro, *Fertil Steril.* 58:465-483.

Liu, D.Y. and Baker, H.W.G., 1992b, Sperm nuclear chromatin normality: relationship with sperm morphology, sperm-zona pellucida binding, and fertilization rates in vitro, *Fertil Steril.* 58:1178-1184.

Lolis, D., Georgiou, I., Syrrou, M., Zikopoulos, Kl, Konstantelli, M., and Messinis, I., 1996, Chromomycin A$_3$-staining as an indicator of protamine deficiency and fertilization, *Intern J Androl.* 19:23-27.

Lopes, S., Jurisicova, A., and Casper, R.F., 1998a, Gamete-specific DNA fragmentation in unfertilized human oocytes after intracytoplasmic sperm injection, *Hum Reprod.* 13:703-708.

Lopes, S., Sun, J.G., Jurisicova, A., Meriano, J., and Casper, R.F., 1998b, Sperm deoxyribonucleic acid fragmentation is increased in poor-quality semen samples and correlates with failed fertilization in intracytoplasmic sperm injection, *Fertil Steril.* 69:528-532.

Lowe, X.R., de Stoppelaar, J.M., Bishop, J., Cassel, M., Hoebee, B., and Wyrobek A.J., 1998, Epididymal sperm aneuploidies detected in three genetic strains of rats by multi-color FISH, *Env Mol Mut.* 31:125-132.

Lowe, X., O'Hogan, S., Moore, D. II, Bishop, J., Wyrobek, A.J., 1996, Aneuploid epididymal sperm detected in chromosomally normal and Robertsonian translocation-bearing mice using a new three-chromosome FISH method, *Chromosoma* 105:204-210.

Manicardi, G.C., Bianchi, P.G., Pantano, S., Azzoni, P., Bizzaro, D., Bianchi, U., Sakkas, D., 1995, Presence of endogenous nicks in DNA or ejaculated human spermatozoa and its relationship to chromomycin A3 accessibility, *Biol Reprod.* 52:864-867.

Manicardi, G.C., Tombacco, A., Bizzaro, D., Bianchi, U., Bianchi, P.G., Sakkas, D., 1998, DNA strand breaks in ejaculated human spermatozoa: comparison of susceptibility to the nick translation and terminal transferase assays, *Histochem J.* 30:33-39.

McInnes, B., Rademaker, A., Greene, C.A., Ko, E., Barclay, L., and Martin, R.H., 1998, Abnormalities for chromosome 13 and 21 detected in spermatozoa from infertile men, *Hum Reprod.* 13:2787-2790.

Nasr-Esfahani M.H., Razavi, S., Mardani, M., 2001, Relation between different human sperm nuclear maturity tests and in vitro fertilization, *J Assist Reprod Genet.* 18:219-225.

Olshan, A.F. and Mattison, D.R. 1994, *Male-Mediated Developmental Toxicity*, Plenum Press, New York.

Perreault, S.D., Rubes, J., Robbins, W.A., Evenson, D.P. and Selevan, S.G., 2000, Evaluation of aneuploidy and DNA damage in human spermatozoa: applications in field studies, *Andrologia* 32:247-254.

Robbins, W.A., Witt, K.L., Haseman, J.K., Dunson, D.B., Troiani, L., Cohen, M. S., Hamilton, C.D., Perreault, S.D., Libbus, B., Beyler, S.A., Raburn, D.J., Tedder, S.T., Shelby, M.D., and Bishop, J.B., 2001, Antiretroviral therapy effects on genetic and morphologic end points in lymphocytes and sperm of men with human immunodeficiency virus infection, *J Infectious Dis.* 184:127-135.

Sakkas, D., Mariethoz, E., Manicardi, G., Bizzaro, D., Bianchi, P.G. and Bianchi, U., 1999, Origin of DNA damage in ejaculated human spermatozoa, *Rev Reprod.* 4:31-37.

Sakkas, D., Urner, F., Bianchi, P.G., Bizzaro, D., Wagner, I., Jaquenoud, N., Manicardi, G., and Campana, A., 1996, Sperm chromatin anomalies can influence decondensation after intracytoplasmic sperm injection, *Hum Reprod.* 11:837-843.

Sakkas, D., Urner, F., Bizzaro, D., Manicardi, G., Bianchi, P.G., Shoukir, Y., and Campana, A., 1998, Sperm nuclear DNA damage and altered chromatin structure: effect on fertilization and embryo development, *Hum Reprod.* 13 (Suppl 4):11-19.

Schmid, T.E., Wang, X., and Adler, I.-D., 1999, Detection of aneuploidy by multicolor FISH in mouse sperm after in vivo treatment with acrylamide, colchicines, diazepam, or thiabendazole, *Mutagenesis* 14:173-179.

Selevan, S.G., Borkovec, L., Slott, V.L., Zudova, Z., Rubes, J., Evenson, D.P., and Perreault, S.D., 2000, Semen quality and reproductive health of young Czech men exposed to seasonal air pollution, *Environ Health Perspect.* 108:887-894.

Shi, Q.H. and Martin, R.H., 2000, Aneuploidy in human sperm: a review of the frequency and distribution of aneuploidy, effects of donor age and lifestyle factors, *Cytogenet Cell Genet* 90:219-226.

Singh, N.P., Danner, D.B., Tice, R.R., McCoy, M.T., Collins, G.D., and Schneider, E.L., 1989, Abundant alkali-sensitive sites in DNA of human and mouse sperm, *Exp Cell Res.* 184:461-470.

Singh, N.P., and Stephens, R.E., 1998, X-ray-induced double-strand breaks in human sperm, *Mutagenesis* 13: 75-79.

Spano, M., Kolstad, H., Larsen, S.B., Cordelli, E., Leter, G., Giwercman, A., Bonde, J.P., and Asclepios, 1998, The applicability of the flow cytometric sperm chromatin structure assay in epidemiological studies, *Hum Reprod.* 13:2495-2505.

Sloter, E., Lowe, X., Moore, D. II, Nath, J., Wyrobek, A., 2000, Chromosomal breaks, duplications, deletions, and aneuploidy in the sperm of healthy men., *Am J Hum Genet.* 67:862-872.

Spano, M., Bonde, J.P., Hjollund, H.I., Kolstad, H.A., Cordelli, E, and Leter, G., 2000, Sperm chromatin damage impairs human fertility. The Danish Pregnancy Planner Study Team, *Fertil Steril.* 73:43-50.

Steger, K., Failing, K., Klonisch, T., Behre, H.M., Manning, M., Weidner, W., Hertle, L., Bergmann, M., and Kliesch, S., 2001, Round spermatids from infertile men exhibit decreased protamine-1 and -2 mRNA, *Hum Reprod.* 16:709-716.

Sun, J.G., Jurisicova, A., and Casper, R.F., 1997, Detection of deoxyribonucleic acid fragmentation in human sperm: correlation with fertilization in vitro, *Biol Reprod.* 56:602-607.

Tice, R.R., Agurell, E., Anderson, D., Burlinson ,B., Hartmann, A., Kobayashi, H., Miyamae, Y., Rojas, E., Ryu, J.C., and Sasaki, Y.F., 2000, Single cell gel/comet assay: guidelines for in vitro and in vivo genetic toxicology testing, *Environ Mol Mutagen.* 35:206-221.

Twigg, J., Fulton, N., Gomez, E., Irvine, D.S. and Aitken, R.J., 1998a, Analysis of the impact of intracellular reactive oxygen species generation on the structural and functional integrity of human spermatozoa: lipid peroxidation, DNA fragmentation and the effectiveness of antioxidants, *Hum Reprod.* 13:1429-1436.

Twigg, J., Irvine, D.S., Houston, P., Fulton, N., Michael, L., and Aitken, R.J., 1998b, Iatrogenic DNA damage induced in human spermatozoa during sperm preparation: protective significance of seminal plasma, *Molec Human Reprod.* 4:439-445.

Van Hummelen, P., Lowe, X.R., and Wyrobek, A.J., 1996, Simultaneous detection of structural and numerical chromosome abnormalities in sperm of healthy men by multicolor fluorescence in situ hybridization, *Hum Genet.* 98:608-615.

Wyrobek, A.J., Anderson, D., Lewis, S., Nagao, T., Perreault, S., Robaire, B., and Schrader, S., 1994, Biomarkers and health endpoints of developmental toxicology of paternal origin: Summary of working group discussion, in: *Male-Mediated Developmental Toxicity*, A.F. Olshan and D.R. Mattison, eds., Plenum Press, New York, p. 359.

Wyrobek, A. J., Marchetti, F., Sloter, E., Bishop, J., 2000, Chromosomally defective sperm and their developmental consequences. in: *Human Monitoring after Environmental and Occupational Exposure to Chemical and Physical Agents*, D. Anderson, A.E. Karakaya, and RJ Sram, eds., NATO Science Series, Vol. 313, IOS Press, p. 134.

RISK ASSESSMENT

Andrew F. Olshan, [1] Gary Kimmel, [2] and Donald Mattison[3]

[1]Department of Epidemiology
University of North Carolina
Chapel Hill, NC 27599
[2]United States Environmental Protection Agency
National Center for Environmental Assessment
Washington, DC 20004
[3]March of Dimes
White Plains, NY 10605

INTRODUCTION

The Risk Assessment breakout group was chaired by D. Mattison and A. Olshan, with approximately twenty conference participants in attendance. A wide range of interests and expertise were represented. G. Kimmel was asked to provide some background in the area of general risk assessment and how it might apply to male-mediated effects.

The breakout group felt that a primary issue is the lack of experimental and human research, a prerequisite that must be addressed to facilitate the risk assessment process. The fact that male-mediated developmental toxicity (MMDT) is a relatively understudied area is due, in part, to what might be termed a perception problem. That is, researchers in reproductive toxicology, teratology, reproductive and perinatal epidemiology, and genetics often overlook the potential importance of MMDT. This may be due in part to a lack of a clear definition of what constitutes male-mediated "developmental toxicity." Developmental toxicity, as often defined, does not include many of the sperm/semen endpoints that were subjects of this conference. While expanding the definition may be appropriate, making this clear is imperative, so other researchers and practioners of risk assessment will have a clear understanding of what is included.

The breakout group felt that there is an acute lack of awareness of the promising leads generated in previous research and the growing mechanistic framework for male-mediated effects. The breakout group suggested that efforts be made to communicate more effectively with various organizations including: funding agencies, federal and international regulatory bodies, and scientific societies. In addition, meetings such as the present one are effective means to provide useful summaries of the state of knowledge. In order to do this, there may

Advances in Male Mediated Developmental Toxicity, edited by Bernard Robaire and Barbara F. Hales.
Kluwer Academic/Plenum Publishers, 2003.

need to be a more formal organization that speaks as a collective body, as opposed to a group of individual scientists that meet occasionally.

ANIMAL STUDIES

With respect to the formal risk assessment process, MMDT has not been well established in the hazard identification phase. While dominant-lethal studies in animal testing have been used for a long time to identify the influence of stress on genetic integrity, there has only recently been a focus on other sperm and semen parameters in toxicity testing. Two issues within animal test systems were discussed by the breakout group: 1) the apical nature of toxicity testing, and 2) the need for dose-response information. Traditionally, toxicity testing is apical, i.e., many potential modes of toxicity are brought together in the collective endpoints that are measured in the standard whole animal tests. As we move to incorporating newer study designs, we will have to be able to define how the endpoints measured provide significant information and overall toxic potential of a particular stress/exposure. This will require an understanding of the dose-response nature of the effect, otherwise, the information will have limited application in the risk assessment process.

The breakout group discussed the potential short-term and long-term application of information on male-mediated effects in risk assessment. Short-term, there was the feeling (as noted above) that risk assessors may be missing important endpoints of toxicity that may be particularly responsive to specific stressors. There was also the feeling that a stronger data base in this area could help support or refute some of the assumptions that must be made in the current risk assessment process. From a long-term perspective, data on male-mediated effects should be very useful in the ongoing efforts at harmonizing various risk assessment approaches and developing methods for using mechanism of action data and for assessing structure-activity relationships, mixtures, and different exposure paradigms.

HUMAN STUDIES

For some human endpoints such as miscarriage, birthweight, preterm birth, and neurobehavioral parameters, there clearly has been insufficient evaluation of male mediated effects for a broad array of exposures, even for known mutagenic exposures such as ionizing radiation and cigarette smoking. For other endpoints such as birth defects and childhood cancer there have been more epidemiologic studies, but the studies have not provided consistent and valid evidence for a causal assessment. There are several central methodological limitations of most of these studies that have severely limited interpretation and impeded hazard identification. One of the most important is the lack of accurate exposure information collected in the majority of the studies. Often this is because the study was not designed to investigate paternal exposures, and therefore only limited information was obtained. The group recognized the critical need for more informative epidemiologic studies to guide the risk assessment process. Greater emphasis should be placed on studies designed with a sufficiently large sample size and collecting high quality data on exposures and potentially confounding factors. These studies should collectively examine a diverse array of exposures and endpoints. Greater emphasis should be placed on increasing communication between epidemiologists and laboratory investigators. The laboratory work can better inform epidemiologic studies as to candidate toxicants and mechanisms; epidemiologic findings can provide leads for further evaluation in the laboratory setting.

STUDY DESIGNS FOR THE ASSESSMENT OF MALE-MEDIATED DEVELOPMENTAL TOXICITY

Barbara F. Hales[1] and Daniel G. Cyr[2]

[1]Department of Pharmacology and Therapeutics
McGill University, Montreal, QC, H3G 1Y6, Canada
[2]Institut National de Recherche Scientifique-Institut Armand Frappier
Pointe Claire, Quebec H9R 1G6, Canada

INTRODUCTION

It is clear from many of the chapters in this volume that exposure of males to toxic chemicals or radiation can adversely affect subsequent generations. The endpoints for these adverse effects on progeny outcome have varied from pre-implantation and/or post-implantation death, to malformations, altered behaviour, or cancer. Effects may not be manifested until adulthood. Furthermore, several studies have shown that even "apparently normal" F_1 offspring can transmit adverse effects to the F_2 generation (Auroux et al., 1988; Hales et al., 1992)

How widespread are the agents that produce male mediated developmental toxicity? Men exposed to toxic chemicals need to know if the health of their unborn children will be affected. To extrapolate potential risks to humans, reproductive and developmental toxicology risk assessors need information about the chemicals identified as male mediated developmental toxicants in animal models. Epidemiology studies have identified occupational and life style exposures that are associated with adverse progeny outcome; these exposures need to be pursued in animal models to identify the specific chemicals and exposure paradigms that are involved. Elucidation of the molecular mechanisms that are responsible for male-mediated developmental toxicity is critical for the development of new endpoints and approaches to assess male mediated developmental toxicity.

The focus of this breakout group was on the design of studies to assess male-mediated developmental toxicity. Discussions were broadly aimed at addressing the following questions: How should we design studies to assess male-mediated developmental toxicity? Are there critical windows of exposure? Can we do cross species comparisons? What are the specific mechanisms underlying male-mediated developmental toxicity? Elucidation of the underlying mechanisms will permit the testing and validating of new endpoints for the analysis of male mediated developmental toxicity.

Advances in Male Mediated Developmental Toxicity, edited by Bernard Robaire and Barbara F. Hales.
Kluwer Academic/Plenum Publishers, 2003.

From the human perspective, epidemiology studies have focused largely on a case-control approach (see the breakout group report from Olshan et al., this volume), with the need for careful evaluation of exposure. Our increasing ability to characterize the genomic integrity of male germ cells (see the breakout group report from Perreault et al., this volume) will impact on the information that can be obtained from human studies and should shed light on the types of damage associated with specific exposures and pregnancy outcomes.

EXPERIMENTAL APPROACHES

The current study designs that are most appropriate to characterize the effects of specific chemicals or exposures of the male on his progeny outcome in animal models are: the two- or three-generation reproduction and fertility test, and the dominant lethal and extended dominant lethal test. Other tests that focus on the effects of toxicants on the male germ cell, but were not addressed in the breakout group discussion, include the heritable translocation test (detects transmitted chromosomal damage manifested as balanced reciprocal translocations in the progeny of mutagen-exposed males) (Bishop and Kodell 1980) and the specific locus mutation test (detects the mutation rate at 7 specific loci as a result of mating treated males to test females homozygous for 7 recessive markers for visible parameters, such as coat or eye colour) (Russell et al., 1981).

Two-Generation Reproduction and Fertility Test Guidelines

The Reproduction and Fertility Effects Test aims to assess the ability of drugs and chemicals to disrupt reproduction (U. S. EPA Health Effects test guideline 870.3800; Clegg et al., 2001). This test provides information on the effects of a substance on various aspects of reproduction, including the estrous cycle, mating behaviour, conception, gestation, parturition, lactation, weaning, growth, and development of the offspring. The data that may be obtained on the second and/or third generations include morbidity, mortality and developmental toxicity.

Both the male and female parental animals, usually rats, are exposed to the test substance before and during mating, gestation and lactation. Males of the parental generation should be dosed for at least one complete spermatogenic cycle (approximately 56 days in the mouse and 70 days in the rat) prior to mating. Some of the resulting F_1 offspring are exposed until adulthood and then throughout mating and pregnancy. Subsequently, the F_1 animals are mated with animals in the same dose group and their offspring (F_2) are then exposed to the chemical until weaning. The test relies heavily on breeding success in each generation, with the consequence that the animal model is chosen for its fecundity. The test includes histological examination of the reproductive organs, and measures of sperm production, motility and morphology that may help to localize adverse effects to the male germ cells. Nevertheless, since both sexes are dosed, it may not be easy to identify the affected sex, or to distinguish effects that are mediated via an exposure that is prior to conception from one that is due to in utero or post-natal exposure. As reproductive success is measured by the litter size at birth, the standard multi-generation test does not distinguish between reduced litter size occurring secondary to failed fertilization or as a consequence of post-fertilization embryo or fetal loss. The incidence of malformed fetuses that is observed will be reduced by any peri-natal death and cannibalism that occurs. The likelihood of being able to observe any functional or latent deficits or cancer in the second generation is also very low. Thus, while reproductive abnormalities that appear in the first or second generation offspring should be detected, this test does have shortcomings as a screen for male-mediated developmental toxicity.

Dominant Lethal/Extended Dominant Lethal Test Guidelines

The "dominant lethal" test is a relatively non-specific test, designed to detect suspected mutagens and DNA-reactive chemicals that induce genetic damage in sperm (U. S. EPA Health Effects Test Guideline 870.5450; Rohrborn, 1970; Ehling U.H. et al., 1978; Green et al., 1985). The dominant lethal test involves acute or short-term dosing of the males only (typically as a single dose or for 5 daily doses) and subsequently mating each male with one to two untreated females per week, in serial fashion, for about ten weeks, a period of time equivalent to the duration of spermatogenesis plus epididymal transit time. Rats or mice are commonly selected as the test species, and normally three dose levels are used. Females are examined after confirmed mating, usually mid-gestation, to determine the numbers of corpora lutea in the ovaries (an indication of the number of potential embryos) and implantations in the uterus. The implants are classified as normal fetuses, dead fetuses, or early or late resorptions; fetal weights may be monitored. Usually both pre- and post-implantation embryonic losses are considered as indicative of dominant lethality.

The fetal loss detected with the dominant lethal test is usually assumed to be due to chromosomal damage (structural and/or numerical anomalies) or to lethal gene mutations derived from the fertilizing spermatozoa. However, a positive dominant lethal test, as assessed by increased pre-implantation loss, may also be due to failure of the sperm to recognize and/or fertilize the oocyte. For one well characterized male reproductive toxicant, methyl chloride, there were elevated rates of post-implantation loss during the first two weeks post-exposure of male rats, and in pre-implantation loss during weeks 2-8 post-exposure. Subsequent studies revealed that the elevated pre-implantation loss was the result of failure of fertilization due to cytotoxic effects on the sperm (Working and Chellmann, 1989).

The use of sequential mating intervals permits the identification of the stage(s) of spermatogenesis most susceptible to the toxicant. In rats, for example, during the first week after exposure the effects observed indicate damage to spermatozoa during epididymal maturation. Effects evident 2-3 weeks after exposure result from damage to post-meiotic germ cells, round or elongating spermatids. Effects that are manifested after longer periods of time indicate damage to meiotic or pre-meiotic spermatocytes, or to mitotic spermatogonia or stem cells. This test may be added at the end of a subchronic or multi-generation test by serially breeding the males with untreated females when their treatment is finished (e.g. Tyl et al., 2000).

The gestational stage when the females are sacrificed in the dominant lethal test will determine whether it is feasible to examine the fetuses for malformations. In developmental toxicity tests with rats, pregnancy is normally terminated on day 21 of gestation to permit the assessment of external, internal and skeletal malformations. To characterize "male teratogenicity", Knudsen and co-workers (1977) suggested a "modified" or "extended" dominant lethal test, with pregnancy termination and examination of the fetuses on day 21, rather than earlier during gestation. This delay in the time of assessment until the end of gestation and the inclusion of additional "standard teratology" parameters together greatly increase the likelihood for detection of any exposure that causes male mediated developmental toxicity.

OPTIMIZATION OF STUDY DESIGNS TO DETECT MALE-MEDIATED DEVELOPMENTAL TOXICANTS

The extended dominant lethal study design permits the detection of male mediated adverse progeny outcomes visible on or before gestational day 21. Exposure of the adult male to the toxicant throughout a complete cycle of spermatogenesis and spermatozoal maturation in the epididymis ensures that all critical stages of germ cell proliferation and differentiation are exposed to the toxicant. Any abbreviated study design with a shorter period of exposure would NOT evaluate effects at all germ cell stages. The two-generation test design, while not focused on effects on the male reproductive system, does address the possibility that fetal and/or neonatal processes in gonadal development may be targets for the adverse effects of exposure to a drug or toxicant.

As mentioned above, the endpoints for male-mediated developmental toxicity vary from pre-implantation and/or post-implantation death, to malformations, altered behaviour, or second generation cancer. A comprehensive evaluation of progeny outcome in the modified dominant lethal test or a two-generation study design would be expected to incorporate many, but not all, of these parameters. Cannibalism would decrease the numbers and types of malformations visible at birth in the two-generation study design, and also make it difficult to detect early or late resorptions. Many male-mediated developmental toxicants, such as cyclophosphamide, cause pre-implantation death or early resorptions, manifested as a reduction in inner cell mass-derived tissues (Kelly et al. 1992). Severe growth retardation is a frequent outcome of male-mediated developmental toxicants. It is unlikely that "runts" would survive the process of parturition in the two-generation study design, but they would be observed in the modified dominant lethal test.

The functional or latent deficits that are most commonly reported as a consequence of exposure to a male-mediated developmental toxicant are behavioural alterations, especially with respect to learning, and cancer. The addition of behavioural testing to the two generation test paradigm would address this consideration. There was debate about whether the cost of maintenance of the F_1 offspring long enough to determine whether or not there was an increase in cancer in this generation was warranted. The fact that the F_1 generation is maintained and mated ensures that any effects transmitted to the F_2 generation will be observed. An alternate, more focused approach would be to add sufficient females to the extended dominant lethal test paradigm to allow half to be killed on day 21 of gestation and the other half to litter and be maintained for subsequent behavioural and cancer tests.

Currently there is no in vitro test to characterize male-mediated developmental toxicity, or the reproductive or developmental toxicity of chemicals. Furthermore, it is unlikely that it will ever be feasible to capture all the potential sites of action of a male-mediated developmental toxicant in a single mammalian in vitro system. One possibility is the use of other model species such as the fruit fly; importantly, most of the compounds that have been shown to be male-mediated developmental toxicants in mammals also affect male germ cells in *Drosophila* (Vogel and Nivard, this volume). The recent establishment of primordial germ cell-derived permanent female and male embryonic murine cell lines presents an interesting avenue to explore the establishment of an in vitro method to assess germ cell mutagenicity (Klemm et al., 2001). The use of an "isolated germ cell" model system to assess male mediated developmental toxicity would presuppose that the male germ cell is the exclusive target, and that the role of the oocyte is "unimportant" in translating any male germ cell "hit" into developmental toxicity. We know that this is not true. Studies from the lab of Generoso (Generoso, 1980) showed that damage to the male gamete could be repaired in the oocyte, and that murine strain differences in repair capacity in the oocyte were responsible for differences in the incidence of male germ cell mutations.

CROSS-SPECIES COMPARISONS: CAN WE EXTRAPOLATE FROM STUDIES IN RODENTS TO HUMANS?

Rats are the preferred model for most tests of the effects of chemicals on reproduction and fertility, including the two-generation test and the dominant lethal test. Rats have a short gestational period, produce large litters and have relatively low rates of resorptions or malformations. Depending on pharmacokinetics and metabolism, exposure of the germ cell may be similar to that in man, or not. Significantly, rats produce sperm in numbers that greatly exceed the minimum required for fertility; reduction of sperm production by more than 90% does not have any significant effect on fertility (Robaire et al., 1984). Very few rat spermatozoa have abnormal morphology or altered motility parameters. The consequence is that there will be few false positives. This is important. On the other hand, how does this affect our ability to use rats as a model for predicting human risk? Humans are notorious as the species having a very high proportion of abnormal spermatozoa. It is not considered abnormal to have as many as 50% of spermatozoa with an abnormal form or as many as 50% of spermatozoa with reduced or absent motility (World Health Organization, 1999); equivalent numbers for the rat are less than 5% for both parameters. It is likely that the threshold for normal fertility is much lower in man than in rats. Exposure to anticancer drugs which induce azoospermia, oligozoospermia, or aztheno-zoospermia in men do not cause a noticeable decrease in sperm count in rats (e.g., Trasler et al., 1985). Therefore, although a rodent animal model is robust for the identification of chemicals likely to have effects in man (low false positives), it may be weak in identifying chemicals that cause more limited or selective damage (high false negatives).

Some of the concern about the consequences to the progeny of altered or defective male germ cells has arisen from the increasing use of assisted reproductive technologies such as in vitro fertilization and intracytoplasmic sperm injection (ICSI) (see Hewitson et al., this volume). A drawback to the rat as a model for studies of male-mediated developmental toxicity pursuant to these technologies is that few labs have succeeded with rat in vitro fertilization techniques and early embryo culture.

CONCLUSIONS

It is important that toxicity testing paradigms include study protocols that will permit the detection and characterization of male-mediated developmental toxicants. Both the two-generation reproduction and fertility test and the dominant/extended dominant lethal test have the ability to meet this objective. As there is evidence that adverse progeny outcomes may result from damage to the male genome at specific stages of germ cell development, it is important that these tests not be "abbreviated" by too short an exposure period. It is apparent that histopathological examination of the testis does NOT replace these tests for male-mediated developmental toxicants. Cyclophosphamide is well esta-blished as a male-mediated developmental toxicant, yet even after long-term exposure few or no changes are detected in the testis or the orderly process of spermatogenesis (Trasler et al., 1988). While rodents serve as good models for much toxicity testing, the extent to which the robustness of male reproduction in these animals may lead to false negatives is not clear. Finally, it is important that the findings of human epidemiology studies be tested in animal models, with exposure paradigms matched to the best of our ability, and that the hypotheses derived from animal models be tested in controlled human studies.

While some of the "classical" endpoints for male-mediated developmental toxicity, such as pre- or post-implantation death and malformations, can be detected, others, such as behavioural alterations and second generation cancer, are not examined routinely in the two-generation reproduction and fertility and the extended dominant lethal tests. While the addition of these endpoints would enhance our ability to detect and characterize male-mediated developmental toxicants, this is probably not sufficient. Even more important is the development and validation of additional endpoints, preferably at the multiple levels of function, cell and molecule. Towards this goal, we need to elucidate the mechanism(s) underlying male-mediated developmental toxicity. It is widely thought that one of the mechanisms underlying male-mediated developmental toxicity is DNA damage in the male genome. In this respect, it is interesting that exposure to widely different male-mediated developmental toxicants, from irradiation to cyclophosphamide, for example, produces the same types of malformations, i.e. severe growth retardation, edema, and hydrocephaly, regardless of whether rats or mice are used as the test species (see chapters by Robaire and Hales or by Nagao, this volume). This finding, and the studies of Müller and co-workers (this volume) with 'Heiligenberger' mice (registered as 'HLG/Zte') in which exposure of male mice to ionizing radiation induced an increased incidence of gastroschisis, a defect which occurred spontaneously in these mice, suggest that male-mediated developmental toxicants are unlikely to be targeting a specific gene. Random damage to the genome may induce genome instability in male germ cells, without affecting their ability to fertilize an oocyte. Ultimately, the more we understand about the biology underlying male-mediated developmental toxicity, the easier it will be to assess risk, on both an individual and population basis.

ACKNOWLEDGMENTS

This report is a summary by the Chairs based on discussions among participants at the Study Design Workshop. We thank the following participants for their input: Jacques Auger, Janice Bailey, Thomas Collins, Tiaan De Jager, Lori Dostal, Yuri Dubrova, Jan Friedman, Anne Golden, Lhanoo Gunawardhana, Stephen Hooser, Michael Joffe, Carole Kimmel, Mary Lee, Michelle Leonard, Dan Livy, Michaela Luconi, Tetsuji Nagao, Jacqui Piner, Gregorz Szymczynski, Lara Vassilieva, Tacey White.

REFERENCES

Auroux, M., Dulioust, E.J.B., Nawar, N.N.Y., Yacoub, S.G., Mayaux, M.J., Schwartz, D., and David, G., 1988, Antimitotic drugs in male rat. Behavioral abnormalities in the second generation. *J Androl.* 9:153-159.

Bishop, J.B., and Kodell, R.L., 1980, The heritable translocation assay: its relationship to assessment of genetic risk for future generations. *Teratog Carcinog Mutagen.* 1:305-332.

Clegg, E.D., Perreault, S.D., and Klinefelter, G.R., 2001, Assessment of Male Reproductive Toxicity, in: *Principles and Methods of Toxicology*, Fourth Edition, A. Wallace Hayes, ed., Taylor & Francis, Philadelphia, p. 1263.

Ehling, U.H., Machemer, L., Buselmaier, W., Dycka, J., Frohberg, H., Kratochvilova, J., Lang, R., Lorke, D., Muller, D., Peh, J., Rothrborn, G., Roll, R., Schulze-Schencking, M., and Wiemann, H., 1978, Standard protocol for the dominant lethal test on male mice set up by the work group "Dominant Lethal Mutations of the ad hoc Committee Chemogenetics". *Arch. Toxicol.* 39:173-186.

Generoso, W.M., 1980, Repair in fertilized eggs of mice and its role in the production of chromosomal aberrations. *Basic Life Sci.* 15:411-420.

Green, S., Auletta, A., Fabricant, R., Kapp, M., Sheu, C., Springer, J., and Whitfield, B., 1985, Current status of bioassays in genetic toxicology: The dominant lethal test. *Mutation Res.* 154:49-67.

Hales, B.F., Crosman, K., and Robaire, B., 1992, Increased postimplantation loss and malformations among the F2 progeny of male rats chronically treated with cyclophosphamide. *Teratology.* 45:671-678.

Kelly, S.M., Robaire, B., and Hales, B.F., 1992, Paternal cyclophosphamide treatment causes postimplantation loss via inner cell mass-specific cell death. *Teratology.* 45:313-318.

Klemm, M., Genschow, E., Pohl, I., Barrabas, C., Liebsch, M., Spielmann, H., 2001, Permanent embryonic germ cell lines of BALB/cJ mice – an in vitro alternative for in vivo germ cell mutagenicity tests. *Toxicol In Vitro.* 15:447-453.

Knudsen, I., Hansen, E.V., Meyer, O.A., and Poulsen, E., 1977, A proposed method for the simultaneous detection of germ-cell mutation leading to fetal death (dominant lethality) and of malformation (male teratogenicity) in mammals. *Mutat Res.* 48:267-270.

Robaire, B., Smith, S. Hales, B., 1984, Suppression of spermatogenesis by testosterone in adult male rats: Effect on fertility, progeny outcome and pregnancy. *Biol Reprod.* 31:221-230.

Rohrborn, G., 1970, The dominant lethals: Method and cytogenetic examination of early cleavage stages, in: *Chemical Mutagenesis in Mammals and Man*, F. Vogel and G. Rohrborn, eds., Springer-Verlag, Heidelberg, p.148.

Russell, L.B., Selby, P.B., von Halle, E., Sheridan, W., and Valcovic, L., 1981, The mouse specific-locus test with agents other than radiations: interpretation of data and recommendations for future work. *Mutat Res.* 86:329-354.

Trasler, J.M., Hales, B.F., and Robaire, B., 1985, Paternal cyclophosphamide treatment of rats causes fetal loss and malformations without affecting male fertility. *Nature.* 316:144-146.

Trasler, J.M., Hermo, L., Robaire, B., 1988, Morphological changes in the testis and epididymis of rats treated with cyclophosphamide: a quantitative approach. *Biol Reprod.* 38:463-479.

Tyl, R.W., Friedman, M.A., Losco, P.E., Fisher, L.C., Johnson, K.A., Storther, D.E. and Wolf, C.H., 2000, Rat two-generation reproduction and dominant lethal study of acrylamide in drinking water. *Reprod Toxicol.* 14:385-401.

U. S. Environmental Protection Agency, 1998, *Health Effect Test Guidelines OPPTS 870.3800 Reproduction and Fertility Effects*, U.S. Government Printing Office, Washington, D.C.

U. S. Environmental Protection Agency, 1998, *Health Effect Test Guidelines OPPTS 870.5450 Rodent Dominant Lethal Assay*, U.S. Government Printing Office, Washington, D.C.

Working, P.K., and Chellman, G.J., 1989, The use of multiple endpoints to define the mechanisms of action of reproductive toxicants and germ cell mutagens. *Prog Clin Biol Res.* 302:211-227.

World Health Organization: *WHO Laboratory Manual for the Examination of Human Semen and Sperm-Cervical Mucus Interaction,* 34th ed. Cambridge, UK, Cambridge University Press, 1999.

TRANSLATIONAL RESEARCH IN MALE MEDIATED DEVELOPMENTAL TOXICITY

Jacquetta M. Trasler

Departments of Pediatrics, Human Genetics
and Pharmacology & Therapeutics
McGill University and The McGill University-Montreal Children's Hospital
Research Institute
2300 Tupper Street
Montreal, Quebec, Canada H3H 1P3

INTRODUCTION

The task of the Workshop Group on Translational Research was to suggest ways in which basic research findings occurring since 1992 and presented at the Second International Conference on Male-Mediated Developmental Toxicity could be applied to human studies and used to develop new screening tests for male-mediated toxicants. The Group considered the impact on the field of male-mediated developmental toxicology of research in areas such as new genes, epigenetics, prenatal exposures, germ cell protection strategies, apoptosis, cell cycle, and heat shock proteins. A basic premise was that a better understanding of the biological mechanisms underlying male-mediated effects is crucial for the design of new clinical epidemiological studies including research on treatments used in cancer patients and environmental exposures. A key issue was to suggest ways in which we can address differences seen in paternal exposure studies between species and strains.

RESEARCH AREAS

Several promising areas of basic research applicable to translational research in male-mediated toxicology were identified and are discussed below.

Genome Project

Genes in human and other species are now being identified, sequenced and characterized at a rapid rate and cross-species comparisons are being carried out. Several techniques allow large scale screening for alterations in gene expression (e.g., expression

Advances in Male Mediated Developmental Toxicity, edited by Bernard Robaire and Barbara F. Hales.
Kluwer Academic/Plenum Publishers, 2003.

microarrays) and gene sequence (e.g., single nucleotide polymorphisms or SNPs) that are applicable to studies of male-mediated toxicity in humans and model species. Techniques such as laser capture microdissection, that are currently being used extensively in human cancer studies, will permit expression and mutation studies on testicular biopsy samples or archived tissues. In order to take advantage of newer techniques emanating from the human genome project, it will be necessary to collect well characterized samples on exposed populations. Such an approach is being taken with the studies on the childhood cancer survivors, where DNA samples are being collected. Similarly, somatic DNA, sperm and testicular biopsy samples are also being collected in groups of men with idiopathic infertility (K. Jarvi, personal communication).

Animal-Human Comparisons

A large number of studies provide evidence in animal models that paternal treatment is associated with effects in the offspring. The underlying mechanisms are postulated to be genetic or epigenetic alterations. In humans, other than epidemiological data, there is little evidence of transmission to the offspring of drug- or chemical-induced heritable changes. It is clear that well-designed epidemiological studies, with sufficient numbers of defined cases, accurate exposure histories and identification of confounders, are necessary. Prior to embarking on expensive epidemiological studies, animal models may help establish biological plausibility and identify mechanisms. One of the weaknesses of animal studies to date is that they do not often mimic human exposures. Typically, both acute and chronic treatment regimens have been used in animal studies; however, multi-drug/agent regimens that are more similar to human environmental exposures or regimens used in cancer chemotherapy are not commonly used.

Are there techniques available currently that allow genetic or epigenetic studies in exposed human populations? Several promising techniques are being used, including fluorescence in situ hybridization (FISH) on sperm, the sperm chromatin structure assay, and the search for Y chromosome deletions. One of the most developed and promising is FISH (see chapters by Martin and by Robbins, this volume). Some of the most useful information comes from studies where sperm are collected pre-treatment, during treatment and post-treatment (e.g., Robbins et al., 1997).

Ideally, when looking at genetic and epigenetic endpoints, both exposed patients as well as their offspring should be studied. Examples of populations amenable to such studies are cancer patients and patients seen in in vitro fertilization (IVF) clinics. In cancer patients, the exposures are known. Paternal exposure histories can be gathered in IVF patients and pregnancy outcomes followed. A number of countries require that outcome data be reported for IVF clinics; such information may include highly informative data such as pregnancy, spontaneous abortion and congenital malformation rates, ultrasound as well as results of chorionic villus sampling, genetic tests, and postnatal follow-up. DNA-based tests such as FISH still need to become more standardized and automated to allow larger numbers of patients to be studied and reliably compared between studies.

Species and Strain Differences

Various animal models have been used in studies of male-mediated effects. For toxicological studies, there are several advantages to using the rat as a model including its well characterized reproductive and endocrinological systems, large litter size and low, consistent background rates of early and late pregnancy losses and congenital malformations.

The mouse has also been used to study the male-mediated effects of acute and chronic chemical and radiation exposures. Advantages of the mouse as a model include the existence of many well-defined inbred strains as well as more recently developed mice with specific targeted gene mutations. Strain differences in response to paternal exposures may help uncover underlying mechanisms. For example, Dubrova (this volume) described the existence of differences in rates of minisatellite instability between inbred mouse strains.

A very active area of research is the study of induced mutations in mice. Targeted null mutations ('gene knockout') are employed to inactivate specific genes and study effects on mice of deficiencies in key enzymes, structural proteins, signaling molecules, etc. A number of genes important for spermatogenesis such as DNA repair genes and genes encoding proteins involved in meiosis and chromatin remodeling have been inactivated in mice, sometimes with dramatic effects. Wide phenotypic variation is often found between different inbred mouse strains, reminiscent of documented inter-strain and inter-species differences in male-mediated effects; such findings suggest that underlying genetic factors (e.g., modifying genes) should be considered in studies of male-mediated toxicants. Another approach is to combine a specific knockout with a drug or chemical exposure (Bailey et al., 2002). In concert with sequence and SNP data derived from the genome projects, such combined gene/chemical studies have the potential of providing additional information to explain species and strain differences.

Once more mechanistic data is available from rodent studies, an additional animal model that may be useful for modelling human male exposures is the primate.

Prenatal Germ Cell Exposures

Key events in germ cell development occur before birth. In the prenatal male, germ cells proliferate and epigenetic events such as the methylation of repeat sequences and imprinted genes such as H19 are initiated. Data from mouse studies indicate that primordial germ cells or gonocytes are sensitive to the induction of mutations; an example is the drug *N*-ethyl-*N*-nitrosourea (Shibuyu et al., 1996). The *N*-ethyl-*N*-nitrosurea studies provided evidence that the male germline may be vulnerable in utero. In humans, testicular cancer incidence has been on the rise over the last century. One theory is that in utero exposures to toxicants in the environment alters gonocytes, predisposing them to changes that lead to testicular cancer post-puberty. Clearly more studies are needed in this area. The consequences of prenatal exposures can best be tested in animal models and may help pinpoint underlying mechanisms.

Epigenetics

In a number of animal studies, adverse effects on the offspring occur at levels too high to be accounted for by mutagenesis suggesting that other mechanisms may be involved. Much work in the area of epigenetics has been carried out over the last 10 years, including work on enzymes and mechanisms underlying DNA methylation and chromatin structure, both of which have profound effects on gene expression and function. DNA methylation patterns are established in the germline and abnormalities in genomic methylation have been associated with cancer and behavioral abnormalities. Profound changes in chromatin structure occur during spermiogenesis.

Paternal exposures to certain chemicals may alter DNA methylation in spermatozoa. For example, chronic treatment of male rats with 5-azacytidine, a drug that alters DNA methylation, resulted in abnormalities in male germ cells and an increase in preimplantation loss (Doerksen and Trasler, 1996). In particular, there is concern for imprinted genes (genes expressed preferentially from only one of the parental alleles) that pick up a unique epigenetic mark or imprint in the male germline (see Bartolomei, this volume). A failure to be marked in the male germline, perhaps as a result of a drug or chemical exposure to developing germ cells, cannot be compensated for by the maternal genome following fertilization. Imprinting defects have been reported in ICSI patients (Cox et al., 2002) and simple manipulations such as alterations in culture conditions can lead to abnormalities in genomic imprinting (Khosla et al., 2001).

In animal models, alterations in sperm chromatin structure have been associated with abnormalities in pregnancy outcome. Normal alterations in chromatin structure occurring during spermiogenesis are associated with the replacement of histones by transition proteins followed by protamine(s). In mice, targeted disruption of the transition protein 2 gene altered chromatin structure and resulted in smaller litter sizes (Zhao et al., 2002). Drugs and chemicals may also alter sperm chromatin structure. For instance, lead, known to be a male-mediated toxicant (see Silbergeld et al., this volume), causes conformational changes in protamine leading to decreases in protamine-DNA binding; it was proposed that such chromatin changes may alter sperm chromatin condensation or decondensation, leading to altered fertility (Quintanilla-Vega et al., 2000).

Techniques, some of which can be used on relatively small numbers of cells, have been developed to look for alterations in DNA methylation (bisulfite genomic sequencing, restriction landmark genomic scanning and methylation profiling) and chromatin structure (chromatin immunoprecipitation assays). Larger scale genomic methylation assays that can be applied to male-mediated studies are being developed. More animal studies need to be designed to look for epigenetic effects of paternal exposures. Human studies are only warranted once animal exposures are identified that cause epigenetic alterations.

Protection of the Seminiferous Epithelium from Exogenous Substances

Transient or permanent infertility results from a number of male exposures, in particular with anti-cancer regimens. Evidence from studies of sperm chromosomes suggests that residual defects exist 6-18 months after the end of treatment (Martin et al., 1986; De Mas et al., 2001; see Martin, this volume). There has been concern that paternal exposures, especially those that cause DNA damage, may permanently affect long-lived spermatogonial stem cells. Effects of a drug on stem cells may then persist, providing a source of abnormal germ cells throughout the life of an individual. Are there ways to protect the seminiferous epithelium during treatment (e.g., anticancer drugs and radiation) or exposures (e.g., Chernobyl, Gulf War)? New research in three different areas is promising: in vivo protection of germ cells during treatment; removal (pre-exposure), cryopreservation and subsequent post-treatment reintroduction of germ stem cells; harnessing of cellular protection mechanisms such as heat shock proteins and molecules that regulate apoptosis.

Over the last 10 years there have been a number of advances in our understanding of mechanisms of protection of the seminiferous epithelium as presented by Meistrich (this volume). Treatment of rats with various hormonal regimens can protect the testis from cancer chemotherapy-induced damage. Rodent studies have suggested that the infertility seen in patients following radiation or chemotherapy may be due to failure of the differentiation of spermatogonia rather than permanent effects on stem cells and that hormonal treatments can reverse drug-induced infertility, even when hormones are given after the initiation of drug therapy. Such findings should be encouraging for cancer patients where anti-cancer drugs need to be started soon after diagnosis before hormone regimens can be initiated. Hormonal protection of the seminiferous epithelium has been extended to men undergoing chemotherapy. Hormonal treatments are expected to protect spermatogenesis in man; however, it is unclear at present whether the hormonal regimens will also prevent genetic damage.

The powerful technique of germline stem cell transplantation has been pioneered in rodent models over the last 8-10 years (Brinster, 2002). The technique allows spermatogonial stem cells to be removed by testicular biopsy and cryopreserved, and subsequently thawed and reintroduced into the seminiferous epithelium. Clinical trials based on this technique are underway and provide a way for men undergoing irradiation or chemotherapy to preserve their fertility post-treatment. Such an approach might also be useful for other types of paternal exposures.

Much is now known about the pathways and molecules involved in apoptosis and DNA repair, including studies of these processes during spermatogenesis. There is evidence of male-mediated effects resulting from compromised functioning of pathways that play a role in protecting the seminiferous epithelium. For example, in mice, p53 may suppress radiation-induced male-mediated teratogenesis (Armstong et al., 1995). While low levels of spontaneous apoptosis are normally present in the seminiferous epithelium, treatment with cyclophosphamide, well known to produce male-mediated effects, resulted in increased levels of apoptosis as compared to saline-treated animals (Cai et al., 1997; Brinkworth and Neischlag, 2000). Robaire and Hales (1999) have postulated that apoptosis of damaged premeiotic germ cells plays a critical protective role in preventing genetic damage from being propagated to the offspring; in contrast, the more streamlined postmeiotic cells may not have the necessary cellular machinery to permit apoptosis to occur (Hales and Robaire, 2001). Heat-shock proteins, including sperm-specific forms are abundant in the testis and may, like apoptosis, play a role in protecting germ cells from paternal exposures. In the future, being able to detect individual differences in DNA repair and apoptosis capabilities may be useful in tailoring treatment regimens in patients.

FUTURE DIRECTIONS

Much basic research carried out over the last decade holds promise for application to the field of male-mediated developmental toxicity. While translational research is already underway for a number of the areas described above, other areas require more data from testing in animal models. It is encouraging that some very promising techniques such as FISH and germ stem cell transplantation have become available for translational studies soon after their development. Over the next decade, it is likely that, due to the human and animal genome projects, methods will become available to carry out inter-species comparative studies and permit the genome to be probed for subtle genetic and epigenetic alterations that might result from paternal exposures to chemicals and radiation.

ACKNOWLEDGMENTS

This report is the Chair's summary based on discussions among participants at the Translational Research Workshop. I thank the following participants for their input: A. Adeeko, A. Aguilar, L. Anderson, M. Bartolomei, M. Brinkworth, C. Campagna, R. Cheng, F. Eustache, K. Jarvi, T. Kelly, D. Lucifero, S. Maier, C. Oakes, K.L. Park, J. Qiu, N. Salama, and S. Siew.

REFERENCES

Armstrong, J.F., Kaufman, M.H., Harrison, D.J., and Clarke, A.R., 1995, High frequency developmental abnormalities in *p53*-deficient mice. *Curr Biol.* 5:931-936.
Bailey, R.W., Aronow, B., Harmony, J.A.K., and Griswold, M.D., 2002, Heat shock-initiated apoptosis is accelerated and removal of damaged cells is delayed in the testis of clusterin/ApoJ knock-out mice. *Biol Reprod.* 66:1042-1053.

Brinkworth, M.H., and Neischlag, E., 2000, Association of cyclophosphamide-induced male-mediated, foetal abnormalities with reduced paternal germ cell apoptosis. *Mutat Res.* 447:149-154.

Brinster, R.L., 2002, Germline stem cell transplantation and transgenesis. *Science.* 296:2174-2176.

Cai, L., Hales, B.F., and Robaire, B., 1997, Induction of apoptosis in the germ cells of adult male rats after exposure to cyclophosphamide. *Biol Reprod.* 56:1490-1497.

Cox, G.F., Bürger, J., Lip, V., Mau, U.A., Sperling, K., Wu, B-L., and Horsthemke, B, 2002, Intracytoplasmic sperm injection may increase the risk of imprinting defects. *Am J Hum Genet.* 71:162-164.

De Mas, P., Daudin, M., Vincent, M.C., Bourrouillou, G., Calvas, P., Mieusset, R, and Bujan, L., 2001, Increased aneuploidy in spermatozoa from testicular tumour patients after chemotherpy with cisplatin, etoposide and bleomycin. *Human Reprod.* 16:1204-1208.

Doerksen, T., and Trasler, J.M., 1996, Developmental exposure of male germ cells to 5-azacytidine results in abnormal preimplantation development in rats. *Biol Reprod.* 55:1155-1162.

Hales, B.F., and Robaire, B., 2001, Paternal exposure to drugs and environmental chemicals: effects on progeny outcome. *J Androl.* 22:927-936.

Khosla, S., Dean, W., Reik, W., Feil, R., 2001, Culture of preimplantation embryos and its long-term effects on gene expression and phenotype. *Hum Reprod Update.* 7:419-427.

Martin, R., Hildebrand, K., Yamamoto, J., Rademaker, A., Barnes, M., Douglas,G., Arthur, K., Ringrose,T., and Brown, I., 1986, An increased frequency of human sperm chromosomal abnormalities after radiotherapy. *Mutat Res.* 174:219-225.

Quintanilla-Vega, B., Hoover, D.J., Bal, W., Silbergeld, E.K., Waalkes, M.P., and Anderson, L.D., 2000, Lead interaction with human protamine (HP2) as a mechanism of male reproductive toxicity. *Chem Res Toxicol.* 13:594-600.

Robaire, B., and Hales, B.F., 1999, The Male Germ Cell as a Target for Drug and Toxicant Action. In: The Male Gamete: From Basic Science to Clinical Applications (C. Gagnon ed.) Cache River Press, FL, pp 469-474.

Robbins, W.A., Meistrich, M.L., Moore, D., Hagemeister, F.B., Weier, H.-U., Cassel, M.J., Wilson, G., Eskenazi, B., and Wyrobek, A.J., 1997, Chemotherapy induces transient sex chromosomal and autosomal aneuploidy in human sperm. *Nature Genet.* 16:74-78.

Shibuyu, T., Horiya, N., Matsuda, H., Sakamoto, K., and Hara, T., 1996, Dose-dependent induction of recessive mutations with N-ethyl-N-nitrosourea in primordial germ cells of male mice. *Mutat Res.* 357:219-224.

Zhao, M., Shirley, C.R., Yu, Y.E., Mohapatra, B., Zhang, Y., Unni, E., Deng, J.M., Arango, N.A., Terry, N.H.A., Weil, M.M., Russell, L.D., Behringer, R.R., and Meistrich, M.L., 2002, Targeted disruption of the transition protein 2 gene affects sperm chromatin structure and reduces fertility in mice. *Mol Cell Biol.* 21:7243-7255.

Gene sequence microassays, 279–280
Genitourinary anomalies, paternal radiation
        exposure-related, 54
Genome, epigenetic modification of, 23–245
Germ cells
    female, mutagenesis in, 6–8
    male
        apoptosis of, 76
        direct effect of toxicants on, 170
        H19/igf2 gene methylation patterns in, 242
    totipotency of, 248–249
Germ stem cell transplantation, 282–283
Gigantism, paternal cyclophosphamide exposure-
        related, 170–171
Glucose-6-phosphate dehydrogenase, 91, 92
Glutathione
    acrylamide binding with, 193
    acrylamide-related depletion of, 193, 194
Glycidamide, 193
Goats, cloning experiments in, 247
Gonadotrophin depletion, as apoptosis cause, 92–
        93
Gonadotropin-releasing hormone analogs,
        spermatogenesis-protective effects of, 227,
        229–231
    genetic consequences of, 231–234, 235
    in humans, 234–235
Gonocytes, *in utero* toxicant exposure of, 281
Granulosa cells, as cloning donor cells, 248
Greece, decreasing sperm counts in, 85
Gulls, chemical pollutant-induced germline
        mutations in, 125

Hair dye, as childhood cancer risk factor, 148
Hamster oocyte/human sperm fusion technique,
        60, 182
    in cancer patients, 184
Heart malformations, paternal radiation exposure-
        related, 54
Heat-shock proteins, 283
Heavy metals, paternal exposure to, as childhood
        cancer cause, 94
Heiliberger strain mice, 163–168, 276
Hepatoblastoma, in children, paternal occupational
        exposure-related, 95
Herbal remedies, 221
Herbicides, paternal exposure to, as childhood
        cancer cause, 151, 152
Heritable translocation test, of male-mediated
        developmental toxicity, 7, 101, 102
    of acrylamide, 142
    dominant lethal effects of, 132
    of etoposide, 142
    positive results in, 132–133
Hexanedione, 170, 229
H19 gene, 281
    differently methylated domain of, 241–242

H19 gene (*cont.*)
    epigenetic regulation of, 239–242
    hypermethylation of, 240–241, 242
    imprinting of, 240, 241–242
        lability in cell cultures, 242–243
Hippocampal cell cultures, effect of lead exposure
        on, 40–42
Hiroshima atomic bombing survivors, 93
    children of, 12, 93, 115, 222
    radiation dose exposure in, 124
Hormonal treatment, of male chemotherapy
        patients, 282
Hormones: *see also specific hormones*
    as male stem cell protectants, 227–228
HP2, interaction with lead, 44–45
Human chorionic gonadotropin, as pregnancy
        indicator, 25, 26, 27, 30, 67
Human Genome Project, 280
Hycanthone, *in vivo* genetic toxicity patterns of, 7
Hydrocarbons, paternal exposure to, as childhood
        cancer cause, 150, 153, 154, 156–157
Hydrocephaly, paternal cyclophosphamide
        exposure-related, 14, 15, 16, 170
Hydronephrosis, paternal cyclophosphamide
        exposure-related, 15
N-Hydroxymethyl acrylamide, 6
Hypospadias, 86, 87
Hypothalamic pituitary axis, effect of toxicants on,
        170

ICF syndrome, 239
ICSI: *see* Intracytoplasmic sperm injection
Imprinted genes, 240
    clustering of, 240
    H19/igf2 locus, epigenetic regulation of, 240–
        242
Imprinting, 281
    in the developing embryo, 242
    effect of cell cultures on, 242–243
    in intracytoplasmic sperm injection, 282
    parental/genomic, 240
Imprinting control region, 241
Indenopyridine, 229
Industries, occupational exposure within, 149
Infants, male, protamine abnormalities in, 261
Infertility, male
    alcohol-related, 63
    caffeine-related, 33
    chemotherapy-related, 282
    chromosome abnormalities associated with, 185
    genetic/chromosomal disorders-related, 66
    genetic factors in, 88, 95
    idiopathic, 280
    in infertility clinic patients, 73
    intracytoplasmic sperm injection-related, 200
    lead exposure-related, 38, 39–40, 43, 45
    radiation therapy-related, 282